DESIGN OF PHOTOVOLTAIC SYSTEMS

L UMANAND

L Umanand is a professor at the Department of Electronic Systems Engineering, Indian Institute of Science, Bangalore.

Published in 2024

Preface

Planet earth, the place of all human endeavours is a thermodynamic system. If one considers the occasional asteroids entering the earth's atmosphere as insignificant, then there exists only energy exchange between the earth system and its external environment. Earth may be considered as a thermodynamically closed system with only the solar energy entering the boundary as input energy to the earth system. The only source of energy being solar, all other forms of energies are derived from this single input source that is the Sun. Within this Earth system, the solar energy is converted into various forms like the wind energy, hydro energy, thermal energy, light energy, wave energy, geo-thermal energy, storage in fossils beneath the surface of the earth, etc.. As of now, most applications are driven by energy sources that are derived in some way from one of the fossil fuels. In fact fossil fuel based energy sources make up about three quarters of the global energy consumption. Any energy system that derives its operational energy from the fossil fuel source are called *conventional* energy systems. Systems that use energy sources other than the fossil fuel sources for their operation may be classified as *alternate* energy systems or *non-conventional* energy systems.

Literature also classifies the fossil fuel based energy as *non-renewable* form of energy even though it is actually derived from the solar energy. All stored forms of energy beneath the surface of the earth has taken many thousands of years to achieve levels of energy storage that was prevalent a century ago before being almost depleted. The time constant for storing the energy beneath the surface of the earth as fossil fuel is in the order of thousands of years, where as the time constant for utilizing the stored energy based on the current energy consumption levels is in the order of few decades. As the utilization and replenishment time constants are so vastly different, one may consider the stored or fossil fuel energy forms as non-renewable for time spans in the order of centuries. Alternate energy forms like light, thermal, wind, hydro etc. are derived directly from the solar energy that is incident on the earth's surface. Hence these energy forms are called *renewable*. Systems that are based entirely on these renewable forms of energy are called renewable energy systems.

Energy usage is categorized into two broad generic application groups viz. (i) mass transport and (ii) energy transport. Mass transport includes all applications that involve physical movement – either linear movement or rotational movement. Movement of people, animals and goods by means of motorized and non-motorised vehicles are a set of applications that fall in the category of mass transport. Another example is that of pumping up water through a specific height. When water is lifted from an underground sump to an overhead tank, then a specific mass of water is moved from one potential level to another potential level. Likewise, pumping of water from wells and bore wells for purposes of drinking and irrigation will also fall under the category of mass transport. Another example of liquid mass transport is that of lifting oil from oil wells. Rotary motion can also be classified as mass transport. As examples, the movements of a flywheel, the movement of the blades of a windmill, the movements of the blades of a turbine etc. all fall in the category of rotational mechanical movement of masses.

Applications that involve transport of energy by means of wave or electromagnetic propagation, heat and light fall in the energy transport category. Radio waves used for transmission of audio energy for the purposes of sound based communication is an example of energy transport. Application that involves the transmission and reception of audio and video energies for entertainment and communication are also examples of energy transport. Cell phones, local area networks, wireless networks, land and mobile communication devices, satellite communication and control are all examples of applications that fall in the category of energy transport. The main feature of the energy transport based applications is that the core energy for the application is transported without any movement of mass vis-a-vis the mass transport applications that will always involve usage of energy for physical movement of a mass component. Heating is another major application category of energy transport systems that includes fluid heating like water heating, oil heating, air heating; space heating, cooking, drying, steam generation, distillation, industrial furnaces and similar applications wherein the main function is based on heat flow. Lighting is also another core application category of energy transport system that includes home lighting, community lighting, street lighting, sign boards, monitors, scopes, torches, lamps, candles, industrial lighting and similar applications wherein the core function is based on light. This book is based on applications related to photovoltaic source. Photovoltaic sources convert light energy to electric energy and falls in the category of energy transport applications.

Whatever be the energy domain that a system exists in, the thermodynamic laws are active and coexist in every system, affecting their function. Any system needs energy to do work and deliver the expected output. For any energy output delivered from a system, the amount of energy required at the input will always be greater. This implies that, of the energy applied to the input, only a part of it will be delivered to the output and the remaining part is lost as heat. The part of energy that is lost as heat is irrecoverable-it remains lost forever. This energy that is lost as heat into the thermodynamic domain is a universal law occurring in all systems-simple and complex. The ratio of the delivered energy to the input energy is defined as the energy efficiency of a system. This energy efficiency is always less than one for a real system.

The concept of entropy is central to energy efficiency being always less than 100%. Every action in the universe will need energy. A part of the energy will get converted as heat and will be lost forever. Eventually as time progresses, the usable energy decreases with a consequent increase in the irrecoverable heat energy. Ultimately all available energy is converted to dead heat at which time no activity in the universe can be sustained. *The energy that becomes unavailable or irrecoverable* is the *entropy*. It should be observed that as there is always action happening in universe, the irrecoverable energy is monotonically increasing implying that *entropy will always increase* with time.

The irrecoverable energy or the expended energy is nothing but heat having the lowest work or activity potential. The expended energy exists in the thermal energy domain. Every activity in the universe is linked to the underlying thermal domain that is apparently all pervading. The energy in the electrical domain has a 'one-way' window into the thermal domain by means of the i^2R loss in a resistor. The energy in the magnetic domain is linked to the underlying thermal domain by means of the hysteresis and eddy current losses. The energy in the mechanical domain is linked to the underlying thermodynamic domain through the friction loss. Likewise every energy domain has an one-way outlet or window or port that links with the thermal domain. The energy that flows through this outlet into the thermal domain is in general the energy that is lost from that domain in performing a specific activity. This energy that goes out through the loss outlet into the thermal domain is lost forever.

The key take away from the laws of thermodynamics is that *energy can neither be created nor destroyed*; and *entropy monotonically increases*. However, there exists an apparent flexibility on the *entropy rate*. There is no restraint on the rate at which entropy can increase. It is this apparent flexibility that is the root cause of the present energy problems that is being witnessed globally. Over time, humans have gradually increased the rate of energy consumption per capita and now it is at an alarming level. The one aspect that nature has given the freedom of choice i.e. the rate at which energy is used up, is a matter of concern. This has been vastly misused and as a consequence it has lead to an unhealthy situation with pollution, global warming, deteriorating environment and climate change.

It is understandable that humans strive to achieve a state of physical comfort and luxury in their respective living styles. However, luxury and *entropy rate* conflict each other. Improving luxury in most cases would imply a higher entropy rate wherein energy depletion is much faster. At one end of the spectrum would be the tribal society who are hunters and gatherers. They live off the land and energy depletion would probably be at the minimum. At the other end of the spectrum would be the seemingly highly ordered society where physical comfort and luxury takes precedence. Here the energy consumption per capita can be alarmingly high. One must consciously contemplate and strike a balance between the two extremes. More and more one must strive to move from activities that demand concentrated energy use to equivalent activities that demand distributed energy use over time. This will have a tendency to reduce the entropy rate. It is with this attitude and frame of mind that one must read this book.

This book is on applications related to solar photovoltaic cells. It delves into circuits and system aspects of design of photovoltaic systems. This topic is rather broad based and extends from photovoltaic cells to system level design decisions on applications. Effort has been placed to strike a balance between circuits and system design aspects so that the book can cater to a range of photovoltaic based designers and practitioners. The reader should have a background in power electronics or at least electronics. Where necessary the concepts and theory related to circuits have been discussed to aid readers without background in power electronics.

This book is composed of 12 chapters. The first two chapters are related to the photovoltaic (PV) cell as a power source. Chapter 1 describes the characteristics and parameters that are associated with a solar cell. This chapter also describes the model or circuit equivalent of the PV cell. Chapter 2 is important in that it leads to the interconnection of individual cells in a PV panel. Interconnection of cells in series and parallel are discussed in detail. Aspects related to protection of cells, simulation of PV source and the experimental setup to practically measure the characteristics of a solar cell are described. The chapter concludes with a discussion on PV source emulation techniques which is an important requirement for PV application designers.

Chapter 3 relates to the energy from the Sun that is incident on earth. In order to size the solar panels for a given application, it is important to estimate the amount of incident irradiation at the specific place of interest. This chapter describes the terminologies and the solar geometry that is essential to address this aspect. Chapter 4 is on sizing of the photovoltaic panels which is an application of chapter 3. The sizing design of the PV panels for a given application and for a specific place is based on the incident energy estimates obtained using the concepts described in chapter 3.

Chapter 5 discusses the important topic of energy storage. In particular the battery as an energy storage is discussed in detail. The various parameters related to battery storage systems are discussed leading to the selection of batteries. This chapter also sensitizes the reader to other energy storage methods like flywheel, pumped hydro and compressed air storage mechanisms. This is followed by chapter 6 which discusses sizing applications related to PV sizing and battery sizing.

Chapter 7 deals with the topic of maximum power point tracking (MPPT). The characteristics of the solar cell is such that it can deliver the maximum power at only one operating point for a given incident solar power. Therefore, the electronic control should be such that at any given load, the PV source should always operate at the operating point that delivers maximum power. This chapter discusses methods and algorithms that can achieve this tracking of maximum power operating point. This is an essential technique that needs to be incorporated into the design in order to optimally utilise the PV panels. Chapter 8 is another important application that relates to battery charging. Circuits and control aspects related to charging batteries from PV sources are discussed. The issues associated with series interconnection of batteries and parallel interconnection of batteries are discussed. The solutions for these problems are also addressed.

Chapter 9 deals with applications related to refrigeration using Peltier elements. Powering Peltier junctions with PV sources for the applications of cooling and heat pumps are discussed in detail in this chapter. A part of the chapter is also dedicated to thermal aspects related to heat flow for refrigeration. Chapter 10 describes the application of PV sources to water pumping. The design of PV based pumping system is discussed in detail. Chapter 11 treats the injection of PV source power into the power grid in order to supplement the grid power in some detail. The final chapter is not on system design but on life cycle cost. This chapter finds a place in this book due to its importance in bench marking and estimation of life cycle and annual life cycle energy costs in PV based applications.

Though I am the sole author of this book, the content in this book has evolved over many years with constant interaction with my teachers, colleagues and students. My sincere salutations to all of them. I am grateful to NPTEL (national programme for technology enhanced learning) for having given me an opportunity to deliver a MOOC (massive open online course) on this topic which is hosted on the national SWAYAM portal. My students have devoted time and supported as TAs (technical assistants) managing the topics. They have also contributed to the chapter end questions in the book. I sincerely and gratefully acknowledge my students Syam Sundar Nair, Atmanandamaya, Subhabrata Basak, Ruman Kalyan Mahapatra, Divyanshu Bansal, Aravind G, Rohit P, Anchal Singh Thakur, Ashwin Kanavalkadan, Shubham Agarwal, Gautam Anil Raiker, Vaidika Topendra, Naresh Kumar Meena and Joshua Thomas Mathew for having directly or indirectly contributed towards the book.

I must also express my appreciation to my wife Vijayashree and son Tharakeshwar for the cover page design and support on manuscript format conversions. On several occasions during COVID times, I have thought of dropping the idea of writing this book. However, my wife and son have been very supportive with constant encouragement giving me the needed inspiration to complete this book. I sincerely acknowledge my gratitude for their continuous support.

L UMANAND

Contents

1 Photovoltaic Characteristics 1
   1.1 Brief history 1
   1.2 PV Cell 5
   1.3 Model 7
   1.4 I-V Characteristics 8
      1.4.1 Short circuit current point 9
      1.4.2 Open circuit voltage point 10
      1.4.3 Maximum power point 10
   1.5 Efficiency 10
   1.6 Temperature effects 11
   1.7 Temperature coefficients 12
   1.8 Fill factor 13
   1.9 Datasheet 14
   1.10 Example 14
   1.11 Simulation 15
   1.12 Questions 16

2 Cell and Module Interconnections 19
   2.1 Identical cells in series 19
   2.2 Load line 20
   2.3 Non-identical cells in series 21
   2.4 Identical cells in parallel 23
   2.5 Non-identical cells in parallel 24
   2.6 Protecting and Interconnecting modules 25
   2.7 Simulation model of PV source 27
   2.8 Measurement setup for i-v characteristics 30
   2.9 PV source emulation 32
   2.10 Questions 35

3 Sun Power 39
   3.1 Introduction 39
   3.2 Solar radiation terms 40
   3.3 Estimation of daily irradiance 41
   3.4 Irradiance and insolation 41

3.5		Incident energy	42
3.6		Declination	44
3.7		Solar geometry	44
3.8		Insolation on horizontal flat plate collector	47
3.9		Daily energy on a horizontal flat plate	49
	3.9.1	Insolation on earth surface	51
	3.9.2	Sunrise and sunset angles	51
	3.9.3	Ex.3.1	52
	3.9.4	Ex.3.2	52
3.10		Energy on a tilted flat plate	53
3.11		Irradiance simulation in Octave	55
3.12		Large tilt generalisation	57
3.13		Optimum fixed tilt for collectors	60
3.14		Atmospheric effects	62
	3.14.1	Airmass coefficient	64
	3.14.2	Clearness Index	66
	3.14.3	Tilt factor	70
3.15		Energy on tilted surface with atmospheric effects	71
3.16		Energy script in Octave	74
3.17		Irradiance on vertically placed collectors	76
3.18		Questions	77

4 Photovoltaic Sizing — 81

4.1		Sizing for applications without battery	81
	4.1.1	Estimate H_{at}	83
	4.1.2	Estimate peak sun insolation hours	84
	4.1.3	Estimate the intrinsic area	84
	4.1.4	Estimate the estate area	85
4.2		Roof top PV example	86
4.3		Parking lot PV example	87
4.4		1MVA grid connected PV system example	89
4.5		Cleaning and maintenance	92
4.6		Questions	92

5 Energy Storage — 95

5.1		Batteries	96
	5.1.1	Capacity	96
	5.1.2	C-rate	98
	5.1.3	Efficiency	100
	5.1.4	Energy and power densities	101
	5.1.5	Comparison of batteries	102
	5.1.6	Battery selection	104
5.2		Flywheel storage	105
5.3		Pumped hydro storage	106
5.4		Compressed air storage	108
5.5		Questions	109

6 Photovoltaic Application — 113
- 6.1 Load Profile — 113
- 6.2 Days of autonomy and recharge — 115
- 6.3 Battery sizing — 116
- 6.4 PV array sizing — 118
- 6.5 Design toolbox in Octave — 120
- 6.6 Questions — 122

7 Maximum power point tracking — 125
- 7.1 Introduction — 125
- 7.2 Input resistance of boost converter — 126
- 7.3 Input resistance of buck converter — 128
- 7.4 Input resistance of buck-boost converter — 130
- 7.5 Input impedance of isolated converters — 132
  - 7.5.1 Forward converter — 132
  - 7.5.2 Flyback converter — 133
  - 7.5.3 Pushpull converter — 135
  - 7.5.4 Half bridge converter — 136
  - 7.5.5 Full bridge converter — 137
- 7.6 MPPT through input resistance control — 138
  - 7.6.1 Reference cell method - voltage scaling — 139
  - 7.6.2 Reference cell method - current scaling — 141
  - 7.6.3 Sampling method — 142
  - 7.6.4 Power slope - incremental resistance method — 143
  - 7.6.5 Power slope - pulse phase method — 145
  - 7.6.6 Hill climbing method — 146
  - 7.6.7 MPPT with shading — 147
- 7.7 Questions — 148

8 Battery Charging — 151
- 8.1 Introduction — 151
- 8.2 Direct connection — 151
  - 8.2.1 Example — 153
- 8.3 Charge controller — 154
- 8.4 Charge controller circuit — 155
- 8.5 Switched mode battery charger with current control — 156
  - 8.5.1 Slope Compensation — 157
- 8.6 Charging with MPPT — 160
- 8.7 Batteries in Series — 160
  - 8.7.1 Resistive equaliser — 161
  - 8.7.2 Charge pump based active equaliser — 162
- 8.8 Batteries in Parallel — 165
- 8.9 Questions — 168

9 Peltier Cooling with Photovoltaic — 171

- 9.1 Introduction 171
- 9.2 Peltier device 171
- 9.3 Heat pump 173
- 9.4 Datasheet 175
- 9.5 Peltier cooling 177
  - 9.5.1 Heatsink 179
  - 9.5.2 Heatsink with peltier 180
  - 9.5.3 Forced cooling 181
- 9.6 Thermal aspects 182
  - 9.6.1 Conduction 183
  - 9.6.2 Convection 184
    - 9.6.2.1 Nusselt's number for Free convection 186
    - 9.6.2.2 Nusselt's number for forced convection 188
- 9.7 Peltier refrigeration 189
  - 9.7.1 Free convection method 189
  - 9.7.2 Forced convection method 191
- 9.8 Radiation 193
- 9.9 Mass transport 194
- 9.10 Question 195

10 Water Pumping with Photovoltaic — 199

- 10.1 Introduction 199
- 10.2 Water pumping system 199
- 10.3 Pressure heads 200
- 10.4 Hydraulic energy 201
- 10.5 Hydraulic power 202
- 10.6 Friction head determination 203
  - 10.6.1 Moody Chart 204
  - 10.6.2 Numerical solution to Colebrook-White formula 206
- 10.7 Calculation steps for hydraulic power 208
- 10.8 Example 1 210
- 10.9 Example 2 211
- 10.10 Example 3 213
- 10.11 Photovoltaic sizing 215
- 10.12 Pumped hydro application 216
  - 10.12.1 Pico hydel system 218
- 10.13 Questions 221

11 Grid Interaction — 225

- 11.1 Introduction 225
- 11.2 Interconnection principle 225
- 11.3 Controlled inverter source 228
- 11.4 T-Network 229
  - 11.4.1 Characteristic impedance 230
  - 11.4.2 Propagation constant 231

CONTENTS

    11.5 L-C-L interface 231
    11.6 Transformer-less versus galvanic isolation 233
    11.7 Single phase grid interface 235
        11.7.1 Open loop plant 236
        11.7.2 Gate drive 237
        11.7.3 PWM 237
        11.7.4 α, β currents 239
        11.7.5 Stationary to rotating frame transformation 240
        11.7.6 Frequency and angle estimation 244
        11.7.7 Feedforward components 244
        11.7.8 MPPT integration 245
    11.8 Three phase grid interface 246
    11.9 Questions 249

12 Life Cycle Cost   **253**
    12.1 Introduction 253
    12.2 Growth models 253
        12.2.1 Linear growth 254
        12.2.2 Compound growth 255
        12.2.3 Exponential growth 257
    12.3 Inflation 258
    12.4 Time frame transformation 259
    12.5 Annual payments 260
    12.6 Life cycle costing 262
    12.7 Example 1 - Estimation of LCC 263
    12.8 Example 2 - Estimation of ALCC 266
    12.9 Example 3 - Estimation of break even point 268
    12.10 Questions 271

Appendix-A : Typical Datasheet of 308W PV module   **273**

Appendix-B : Typical Datasheet of 240W PV module   **275**

Appendix-C : Simulation with KiCAD and NgSPICE   **277**

Appendix-D : Octave scripts   **283**

Appendix-E : Answers to chapter end questions   **291**

Bibliography   **293**

Chapter 1

Photovoltaic Characteristics

1.1 Brief history

This book deals with the topic of interfacing solar photovoltaic cells to various applications and loads. The power electronic interface will be the primary topic of discussion here. It will comprise of power electronic circuits that will interface the solar cell to various loads. However, the central component in all these circuits and applications is the solar cell or the photovoltaic cell. It will be considered as a black box with two power terminals. The characteristics and the parameters of this device will be studied just like any other component that would be bought off the shelf like a diode or a MOSFET or a BJT. This book will not discuss on the physics of the photovoltaic cell or its fabrication. None the less, it will be interesting to look back and see how this photovoltaic cell came about. A historical perspective about the photovoltaic effect and the photovoltaic cell, will be in keeping with the discussions to follow and it will give an insight into how the 19th century scientists discovered and thought about the principles behind the photovoltaic cell and gave such a wonderful gift.

It began in the early 19th century. In 1839, a young boy around 19 years old, Edmond Becquerel discovered the photovoltaic effect and it came to be known as the Becquerel effect. He was working in his father's lab, wherein he discovered that when light was shown upon an electrode dipped in an acidic medium, current flows through the electrode. Later on, he published the results in his memoirs which were the very first publication on photovoltaic effect. He came to be known as the father of photovoltaics. Early photovoltaic cells made by Edmond Becquerel consisted of a container which was painted black so that it traps light. It was filled with an acidic solution and there were two electrodes coated with silver chloride or silver bromide. A membrane is placed in between and connected the external circuit to the two electrodes as shown in figure 1.1.1. Light incident on the electrode results in flow of current. This is the earliest photovoltaic cell. From 1839 to 1876 there was not much reported on this topic.

The first publication after 1839 was in 1876, more that three decades after the discovery of the photovoltaic effect. Adams and Day published in the Philosophical

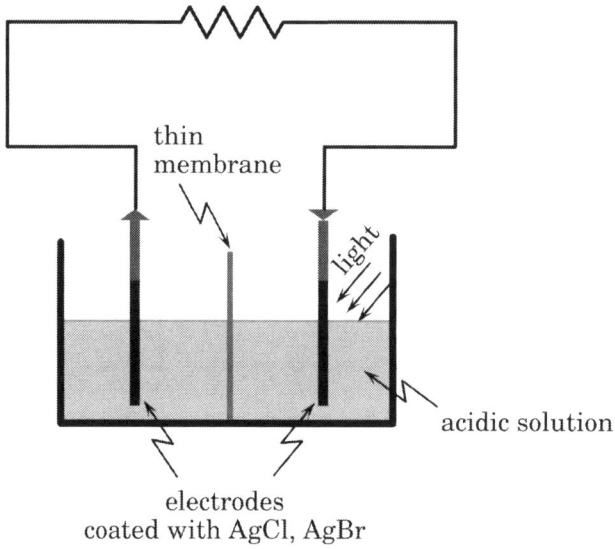

Figure 1.1.1: Early photovoltaic cell

Transactions of the Royal Society of London, where the selenium photovoltaic cell is described. They actually borrow on the discovery of Willoughby Smith who found that the conductivity of selenium is dependent on exposure to light. Based on this principle Adams and Day made the selenium photovoltaic cell. They used vitreous selenium and to that, platinum electrodes were attached. This is encased in a glass tube as shown in figure 1.1.2 and light is allowed to be incident on the selenium through the glass tube. It is observed that a current flows through the electrodes and external circuit. This is the selenium photovoltaic cell.

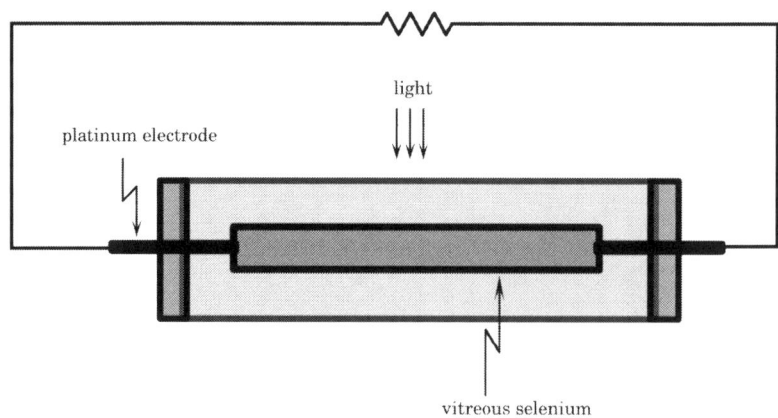

Figure 1.1.2: Selenium photovoltaic cell

1.1. BRIEF HISTORY

Few years later in 1883, Fritts published in the American Journal of Science on a new form of selenium photocell, the first thin film selenium solar cell. It consists of a metal substrate like brass upon which a 25 micron selenium layer is placed. Molten selenium is pressed in between two metal sheets. The top layer is a gold leaf, a very thin semi-transparent gold layer. Contacts are attached to the top gold layer and the bottom brass layer to which the external circuit is connected as shown in figure 1.1.3. When light falls on the selenium through the semi-transparent metal, there is electric current flow through the contacts and the external circuit. The Fritts thin film selenium solar cell was not well received by the researchers at that time, even though it was demonstrated and performed well. Apparently there was some skepticism about the thin film solar cell. The concept of light as a fuel source and the quantum nature of light were not yet clear. Probably, at that time, classical physics was unable to answer the underlying theoretical principles of operation of the solar cell, at least not until Einstein was able to explain it much later.

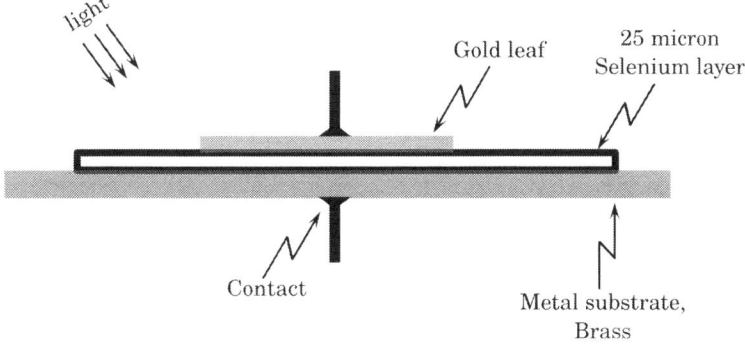

Figure 1.1.3: Thin film selenium photovoltaic cell

In 1900, something remarkable and revolutionary happened. Not only was it the birth of a new century, it also marked the birth of quantum physics. On December 14, 1900, a german physicist Max Planck, reported in his talk at a meeting of the Berlin Academy of Sciences' Physical Society, on the problem that he was facing in the thermal radiation experimental results. He observed in black body radiation that certain discrete amounts of energy only were emitted, contrary to the classical notion that any value of energy could be emitted. He could not find any solution to explain the results by traditional classical physics. He tried to introduce a totally new concept, the concept of quanta. Max Planck introduced the now very famous relation, $E = h\nu$, wherein E is the quantized energy which is proportional to the frequency, ν. h is a proportionality factor which turns out to be same for all types of energy and is called the "planck's constant" or the "quantum of action". At that point in time, discretization of energy was unimaginable. Classical physics would not allow it but Max Planck's seminal ideas which introduced the concept of energy packets or the quanta was the starting point into a totally new realm of the ultra small.

Sometime in 1905, at the time a relatively unknown person of the Swiss patent office, Albert Einstein, published an article in the *Annalen der Physik* wherein he explains the concept of light packets or light quanta called photons. Through this principle, he explains the photovoltaic principle or the photovoltaic mechanism completely and thus laid the foundation for even the present day semiconductor industry.

In 1933, Grondahl worked and published several articles on copper-cuprous oxide solar cell. Among his several publications, the particular article in 1932 on copper cuprous oxide rectifier and photoelectric cell, provides specific details into the manufacturing process of the photoelectric or the solar cell. This copper-cuprous oxide solar cell became quite popular because of its low-cost of production.

Much later in 1941, Ohl came out with the first patent for a silicon solar cell. It is for a light sensitive electric device including silicon. This silicon solar cell had an efficiency of much less than 1%. It did not have much commercial value or commercial implication but it is a landmark milestone point in history as far as the evolution of solar cell or the photovoltaic cell is concerned. In 1954, Chapin, Fuller and Pearson published in the Journal of Applied Physics, a silicon semiconductor solar cell having an efficiency of 6%. Fuller and Pearson were employees of the Bell labs. This is the result of an observation that the Bell labs silicon diode produced significant amount of current and voltage in the presence of light. It is the begining of the modern day silicon semiconductor photovoltaic cells. From here on, there were rapid improvements in the semiconductor industry which also significantly benefited the development of the modern day solar photovoltaic cells and modules. Today, the top three technologies that are commercially most popular in use are the monocrystalline silicon solar cells, the polycrystalline (multi-crystalline) and the thin film silicon solar cells.

The monocrystalline silicon solar cells have an efficiency of nearly 20%. Polycrystalline or Multi-crystalline solar cells are slightly lower in efficiencies and reach about 18% in most commercial products. The thin film solar cells also have efficiencies about 18%. There are several other technologies that are popular at the research and laboratory experimental works but not yet commercially viable as yet. Gallium arsenide germanium solar cell having efficiency of up to 30% have been developed in the research laboratories. Copper indium gallium selenide solar cell with efficiency of 21%, cadmium telluride solar cell with 21% efficiency, amorphous silicon solar cells having efficiency of 10% but very low production costs, dye-sensitized solar cell around 11% efficiency are possible technologies that are being researched. There are also organic solar cells having efficiency of 8%. Multi-junction solar cell, indium gallium phosphide, gallium arsenide, indium gallium arsenide with very high efficiency of 37% have been achieved at research labs. The perovskite solar cells are showing lot of promise. Perovskite is a material which is coated on top of silicon solar cells to improve the efficiency. Quantum dot solar cells are something that will start coming in the near future and they are supposed to yield very high-efficiency.

The early solar cells were used for telephone repeaters and one of the earliest popular application was in space. The Vanguard 1 satellite in 1958 was launched with six silicon solar cell panels having the power of 5 mW. When the battery stopped operating after few months, the photovoltaic (PV) source continued to operate the transmitter for years, even after the mission life of the satellite. This in fact was a very great success for photovoltaic cell application. Now a days, the photovoltaic cell is being populated on rooftops to power the homes, buildings and communities.

1.2. PV CELL

There are several solar fields that convert the solar power and inject the power into the power grid to supplement the load requirements. In Kamuthi, a place in Tamil Nadu, India, you can find one of the earliest large solar array fields. It is a mega solar power plant which delivers about 650 MW of peak power. There are several such mega plants that have been commissioned in India and also around the world. Another fully operational solar park that is capable of generating 2GW of peak power is the Shakti Sthala solar power project situated in Pavagada taluk, Tumakuru district, Karnataka, India. The Bhadla Solar Park located in Bhadla village, in Rajasthan's Jodhpur district, is probably one of the largest solar power plant in the world. It is capable of generating a massive 2.5GW of peak power.

It has been a long journey for the solar photovoltaic cell. About 180 years, starting from 1839 when Edmond Becquerel discovered photovoltaic effect till today, when we see high-efficiency semiconductor silicon solar cells powering several communities. The photovoltaic cell has come to stay and is now poised to become the main electric power supply for this planet.

1.2 PV Cell

In this section, the photovoltaic cell will be discussed from the perspective of electrical design engineers. There are several parameters that have a bearing on the operation of the photovoltaic source. They need to be understood in detail for good design of PV based applications. Figure 1.2.1 shows a picture of a polycrystalline photovoltaic cell and an encapsulated monocrystalline cell. The top layer that is visible is the N type part of the p-n junction. The bottom layer is the p-type substrate. The n-type metallization are joined together and are connected to the minus terminal. The p-type metalization are joined together and connected to the plus terminal.

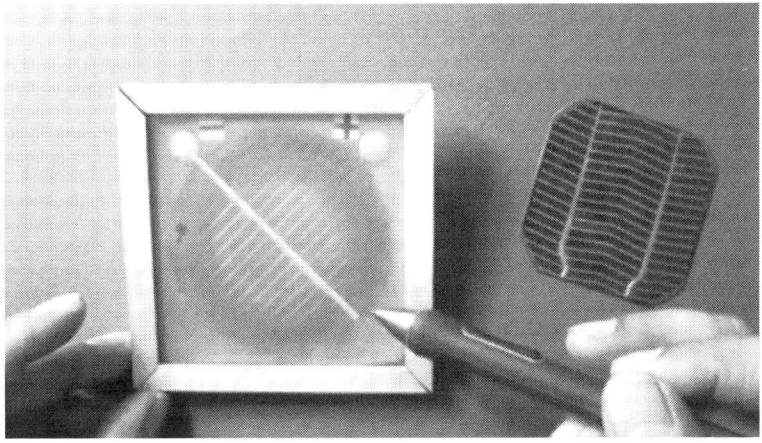

Figure 1.2.1: Typical photovoltaic cell

The PV cell is nothing but a p-n junction just like a diode. The anode and cathode are brought out as the two terminals of the PV cell. The working of the cell is pretty simple. When light falls on this glass surface, i.e. on the n-type surface of the p-n junction, the valence electrons get excited after having acquired photon energy packets. They then move into the conduction band and through the top metallization into the negative terminal. From the negative terminal, the electrons flow through the external circuit and reaches the positive terminal and comes back to the p-type substrate through the bottom metallization. It is important that the exposed part covered by glass here, has to be perpendicular to the incident radiation from the sun. Several PV cells are connected in series and parallel combination to form a module. Severals modules may be arranged as a panel and several panels form a solar array.

The two most common types of sources are viz. (i) voltage source wherein the voltage is constant irrespective of load and (ii) current source wherein the current is constant irrespective of the value of load voltage. The PV cells or the PV source is very much different in character. It behaves like a current source for a range of loads and behaves more like a voltage source for some other range of loads. The i-v characteristics of the PV source is shown in figure 1.2.2 and compared with those of current and voltage sources. Figure 1.2.2(a) gives the characteristics of a voltage source where v_T is the terminal voltage and i_T is the terminal current. The voltage across the terminal is V_C independent of current. Likewise, figure 1.2.2(b) represents the characteristics of a current source wherein the current is fixed and independent of the terminal voltage. In figure 1.2.2(c), the characteristics of the PV source is shown in comparison with the voltage and current sources. Here one can observe that it behaves as both current source and voltage source for specific ranges of the terminal load.

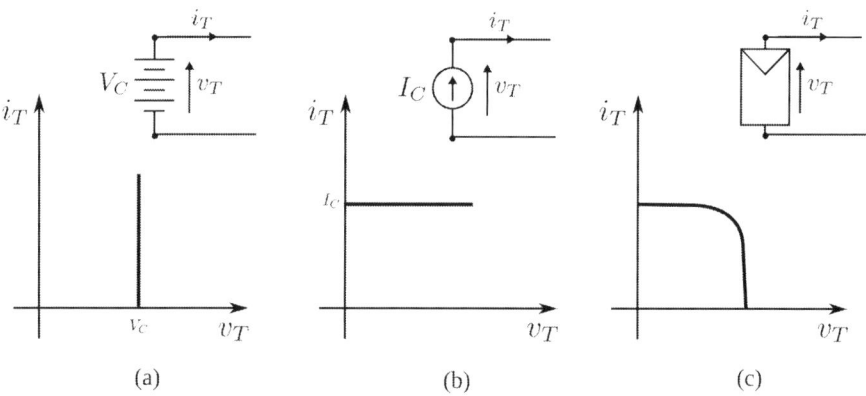

Figure 1.2.2: i-v characteristic of (a) voltage source, (b) current source, (c) PV source

The PV source operates by virtue of light/solar energy being incident on its surface. A photocurrent, i_p is produced which is directly proportional to the solar power that is incident normal to the surface of the panel.

If i_p is zero, then it represents dark condition wherein no light is incident on the solar cell. The characteristic marked '1' in figure 1.2.3 represents this condition. Observe that it is almost identical with the i-v characteristic of a p-n junction diode. As the light intensity increases, there will be a resultant increase in the photocurrent i_p.

1.3. MODEL

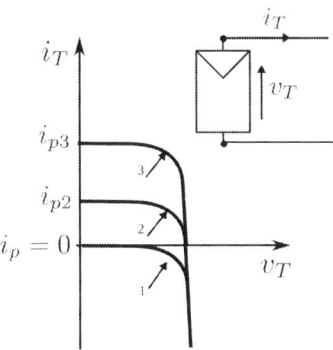

Figure 1.2.3: i-v characteristic variation with solar intensity

This will make the entire i-v characteristic to shift up as shown in figure 1.2.3. It can be noted that $i_{p3} > i_{p2} > 0$. The PV cell or the solar cell is symbolically represented as shown in figures 1.2.2(c) and 1.2.3. It looks similar to an envelope. There are two terminals, the anode terminal and the cathode terminal. v_T is the voltage across the anode-cathode terminals and i_T is the current flowing from the anode terminal or positive terminal to the external circuits.

1.3 Model

The photocurrent, i_p is the current at zero terminal voltage across the PV source. This can be modeled as a current source. A voltage source representing the p-n junction potential can be connected across the current source. Instead an ideal junction diode can replace the shunt voltage source in order to model the junction potential that appears across the photo current source. A non-ideal current source will have a high impedance, R_{sh}, across it. This will make the constant current line in the i-v characteristic to have a small negative slope to represent the finite shunt resistance, R_{sh} across the photo current source. In order to represent the non-idealness of the constant voltage part of the i-v characteristic, a small series source resistance, R_s is included. This would form the first level model of the PV source which is shown in figure 1.3.1(a). If a junction capacitance is considered across the p-n junction diode, then one obtains the first order dynamic equivalent circuit model of the PV source as shown in figure 1.3.1(b).

Referring to figure 1.3.1(a), the model is as follows:

$$i_p = i_d + i_{sh} + i_T \quad (1.3.1)$$

The current through the diode, i_d is the p-n junction current which is given as,

$$i_d = I_o \left(e^{\frac{v_T + i_T R_s}{n V_T}} - 1 \right) \quad (1.3.2)$$

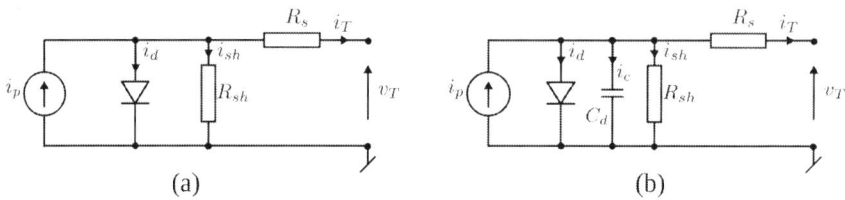

Figure 1.3.1: (a) First level equivalent circuit model of PV source (b) Dynamic equivalent circuit model of PV source

where V_T is the voltage equivalent of temperature and is given by $V_T = \frac{kT}{q} = \frac{T}{11600}$; k is the Boltzmann constant, q is the electronic charge in eV. The diode reverse saturation current is given as $I_o = KT^m e^{-\frac{V_{GO}}{nV_T}}$ wherein K is a constant, $m = 1.5$ and $n = 2$ for silicon material. The voltage equivalent of the forbidden band gap energy, V_{GO} is a value between 1.16V to 1.21V depending on the grade of purification for silicon. From equations 1.3.1 and 1.3.2, the terminal current of the PV source can be written as

$$i_T = i_p - I_o \left(e^{\frac{v_T + i_T R_s}{\gamma}} - 1 \right) - \left(\frac{v_T + i_T R_s}{R_{sh}} \right) \quad (1.3.3)$$

where $\gamma = n \cdot V_T = \frac{nkT}{q}$. The model of equation 1.3.3 is a steady state relationship which is an algebraic equation. This is sufficient for studying the static characteristics of the source and also for steady state system design. However, it cannot be used for simulation and dynamic analysis. The figure 1.3.1(b) introduces a dynamic element in the form of the junction capacitance. This will lead to a more realistic model which is amenable for time domain analysis. The dynamic model is given in equation 1.3.4. Observe that it is a non-linear equation. There is an extra current term representing the capacitor current, i_c in the photo-current equation. The voltage across the capacitance, v_c is the state variable for the dynamic model. It is the sum of the terminal voltage and drop across the series resistance, R_s.

$$i_p = i_d + i_c + i_{sh} + i_T$$
$$v_c = v_T + i_T R_s$$
$$C_d \frac{dv_c}{dt} = i_p - I_o \left(e^{\frac{v_c}{\gamma}} - 1 \right) - \left(\frac{v_c}{R_{sh}} \right) - \left(\frac{v_c - v_T}{R_s} \right) \quad (1.3.4)$$

1.4 I-V Characteristics

The I-V characteristics is an useful tool for analysis and design of PV based systems. The I-V characteristics provides a visual representation of the terminal voltage and current of the PV source. The model of the PV source as given in equations 1.3.3 and 1.3.4 together with the graphical I-V characteristic will help in better understanding

1.4. I-V CHARACTERISTICS

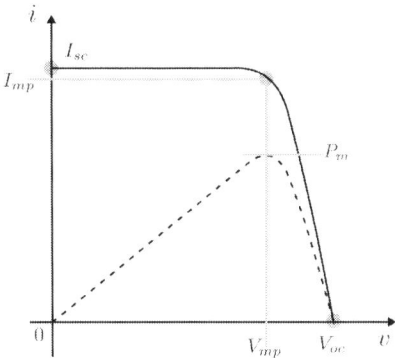

Figure 1.4.1: I-V characteristics of PV source

the PV source. Figure 1.4.1 shows the typical I-V characteristics of the PV source. The x-axis is represented by the terminal voltage, v_T and the y-axis is represented by the terminal current, i_T.

From the I-V characteristics shown in figure 1.4.1, one can observe that there are three significant operating points on the curve, viz. (i) the short circuit current operating point, $(0, I_{sc})$ on the y-axis, obtained when the terminal voltage is zero by short-circuiting the terminals, (ii) the open circuit voltage operating point, $(V_{oc}, 0)$ on the x-axis, obtained when the terminal current is zero by open-circuiting the terminals, and (iii) the peak power operating point, (V_{mp}, I_{mp}) on the I-V curve where the power obtained from the PV source is maximum.

1.4.1 Short circuit current point

When the terminals of the PV source are short-circuited, it means that the terminal voltage, $v_T = 0$. At this time, let the current flowing through the terminals of the PV source, $i_T = I_{sc}$. Applying these constraints to the PV source model given in equation 1.3.3, one obtains,

$$I_{sc} = i_p - I_o \left(e^{\frac{I_{sc}R_s}{nV_T}} - 1 \right) - \left(\frac{I_{sc}R_s}{R_{sh}} \right) \tag{1.4.1}$$

A better visualisation can be obtained when one applies the constraints of ideality to equation 1.4.1. For an ideal PV source, $R_s = 0$ and $R_{sh} \to \infty$. Applying these constraints in equation 1.4.1, one sees that the short circuit current turns out to be equal to the photo current, i_p.

$$I_{sc} = i_p \tag{1.4.2}$$

Equation 1.4.2 implies that the short circuit current is equal to the photo current which in turn is proportional to the incident solar power. The short circuit current, I_{sc} is a measurable electrical parameter which provides a direct measure of the incident solar radiation.

1.4.2 Open circuit voltage point

When the terminals of the PV source are open-circuited, it implies that the terminal current, $i_T = 0$. At this time, let the terminal voltage, $v_T = V_{oc}$. Applying these constraints to the PV source model given in equation 1.3.3, one obtains,

$$0 = i_p - I_o \left(e^{\frac{V_{oc}}{nV_T}} - 1 \right) - \left(\frac{V_{oc}}{R_{sh}} \right) \quad (1.4.3)$$

Applying the ideality constraint for an ideal PV source, $R_s = 0$ and $R_{sh} \to \infty$. Equation 1.4.3 can be re-arranged to give,

$$V_{oc} = nV_T \cdot \ln\left(1 + \frac{i_p}{I_o}\right) \quad (1.4.4)$$

If one considers the saturation current, $I_o \ll i_p$, then the open circuit voltage V_{oc} is proportional to the natural logarithm of the photocurrent i_p. If there is a change in the incident solar power on the PV source, the open circuit voltage varies logarithmically with the incident solar power, whereas the short circuit current would vary proportionately with the incident solar power.

1.4.3 Maximum power point

A third important characteristic operating point on the I-V characteristic of the PV source is the maximum power point. In figure 1.4.1, apart from the I-V curve, there is also a (power versus voltage) P-V curve which is superimposed on the characteristic. Observe that when the terminal voltage, $v_T = 0$ the power is zero. When the terminal voltage, $v_T = V_{oc}$, then the terminal current, $i_T = 0$. At this point also, the power is zero as indicated by the P-V curve. There will be a point $0 < v_T < V_{oc}$ on the x-axis where the power is maximum. The terminal voltage at this point is $v_T = V_{mp}$ and the terminal current at this point is $i_T = I_{mp}$. The maximum power that the PV source can provide to the load is $P_m = V_{mp} \cdot I_{mp}$ and this can happen only if the operating point on the I-V curve corresponds to the maximum power point. One of the desirables in PV system design is to always ensure that the PV source is operating at the maximum power operating point of its I-V characteristic.

1.5 Efficiency

The efficiency of the PV source is considered at the operating point corresponding to maximum power point as indicated in figure 1.4.1. The output power, P_o is

$$P_o = V_{mp} \cdot I_{mp} \quad (1.5.1)$$

The input power is actually the incident solar power. The input power depends on the solar insolation, L, which is the amount of solar power per unit area incident normally on the solar panels. It also depends on the area of exposure, A_e of the solar cells to the solar radiation. The input power, P_{in} is therefore given as,

$$P_{in} = L \cdot A_e \tag{1.5.2}$$

The efficiency of the solar cell, η_{cell} is defined as the ratio of output power to input incident solar power. It is given as,

$$\eta_{cell} = \frac{P_o}{P_{in}} = \frac{V_{mp} \cdot I_{mp}}{L \cdot A_e} \tag{1.5.3}$$

The insolation, L will vary with geographical location, time of day, time of year and atmospheric conditions at a place. In a later chapter, we will discuss in detail about insolation and its estimation at a place. The standard insolation is however set as $1kW/m^2$. This standard value is used as a reference or bench mark value for all comparisons. The value of efficiency will also vary with the area of exposure. If the area of exposure is the actual area of the cell, then the efficiency obtained is the cell efficiency. If the area of exposure is the area of a PV module, then the efficiency obtained is the module efficiency. The module efficiency will in general be slightly lower than the cell efficiency as actual cell area with in a module is lesser than the overall module area.

1.6 Temperature effects

Temperature changes have a significant effect on the parameters of the PV source. It affects the short circuit current, open circuit voltage and the maximum power point values. From equation 1.4.2, it can be seen that the short circuit current is equal to the photo current. The photo current is directly dependent on the solar radiation energy that is incident on the PV cell. As the temperature increases, more electrons reach the conduction band resulting in increase of the photo current. The short circuit current therefore will increase with increase in temperature. However, this increase is small, around 0.1% per degree K for silicon PV cells.

The effect of temperature on the open circuit voltage works differently. As the temperature increases, the open circuit voltage value decreases. It shows a negative temperature coefficient. The reverse saturation current is given as,

$$I_o = KT^m e^{-\frac{V_{GO}}{nV_T}} \tag{1.6.1}$$

Taking the natural logarithm on equation (1.6.1), one gets

$$ln\{I_o\} = ln\{KT^m\} + ln\left\{e^{-\frac{V_{GO}}{nV_T}}\right\}$$

$$= m \cdot ln\{KT\} - \frac{V_{GO}}{nV_T}$$

$$ln\{I_o\} = m \cdot ln\{K\} + m \cdot ln\{T\} - \frac{V_{GO}}{n\frac{T}{11600}} \tag{1.6.2}$$

Differentiating equation 1.6.2 with respect to T, one obtains

$$\frac{d}{dT}\ln\{I_o\} = \frac{m}{T} + \frac{V_{GO}}{nTV_T} \tag{1.6.3}$$

From equation 1.4.4, and taking $I_o \ll i_p$

$$\frac{V_{oc}}{nV_T} = \ln\left\{\frac{i_p}{I_o}\right\} = \ln(i_p) - \ln(I_o)$$

$$\frac{d}{dT}\left\{\frac{V_{oc}}{nV_T}\right\} = \frac{d}{dT}\ln(i_p) - \frac{d}{dT}\ln(I_o) \tag{1.6.4}$$

Using equation 1.6.3 in 1.6.4, one obtains

$$-\frac{V_{oc}}{nV_T} + \frac{1}{nV_T}\frac{d}{dT}\{V_{oc}\} = -\left(\frac{m}{T} + \frac{V_{GO}}{nTV_T}\right)$$

$$\frac{d}{dT}\{V_{oc}\} = \frac{V_{oc}}{T} - \left(\frac{mnV_T}{T} + \frac{V_{GO}}{T}\right)$$

$$= \frac{V_{oc} - V_{GO} - mnV_T}{T} \tag{1.6.5}$$

for silicon the following typical values can be taken: $m = 1.5$, $n = 2$, $V_{GO} = 1.16$, $V_{oc} = 0.6$. Substituting these values in equation 1.6.5, one obtains at $T = 300°K$

$$\frac{dV_{oc}}{dT} = -2.12 mV/°K$$

This implies that for every degree Kelvin increase in temperature, the open circuit voltage will reduce by -2.12 mV. As open circuit voltage decreases with increase in temperature, one can say that V_{oc} has negative temperature coefficient. On the other hand, the short circuit current, I_{sc} has positive temperature coefficient. This also implies that power delivered by the PV source will have a negative temperature coefficient.

1.7 Temperature coefficients

Consider the i-v characteristic of a PV cell shown in figure 1.7.1. Let the i-v curve (dark), be the characteristics at standard insolation of $1kW/m^2$ and standard temperature of $25°C$. For a temperature increase from $25°C$ to $40°C$, the i-v characteristic changes. The i-v curve (dashed) shows the characteristics with increased temperature.

The parameters of the PV source can be expressed with respect to the parameter at standard insolation and temperature as,

$$I_{sc}|_{40°C} = I_{sc}|_{25°C} + \triangle i$$
$$V_{oc}|_{40°C} = V_{oc}|_{25°C} + \triangle v$$
$$P_m|_{40°C} = P_m|_{25°C} + \triangle p \tag{1.7.1}$$

1.8. FILL FACTOR

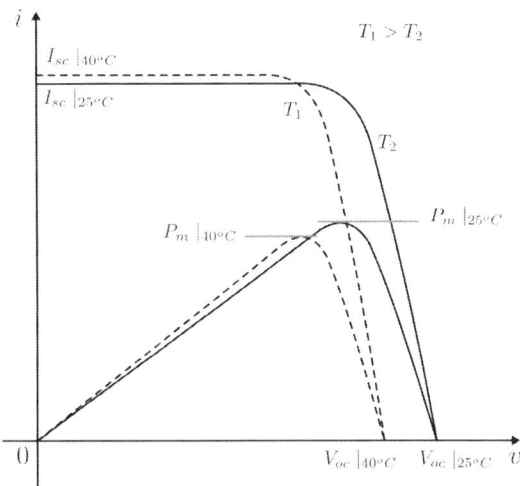

Figure 1.7.1: Effect of temperature on PV cell characteristics

Let the temperature coefficient of a parameter, α_{par} be defined as follows:

$$\alpha_{par} = \frac{\left(\frac{\triangle par}{Par|_{25°C}} \times 100\right)}{\triangle T} \%/K \quad (1.7.2)$$

where the parameter, Par can be I_{sc}, V_{oc} or P_m. Using equation 1.7.2 in equation 1.7.1 one obtains,

$$I_{sc}|_{40°C} = I_{sc}|_{25°C} + \left(\frac{\alpha_I \cdot I_{sc}|_{25°C} \cdot \triangle T}{100}\right)$$
$$V_{oc}|_{40°C} = V_{oc}|_{25°C} + \left(\frac{\alpha_V \cdot V_{oc}|_{25°C} \cdot \triangle T}{100}\right)$$
$$P_m|_{40°C} = P_m|_{25°C} + \left(\frac{\alpha_P \cdot P_m|_{25°C} \cdot \triangle T}{100}\right) \quad (1.7.3)$$

where α_I is the temperature coefficient of I_{sc}, α_V is the temperature coefficient of V_{oc} and α_P is the temperature coefficient of P_m. Using equation 1.7.3, one can estimate the value of a parameter at any given temperature with reference to the value at standard temperature.

1.8 Fill factor

Fill factor is an important benchmark factor that enables comparison of various PV sources. It is in some sense a figure of merit for the PV cell. It tells how good or bad a PV cell is. It is a unit less ratio. It is a ratio of two areas of rectangles. The reference rectangle is the one bounded by I_{sc} on one side and V_{oc} on the other. The area of this rectangle is $V_{oc} \cdot I_{sc}$. This is the largest rectangle that can be obtained from a given PV cell characteristics. For this PV cell, the maximum power point is given at V_{mp} and I_{mp}. The rectangle bounded by I_{mp} on one side and V_{mp} on the other is the maximum power point (MPP) rectangle. The area of this rectangle is $V_{mp} \cdot I_{mp}$. The MPP rectangle area is always less than the reference rectangle area. Closer the area of MPP rectangle is to the area of the reference rectangle, better is the PV cell. The ratio of the MPP rectangle area to the area of the reference rectangle is defined as the fill factor and gives a measure of the goodness of the PV cell. The fill factor, FF is given by

$$FF = \frac{V_{mp} I_{mp}}{V_{oc} I_{sc}} \qquad (1.8.1)$$

Most commercial PV cells have a fill factor around 0.7 to 0.8.

1.9 Datasheet

The datasheet of a PV source is an important document that should be studied in order to understand the various parameters of the PV source. Appendix-A gives the data of a PV module. The PV module is of the mono-crystalline type. All parameters are measured under standard insolation of $1kW/m^2$ and standard temperature of $25°C$. The maximum power obtainable from this module at standard insolation is 308.5W. The y-intercept of the i-v curve, ie. the short circuit current I_{sc} is 9.22A. The x-intercept of the i-v curve, ie. the open circuit voltage V_{oc} is 44.9V. It can also be seen from the datasheet that the current at peak power and the voltage at peak power are 8.51A and 36.2V respectively.

1.10 Example

For another sample PV source, the datasheet is given in Appendix-B. The temperature coefficient of I_{sc} is $0.045\%/°K$. The temperature coefficient of V_{oc} is $-0.34\%/°K$. The temperature coefficient of P_m is $-0.47\%/°K$.

Consider the i-v characteristic shown in figure 1.7.1. Let the i-v characteristic (dark) represent the characteristic at temperature $25°C$. What happens to the i-v characteristics at $40°C$? From the data sheet given in Appendix-B, the open circuit voltage and the short circuit current at standard temperature and insolation are 36.72V and 8.99A for the module. The peak power for the module is 240W at standard temperature and insolation.

In order to find V_{oc}, I_{sc} and P_m at $40°C$ one needs to estimate using equation 1.7.3. From the datasheet $\alpha_I = 0.045\%/°K$, $\alpha_V = -0.34\%/°K$ and $\alpha_P = -0.47\%/°K$. $\triangle T = 15°K$. Substituting these values in equation 1.7.3, the parameters at $40°C$ are obtained.

$$I_{sc}\big|_{40°C} = 9.05A$$
$$V_{oc}\big|_{40°C} = 34.847V$$
$$P_m\big|_{40°C} = 223.08W$$

Observe that the short circuit current parameter increases with temperature. However, both open circuit voltage and peak power points decrease with temperature due to the negative temperature coefficients for these parameters.

1.11 Simulation

Time domain simulation of circuits will provide lot of insights into the operation of the devices and circuits. SPICE is an excellent tool for simulating circuits. There are several open-source and proprietary versions of SPICE available that can be used for this purpose. Among the proprietary versions, PSPICE is a very popular SPICE based tool that includes models for several components. LTSPICE is another proprietary SPICE based circuit simulation tool that is a freeware and quite popular. NgSPICE is an excellent open-source simulation environment that includes the original Berkeley SPICE, XSPICE and CIDER. This makes the NgSPICE environment very powerful and even permits analog behavioural modelling of components. This books will use the open-source NgSPICE as the primary tool for performing circuit simulation. For circuit schematic entry, one can use any plain text editor wherein the SPICE circuit syntax can be typed out or a GUI based schematic entry software like KiCAD can be used to output the circuit netlist that is compatible with NgSPICE. Appendix-C gives information on NgSPICE related setup and *getting started*.

Figure 1.3.1(a) and (b) give the equivalent circuit of the PV source. This circuit can be used as a PV cell sub-circuit. The code snippet for PV cell sub-circuit is given below

Algorithm 1.1 PV cell sub-circuit
```
.SUBCKT PVSOURCE 1 2 Isc=1
VIsc nsc 0 {Isc}
VIo nIo 0 1e-7 0
Vnvt nvt 0 0.05
Bpv 1 2 ipv=-(v(nsc)- v(nIo)*(exp(v(1,2)/v(nvt))-1)>0 ?   (v(nsc)-
v(nIo)*(exp(v(1,2)/v(nvt))-1)) :    0)
.ENDS
```

This PV cell sub-circuit is for a single cell wherein the open circuit voltage swings from 0 to around 0.7V. The diode reverse saturation current is taken as 100nA and $nV_T \approx 0.05$ at $300°K$. The *Bsource* of NgSPICE is used to define the terminal current of the PV source. NgSPICE supports the powerful ternary function as an expression for defining the *Bsource* parameter. A ternary function is defined as a ? b: c. This means that IF a THEN b ELSE c where a, b and c are expressions. Only positive current out of the PV cell terminal is permitted. Any current that sinks into the PV can damage the PV cell. Therefore, any negative current that may flow into the PV cell is blocked and is considered zero. It is equivalent to placing an ideal diode in series with the PV cell. This is achieved by using the ternary function of NgSPICE. The above model of the PV cell may be used to obtain the i-v characteristics. Consider an external load R_o connected across the terminals of the PV cell. If R_o is zero, then it represents the short circuit operating point. If R_o is not connected, then it represents the open circuit operating point. R_o can be swept between these two limits to obtain several operating points and to plot the i-v characteristics. This can be automated by making R_o as a variable load or one can provide an x-axis sweep voltage by replacing R_o with a source incorporating a voltage source having time evolution like a saw-tooth. This would provide the sweep for the x-axis of the i-v characteristic. It can be noted that the wave shape of sweep voltage source need not be sawtooth. It could be triangle or it could also be sinusoidal as long as it is a monotonic function of time. One can get insight into the PV cell equivalent circuit by simulating it and obtaining the i-v characteristics and p-v characteristics at various ambient temperatures.

1.12 Questions

1. Photovoltaic effect was discovered by
    a) Albert Einstien
    b) Adams and Day
    c) Edmond Becquerel
    d) Fritts C E

2. As the temperature increases, the efficiency of PV cell will
    a) improve
    b) decrease
    c) increase
    d) remain unchanged

3. As the incident insolation on a PV cell increases,
    a) both short circuit current and open circuit voltage increase linearly
    b) open circuit voltage increases logarithmically
    c) short circuit current increases logarithmically
    d) both short circuit current and open circuit voltage increase logarithmically
    e) open circuit voltage increases linearly

4. As the temperature increases
    a) the peak power of PV cell decreases
    b) the peak power of PV cell increases
    c) the short circuit current value decreases
    d) the open circuit voltage value increases

5. Which among the following conditions lead to a PV cell having poorest fill factor? (R_s is the series resistance and R_{sh} is the shunt resistance in the PV

1.12. QUESTIONS

model)
a) R_s and R_{sh} are low
b) R_s is low and R_{sh} is high
c) R_s and R_{sh} are high
d) R_s is high and R_{sh} is low

6. As the temperature increases
   a) the open circuit voltage value increases
   b) the peak power of PV cell increases
   c) the short circuit current value decreases
   d) the short current value increases

7. Which of the following statements are true about the PV module?
   a) change in short circuit current of PV cell varies linearly with temperature change
   b) module efficiency is lower than cell efficiency
   c) change in open circuit voltage of PV cell varies logarithmically with temperature change
   d) module efficiency is higher than cell efficiency
   e) temperature coefficient of open circuit voltage is positive
   f) temperature coefficient of short circuit current is negative

8. Consider a 12V battery with a 10 ohm series resistance. The battery-resistance combination is required to emulate a PV cell. What is the fill factor?

9. A PV panel has an open circuit voltage of 40V and a short circuit current of 8A at 25 deg C. The temperature has become 40 deg C. What is the open circuit voltage at 40 deg.C?

10. A PV panel having an area of $1.5m^2$, gives the following readings under standard test conditions. The short circuit current is 8A, the open circuit voltage is 40V, the voltage at peak power is 36.5V and the current at peak power is 7A. The fill factor of the PV panel is found to be 0.72. What is the efficiency of the PV panel?

11. Consider a 12V battery with a 10 ohm series resistance. This battery-resistance combination emulates a PV cell. What is the peak power that can be delivered by this PV emulator.

12. A PV module datasheet provides the following parameter values at standard insolation and temperature,

Parameters	Value
Module efficiency	16.7%
Maximum power	275 W
Voltage at maximum power	31.3 V
Current at maximum power	8.79 A
Open circuit voltage	38.6 V
Short circuit current	9.48 A
Temperature coefficient of V_{oc}	$-1.37 \times 10^{-1} V/^{\circ}C$
Temperature coefficient of I_{sc}	$5.56 \times 10^{-3} A/^{\circ}C$

What is the fill factor of the solar PV module at $25^{\circ}C$.

13. For a PV module which has a datasheet similar to that given in the previous question, calculate the open circuit voltage and short circuit current at a temperature of $45^{\circ}C$.

14. A PV module has a length of 1.662m and width of 0.99m. It has an efficiency of 15% and the maximum output power is 200 W. What is the insolation that is incident normal to the PV module?
Ans: eff=Po/(L.A)

15. A PV source has a junction reverse saturation current of 1nA. The voltage equivalent of temperature is $\frac{T}{11600}$, where T is the temperature in Kelvin. The photocurrent is 8.6A and there are 60 cells connected in series in one module. What is the open circuit voltage of this module?

Chapter 2

Cell and Module Interconnections

The PV cell or solar cell has a single p-n junction. The voltage that is developed across the solar cell is around 0.6V to 0.7V. In order to achieve higher voltages, PV cells must be interconnected in series. In order to achieve higher currents, the PV cells must be interconnected in parallel. Such series and parallel interconnection arrangement is necessary to achieve required voltage and current ratings. The cells that are interconnected in series and parallel arrangement may not be identical in nature. Some of the cells may have partial shading and the i-v characteristics of these shaded cells will be significantly different from that of the unshaded ones. This chapter discusses issues related to series and parallel interconnection of PV cells and the manner in which they can be be resolved.

2.1 Identical cells in series

Consider two identical cells and connect them in series. The series connection and the i-v characteristic of the series combination are shown in figure 2.1.1(a) and (b).

PV cell-1 and PV cell-2 are connected in series. Across the terminals of the series combination, a load is connected as shown. Let v_T and i_T be the terminal voltage and the terminal current of the series combination system. Likewise, v_{T1} and v_{T2} are the terminal voltages of the two cells respectively. i_{T1} and i_{T2} are respectively the terminal currents of the two cells that make up the series system. When two cells are connected in series, the following constraints apply,

$$i_T = i_{T1} = i_{T2}$$
$$v_T = v_{T1} + v_{T2} \tag{2.1.1}$$

Figure 2.1.1(b) shows the i-v characteristic of the combination with v_T as the x-axis variable and i_T as the y-axis variable. The i-v characteristic shown in dotted lines

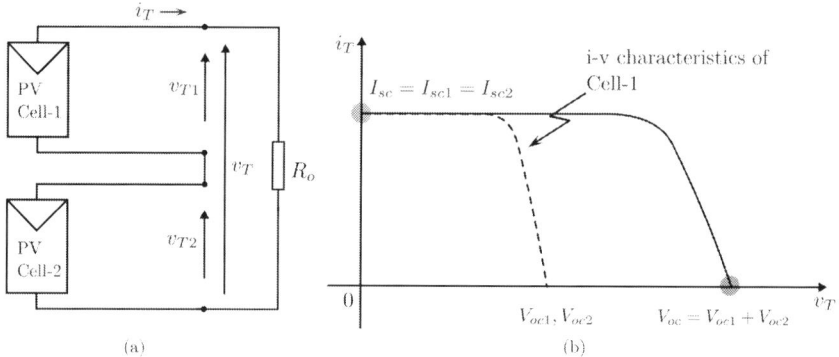

Figure 2.1.1: Series connection (a)Two identical cells in series (b) i-v characteristics of the series combination

is that of PV cell-1 or cell-2 as both are identical. One characteristic point is on the x-axis where the open circuit voltage is the sum of the open circuit voltages of the two cells. By constraints given in equation 2.1.1, $V_{oc} = V_{oc1} + V_{oc2} = 2V_{oc1}$. Another characteristic point in on the y-axis wherein $I_{sc} = I_{sc1} = I_{sc2}$. Using the constraints of equation 2.1.1, the points of the series combination can be visualised as shown in figure 2.1.1(b).

2.2 Load line

The concept of load line is central to analysing and designing systems where the PV source is interfaced to passive and active loads. Consider the PV cell connected to an external load resistance R_o that can vary from 0 to ∞. $R_o = 0$ implies that the PV source is short circuited and $R_o = \infty$ implies that the PV source is open circuited. In the i-v characteristic of the PV source, the x-axis represents the terminal voltage and the y-axis represents the current flowing out of the terminals and into the load impedance. Let an arbitrary straight line passing through the origin be drawn on this i-v coordinate system. The slope of this line gives the ratio of the change in current to the change in voltage or in other words $\frac{1}{R_o}$. Such a line is called the load line. $\frac{\Delta i}{\Delta v} = \frac{1}{R_o}$ is the slope of the load line. If $R_o = \infty$ then the slope of the load line, $\frac{\Delta i}{\Delta v} = 0$ that is the x-axis or the voltage axis. If $R_o = 0$ then the slope of the load line, $\frac{\Delta i}{\Delta v} = \infty$ that is the y-axis or the current axis. Any other finite value of load resistance will result in a load line having slope in between these two extremes. Figure 2.2.1 shows the i-v characteristics of a PV source with the load line superimposed on it for a load resistance of R_o connected across the terminals of the PV source.

Referring to figure 2.2.1, it can be seen that the point of intersection of the load line with the i-v characteristic of the PV source, is the operating point. This point indicates the terminal voltage and the current of the PV source for the given load resistance. When $R_o = 0$, the load line is along y-axis and the point of intersection with the i-v curve gives the operating point as $(0, I_{sc})$. When $R_o = \infty$, the load line is along x-axis

2.3. NON-IDENTICAL CELLS IN SERIES

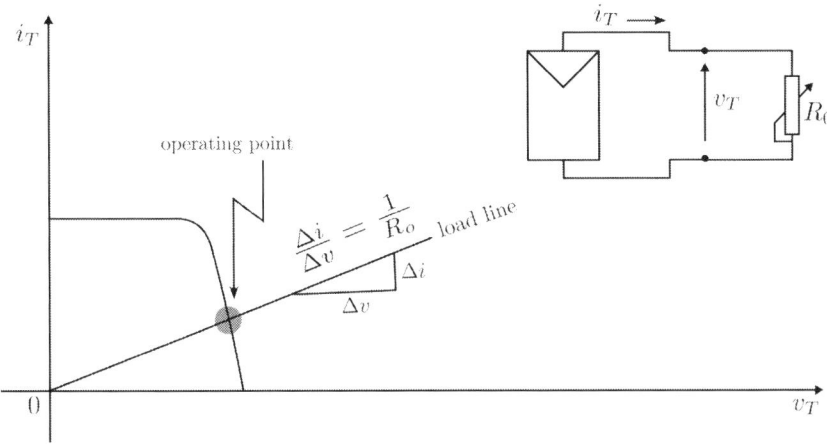

Figure 2.2.1: Load line and operating point on i-v characteristics

and the point of intersection with the i-v curve gives the operating point as $(V_{oc}, 0)$. If the load line intersects the i-v curve at a point where maximum power is being supplied by the PV source, then the operating point is (V_{mp}, I_{mp}) that corresponds to maximum power point (MPP).

2.3 Non-identical cells in series

In a real situation, the PV cells or the solar cells are not identical. Even cells from the same batch of manufacture may have differing i-v characteristics. In applications where the PV modules are used for powering up the buildings and for grid integration, they are mounted on the roof tops of the buildings. Depending on the location of the place, shadows on the PV modules due to trees and buildings in the neighbourhood will have a significant effect on the i-v characteristics of PV cells. Partial shading is a serious and real challenge that needs to be addressed. It is right to conclude that the PV cells having non-identical i-v characteristics is more a norm than otherwise. The objective of this section is to study the i-v characteristics of PV cells in series that have individual i-v characteristics that are not same.

Figure 2.3.1 shows the circuit of two non-identical PV cells connected in series. Alongside in the figure is also shown the i-v characteristics of the individual cells and the series combination. Cell-1 and Cell-2 characteristics are non-identical. Cell-1 has a short circuit current value of I_{sc1} and an open circuit voltage value of V_{oc1}. Likewise, it can be seen that cell-2 has a short circuit current value of I_{sc2} and an open circuit voltage value of V_{oc2}. One can use the same constraints given in equation 2.1.1 to plot the i-v curve of the series combination as given in figure 2.3.1.

The constraints of equation 2.1.1 is applied to obtain the point on the load line that intersects with the i-v curve. The load line is made to swing from 0 slope aligned along x-axis to infinite slope aligned along y-axis. When the load line is along the x-axis,

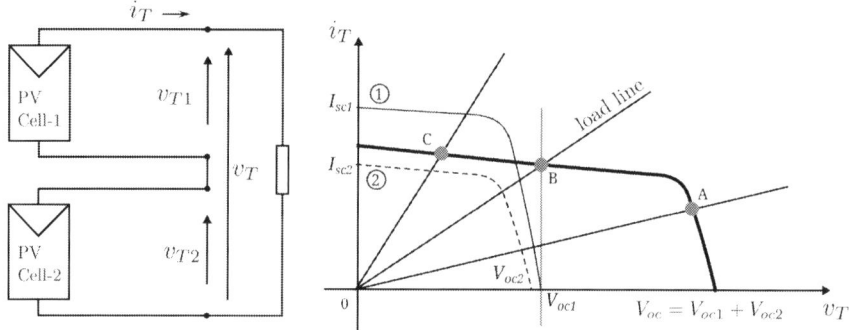

Figure 2.3.1: Non-identical cells in series and its characteristics without protection diodes

and applying equation 2.1.1, the open circuit voltage point for the series combination is obtained. Point A is another point on the characteristics for a load line where the terminal current is less than either of the short circuit currents of the two PV cells. Point B is an interesting operating point where in the load line is such that the terminal current is just equal to the short circuit current of cell with smaller i-v curve. In this case, the cell with the smaller i-v curve is cell-2. Therefore, when $i_T = I_{sc2}$, then $v_{T2} = 0$ and $v_T = V_{oc1}$. If the load line moves up in slope any further like point C, then $v_T < V_{oc1}$ which implies that $v_{T2} < 0$. As the current through both the cells and the load continues to be in the same direction, the power of cell-2 is negative which means that cell-2 is now acting as a sink and not as a source. This will cause cell-2 to run hot and even damage the cell. Therefore, it is essential to ensure that no cell is made to act as a sink and dissipate power. This is a problem that is faced when non-identical cells are connected in series.

The solution to this problem is to protect the cells by placing diodes anti-parallel to the PV cells as shown in figure 2.3.2. Assume for the purpose of this discussion that the diodes are ideal and their forward drop is zero. The diodes ensure that the voltage across the individual cells do not go negative. As a result, each cell in the series combination will either act as source or be bypassed by the anti-parallel diode. None of the cells will go into the sink mode of operation. Consider the adjoining i-v characteristics shown in figure 2.3.2. The load line is made to move from open circuit operating point by increasing the load across the terminals of the series combination. Point A is an operating point wherein the terminal current or the load current is less than either of the short circuit currents of the cells. At point B, the terminal current or the load current is exactly equal to the short circuit current of cell-2. At this operating point and from equation 2.1.1, it is evident that $v_T = V_{oc1}$, as $v_{T2} = 0$. As the load line moves up in slope, the voltage across cell-2 tries to go negative. But the anti-parallel diode across it will turn-on and ensure that the voltage across the cell-2 is clamped to zero. As a result, cell-2 will not dissipate power and will run cool and stay protected. During the portion of the i-v curve from point B onwards till terminal short circuit, cell-2 voltage will be clamped to zero and the terminal voltage of the series combination

2.4. IDENTICAL CELLS IN PARALLEL

will essentially follow the voltage of cell-1 as shown in the i-v characteristics of figure 2.3.2.

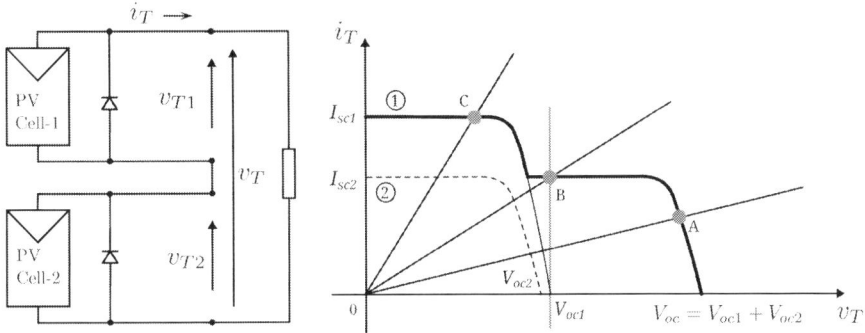

Figure 2.3.2: Non-identical cells in series with protection diode and its characteristics

In the above discussion, the anti-parallel diodes were considered ideal. However, in reality they are also practical p-n junctions like the PV cell. Further, it will not be cost effective to have one protection p-n junction (diode) for every PV cell p-n junction. Therefore, the protection diodes are generally placed anti-parallel across PV modules which contain several PV cells in series. This is a compromise solution that is generally employed.

2.4 Identical cells in parallel

Like cells in series, one can also connect cells in parallel in order to increase the current capability. Consider two PV cells connected in parallel as shown in figure 2.4.1. v_T is the terminal voltage of the parallel combination that is measured across the load. i_T is the terminal current of the parallel combination. i_{T1} and i_{T2} are the terminal currents of cell-1 and cell-2 respectively. Both the cells have identical i-v characteristics. The i-v characteristic of the paralel combination is obtained using the following constraints,

$$i_T = i_{T1} + i_{T2}$$
$$v_T = v_{T1} = v_{T2} \qquad (2.4.1)$$

Note that cells in parallel boost the current rating while cells in series boost the voltage rating of the system. If the load is swept from open circuit to short circuit, the terminal voltage v_T is the terminal voltage of either of the two identical cells in parallel and the terminal current i_T is the sum of the terminal currents of the two cells connected in parallel. Figure 2.4.1 also shows the i-v characteristic of the parallel connected system. Here, the short circuit operating point is at $I_{sc} = I_{sc1} + I_{sc2}$ and the open circuit operating point is at $V_{oc} = V_{oc1} = V_{oc2}$.

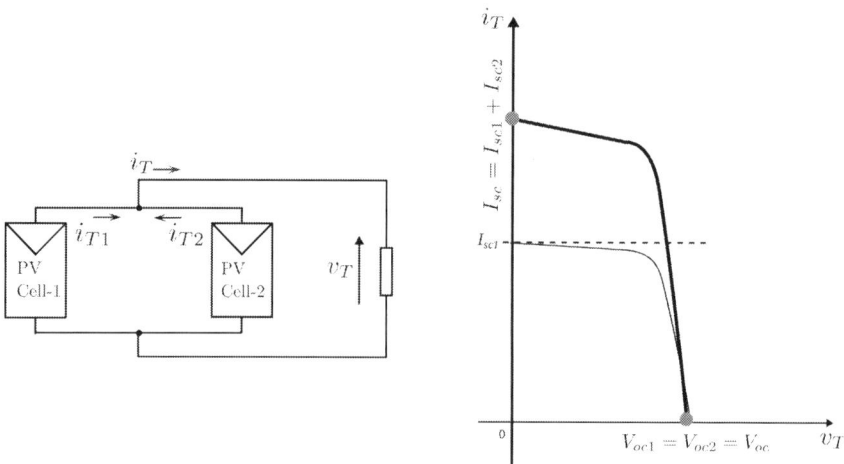

Figure 2.4.1: Two identical cells in parallel and its characteristics

2.5 Non-identical cells in parallel

The PV cells that are connected in parallel will rarely be identical in practice. One needs to understand the i-v characteristics to resolve the problems arising due to the non-identical character of cells in parallel. Consider two PV cells in parallel as shown in figure 2.5.1. It shows the circuit of two non-identical PV cells connected in parallel. Also shown in figure 2.5.1 is the i-v characteristics of the individual cells along with characteristic of the parallel combination. Cell-1 and Cell-2 characteristics are non-identical. Cell-1 has a short circuit current value of I_{sc1} and an open circuit voltage value of V_{oc1}. Cell-2 has a short circuit current value of I_{sc2} and an open circuit voltage value of V_{oc2}. The same constraints given in equation 2.4.1 are used to plot the i-v curve of the parallel combination.

The i-v curve of the combination can be traced by sweeping the operating point from the short circuit operating point to the open circuit operating point. Consider the load is a short circuit. The load line is vertical and will intersect at the short circuit operating point of the parallel combination that is $I_{sc1} + I_{sc2}$ at a terminal voltage of 0 in accordance with the equation 2.4.1. As the load resistance is increased, the operating point which is the intersection of the load line and the i-v curve will move away from the vertical. Point A is an arbitrary operating point wherein the load resistance has just increased from 0 to a finite value. At point B, the terminal voltage of cell-2 has reached open circuit value. At this operating point B, the terminal voltage of the parallel combination is same as the terminal voltage of cell-1. If the load resistance is increased further, the current of cell-2 will become negative. The effective terminal current of the combination would be $i_{T1} - i_{T2}$. At some load, i_{T1} and i_{T2} will be equal in magnitude and opposite in sign resulting in 0 terminal current. This will represent the open circuit operating point of the parallel combination.

As the load resistance is increased and becomes large, the cell will the smaller i-v

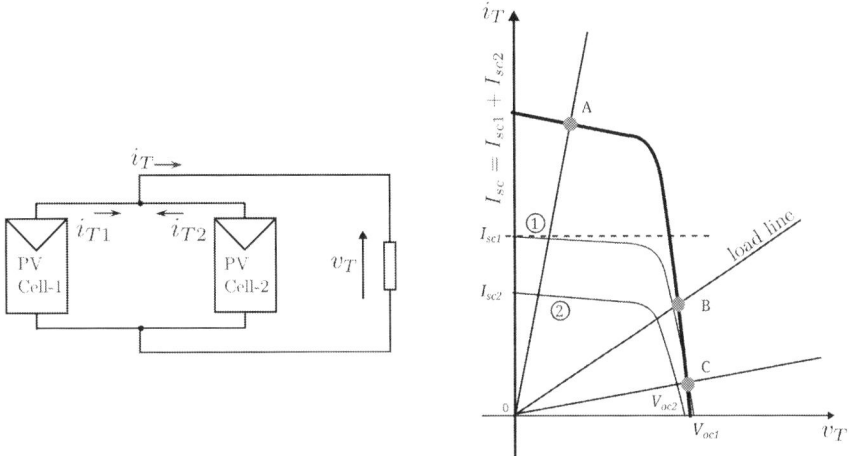

Figure 2.5.1: Two non-identical cells in parallel without protection diodes and its characteristics

characteristic will reach its open circuit point earlier. If the load resistance is increased beyond, the current in the cell will smaller i-v characteristics will become negative with its terminal voltage being still positive. This would mean that this cell would act as a sink rather than a source under such conditions. This would cause heating of the cell and possible damage. In order to resolve this problem, a diode is connected in series along each cell as shown in figure 2.5.2. This will prevent negative current flowing in the cells under any condition and thereby prevent power dissipation in the cells. When the protection diodes are included in the circuit, the i-v characteristic of the combination gets modified. At point B, the cell-2 reaches open circuit voltage point. If the load is increased beyond, due to the presence of diode, there is no current flow in cell-2. Only cell-1 is in play and the terminal characteristics of the combination is same as that of cell-1.

2.6 Protecting and Interconnecting modules

One will need two diodes to protect a PV source. One diode in parallel across the terminals of the PV source that will provide protection for series connection and another diode in series with the positive terminal lead which will be used to provide protection for parallel connection. Though the discussion in the previous sections were for PV cells, it does not make practical sense to use two electronic grade pn junction diodes to protect a single pn junction of solar grade. The protection is generally made at the module level. A solar PV module will consist of several PV cells that are arranged in series and parallel. Cells in series will enhance the terminal voltage of the module and cells in parallel will enhance the terminal current of the module. A module will have one diode in parallel across it to provide series connection protection and another diode in series with the positive terminal lead to provide parallel connection

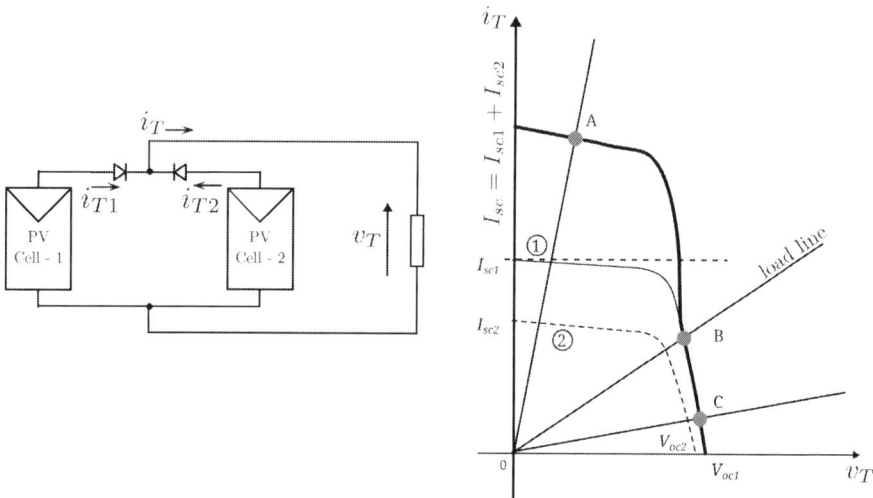

Figure 2.5.2: Two non-identical cells in parallel with protection diodes and its characteristics

protection. Several such modules are connected in series parallel topology to achieve the desired application voltage and power levels. A typical topology is as shown in figure 2.6.1. Here one can see that there are n PV modules in a series string and m such parallel strings. In all there are $n \times m$ modules.

In figure 2.6.1 one can see that each module has a diode in parallel and a diode in series with it. However, all the modules in a series string have the same current flowing through them and therefore just a single diode is sufficient to provide protection for paralleling with similar other series arm. Further, for n modules in series, the n diodes in series will offer n times diode drop and thereby result in lower terminal voltage. The topology of figure 2.6.2 removes the redundant series diodes in each series string and retains only one diode to provide protection for paralleling. This provides a topology that uses lesser number of diodes and also has better terminal voltage performance.

Referring to figure 2.6.2, the terminal current of the interconnected combination is $i_T = i_{T1} + i_{T2} + ... + i_{Tm}$. If one considers all the modules as identical, then $i_T = m \cdot i_{T1}$ or $i_{T1} = \frac{i_T}{m}$. The terminal current of each string is the same as the terminal current of each module in the string. If the series protection diode used for paralleling is considered ideal, then the terminal voltage of each module can be expressed as $v_{T1} = \frac{v_T}{n}$. From equation 1.3.3, the terminal current equation of each module of a string is given as

$$i_{T1} = \frac{I_{sc}}{m} - I_o \left(e^{\frac{v_{T1} + i_{T1} R_s}{\gamma}} - 1 \right) - \left(\frac{v_{T1} + i_{T1} R_s}{R_{sh}} \right)$$
$$\frac{i_T}{m} = \frac{I_{sc}}{m} - I_o \left(e^{\frac{\frac{v_T}{n} + \frac{i_T}{m} R_s}{\gamma}} - 1 \right) - \left(\frac{\frac{v_T}{n} + \frac{i_T}{m} R_s}{R_{sh}} \right) \qquad (2.6.1)$$

2.7. SIMULATION MODEL OF PV SOURCE

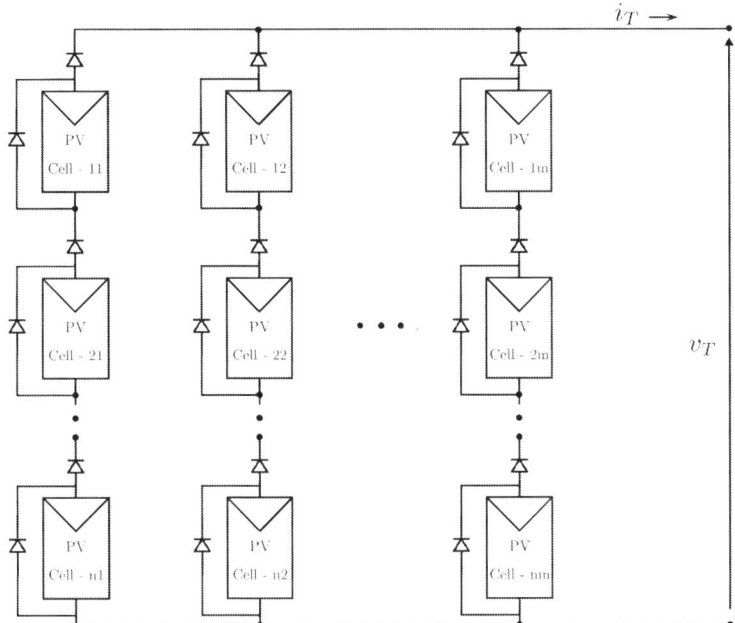

Figure 2.6.1: Interconnection of m parallel set of n modules in series

where n is the number of modules connected in series in each string and m is the number of series string sets in parallel. For the topology as shown in figure 2.6.1, the terminal voltage across each module is given as $v_{T1} = \frac{v_T + n \cdot V_D}{n}$, where V_D is the forward diode drop of each protection diode. In the case of the topology given in figure 2.6.2, there is only one diode drop per string and the terminal voltage across each module is $v_{T1} = \frac{v_T + V_D}{n}$. The junction capacitance based dynamic model can also be used to model the series parallel combination. The dynamic model of a PV cell with junction capacitance is given in equation 1.3.4. This can be applied for each module of a string as

$$v_c = v_{T1} + i_{T1} R_s$$
$$C_d \frac{dv_c}{dt} = \frac{I_{sc}}{m} - I_o \left(e^{\frac{v_c}{\gamma}} - 1 \right) - \left(\frac{v_c}{R_{sh}} \right) - \left(\frac{v_c - v_{T1}}{R_s} \right) \qquad (2.6.2)$$

From this model, the terminal current and terminal voltage of the series parallel combination can be estimated.

2.7 Simulation model of PV source

The SPICE model for a PV source is shown in figure 2.7.1. This is a general purpose SPICE model which can be applied for the $n \times m$ interconnection topology of figure

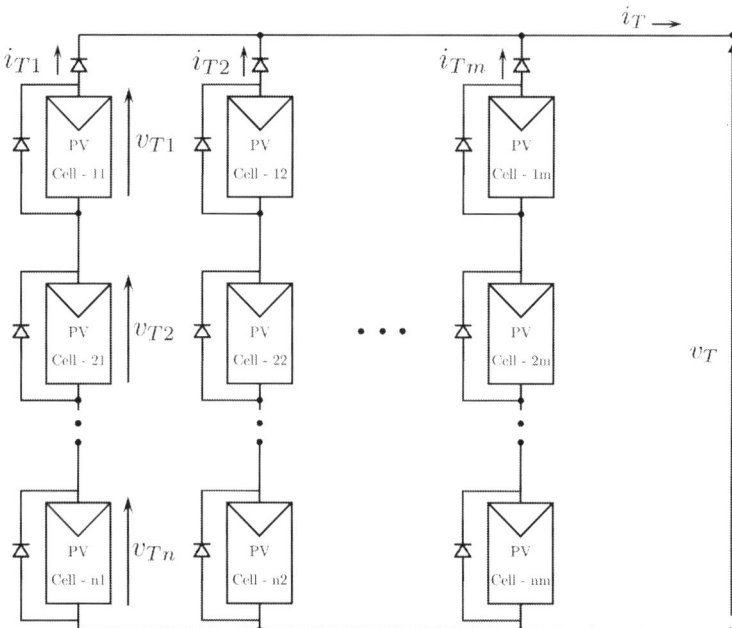

Figure 2.6.2: Interconnection of m parallel set of n modules in series with single parallel protection diode for each series string

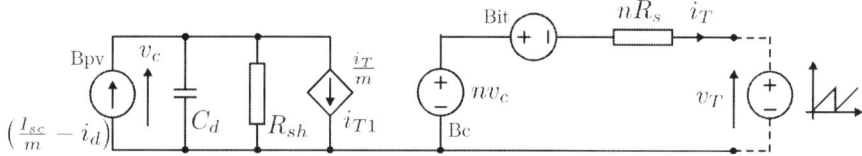

Figure 2.7.1: SPICE model of a PV source

2.6.2. The SPICE model SUBCKT is based on the model equations 2.6.1 and 2.6.2.

```
*----------- * PV SOURCE MODEL*--------------------------
.SUBCKT PV 1 2 Isc=1 nseries=1 nshunt=1
.param Rshunt=1e6 Rseries=0.01
*Using PV model eqn. in B-source
*  (nVT)=0.05 -- n=2 and VT=0.025 at T=290 deg K
*  (ip1-id-iT1) current is modelled as a B source
Bpv nc 2 i=-({Isc/nshunt}- 1e-7*(exp(v(nc,2)/0.05)-1)...
                                  + -i(Bit)/{nshunt})
Cd nc 2 100pF
Rsh nc 2 {Rshunt}
Bc nstr 2 v=(v(nc,2)*{nseries})
Bit nstr nx v=v(nstr,2)<v(1,2) ? v(nstr,2)-v(1,2):0
```

2.7. SIMULATION MODEL OF PV SOURCE

```
Rs nx 1 {Rseries*nseries}
.ENDS
*------------------------------------------------------
```

The ngSpice SUBCKT listing for the PV source is given above. The SUBCKT needs 3 input parameters which are the short circuit current value for the PV source at its terminals (I_{sc}), number of PV blocks in series for a string (*nseries*) and number of strings in parallel (*nshunt*). These input parameters provide details of the topology as per figure 2.6.2. Within the SUBCKT there are 2 parameters that determine the R_{sh} value and the R_s value of each identical PV block. As all PV blocks are considered identical, the SUBCKT simulates for one PV block and then appropriately scales the terminal voltage and current based on the *nseries* and *nshunt* values.

B_{pv} is a non-linear current source that represents $i_p - i_d - i_{T1}$. Across this non-linear current source is connected the equivalent diffusion capacitance, C_d and R_{sh}. The γ value is evaluated at $T = 290°K$. The temperature can also be programmed as a variable parameter. This is left as an exercise to the reader. The voltage that is dynamically built up across the capacitance v_c is scaled by *nseries* to give the voltage source B_c. The series resistance of all the units in a string add up to present ($nseries \cdot R_s$) as the equivalent series resistance at the terminal. The programmable voltage source, B_{it} has the interesting function of preventing negative terminal current. This is achieved by using a conditional statement in the SUBCKT. If the voltage generated at B_c source is less than the terminal voltage v_T, then the correction source B_{it} will produce a drop equal to the difference between the voltage at B_c and that at the terminal. At all other conditions, the voltage across B_{it} is zero. The SUBCKT is placed in a library file with .lib extension. This SUBCKT is called by an ngSpice netlist as given below

```
*--------PV i-v CHARACTERISTICS NETLIST-----------
.include models/luDevices.lib
X1 t 0 PV Isc=2 nseries=2 nshunt=2
V1 t 0 PWL(0 0 0.01 2 0.02 0)
.control
option reltol=0.01 abstol=0.01 vntol=0.01
tran 1us 20ms uic
set color0 = white ; set background as white
set color1 = black; set foreground as black
run plot i(v1) vs v(t)
.endc
.end
*------------------------------------------------------
```

The SUBCKT is placed in the library file luDevices.lib. The PV SUBCKT is called by including a PV source device X1 in the netlist. This is connected to an external voltage source V1 which sweeps the terminal voltage. The source V1 provides a linear sweep. The PV SUBCKT is called with short circuit current value of 2A, two PV cells in series per string and two strings in parallel. The ngSpice simulation commands are given within the *.control* and *.endc* lines. Transient analysis is performed upto 20ms time. The background is set as white and the foreground is black. On execution of the simulation, the i-v characteristic of the PV source is plotted.

2.8 Measurement setup for i-v characteristics

A PV source can be a pv-cell, pv-module, pv-panel or pv-array. PV modules contain cells that are arranged in series and parallel combination to achieve a desired module power, module voltage and module current. A PV panel consists of several modules arranged in either series or parallel to achieve higher voltage capability or higher current capability respectively. Several panels may be arranged in series and parallel combination to form a PV array. PV arrays need large real estate and are generally deployed to feed into the nearby local grid.

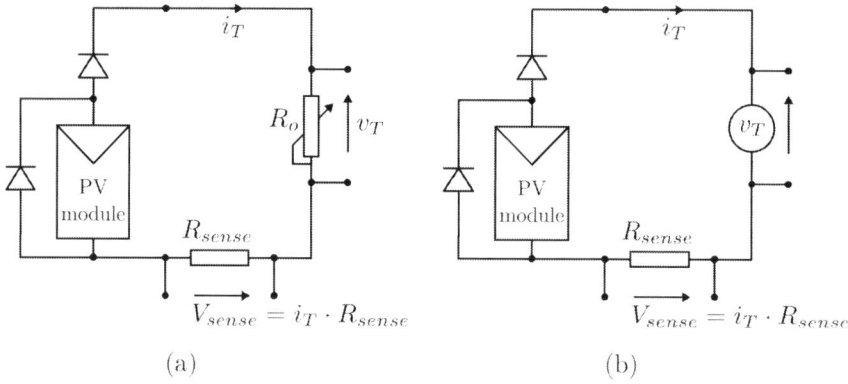

Figure 2.8.1: Measurement of i-v characteristics with (a) rheostat and (b) voltage source

In order to characterize the source, it is important to obtain the i-v characteristics of the given PV source. The simplest method is shown in figure 2.8.1(a). Here the PV source with protection diodes is connected to a load resistor R_o. This load resistor is a rheostat that can be varied from 0 (short circuit) to a large value. The terminal current i_T is measured either by a hall sensor based current probe or by connecting a small value manganin based resistor in series and measuring the voltage across it. Assuming the drop across the current sense resistor R_{sense} to be negligible, the terminal voltage v_T is measured across the load resistor R_o. The terminal voltage and terminal current are measured at a set value of load resistor R_o. Different values of load resistor will lead to different load lines. The point of intersection of the load line, terminal current and terminal voltage will provide a point on the i-v characteristic of the PV source for a given insolation.

The terminal current and voltage values can be obtained and tabulated for various values of the load resistor by sweeping the load resistor from short circuit to open circuit. From the tabulated values, the i-v characteristics can be plotted for a given insolation. In order to perform this experiment, the PV source should be capable of delivering power. This means that a light source is necessary for the PV source to generate power. In the case of PV cells or low power PV sources, a laboratory light source with a spectrum close to white light may be used. This light should be set up such that the light is incident normal to the surface of the PV source. In the case of higher power PV sources, the experiment can be performed outside at noon under clear sky conditions. The measurement and tabulation of the terminal current and voltage for different values of the load resistor should be completed in less than 4 minutes (needed by Sun to cross a longitude) in order to achieve more or less constant insolation condition.

The circuit shown in figure 2.8.1(b) is similar except that the load resistor is replaced by a controlled voltage source. The controlled voltage source, v_T defines the terminal voltage of the PV source. It should also be capable of sinking the current in addition to sourcing current. The controlled voltage source can be set at any value from 0 volts to V_{oc} volts. The terminal voltage and terminal current are tabulated for the various setting of the controlled voltage source. From this data, the i-v characteristic of the PV source can be obtained.

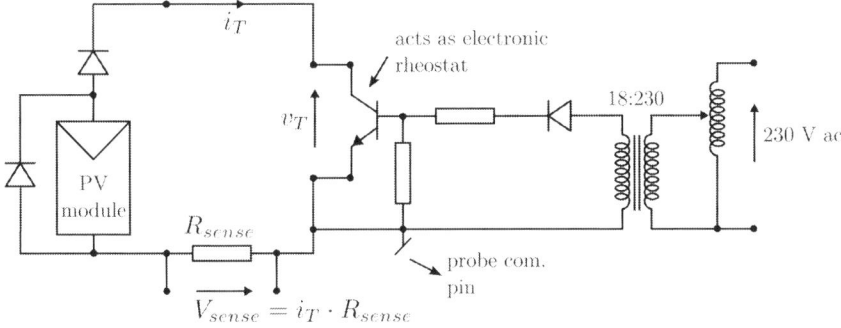

Figure 2.8.2: Measurement of i-v characteristic circuit setup

Both the methods indicated in figure 2.8.1 involves tabulation of the readings in a 4 minute time window during clear sky conditions. The i-v characteristics are obtained off-line by plotting on a graph sheet or in environments like Octave or MATLAB or Python. Figure 2.8.2 gives an improved method. Here there is no need to tabulate and the 4 minute time window is not an issue. The controlled voltage source is made to sweep from 0 volts to V_{oc} volts in a repetitive manner and given to the external trigger of an oscilloscope. The sweep waveform can be of any waveshape. It can be triangle sweep, sawtooth sweep or rectified sinusoidal sweep. The current measurement probe that gives voltage proportional to the terminal current is connected to channel-1 of the oscilloscope. The i-v characteristic can be seen directly on the oscilloscope that gets refreshed in every period of the controlled voltage source period which can be fraction of the 4 minute time window needed by Sun to cross a longitude.

In the circuit of figure 2.8.2, a controlled semiconductor device like a bipolar junction transistor (BJT) is used instead of a load resistor. The BJT acts like a variable resistor which can be swept from near short circuit to open circuit values by controlling the base drive of the BJT. The base drive signal for the BJT is derived from the

230V ac grid. An auto-transformer is used to step down the 230V ac mains voltage. In conjunction with the auto-transformer, a regular transformer is also used as shown. The regular transformer will provide galvanic protection apart from providing a step down ratio. A rectifier diode is used in the secondary of the transformer to provide half wave rectification. The resistive divider network provides the base drive current for the BJT. The auto-transformer together with the regular transformer steps down the main voltage such that the peak mains voltage of $230\sqrt{2}$ will correspond to maximum base drive current which will drive the BJT to saturation. At this conditions, an almost short circuit load is applied across the PV source. As the rectified sinusoidal drive voltage changes, the base drive current also changes making the BJT to transit through its active region and thereby acting as a variable resistor. When the drive voltage is zero, the BJT is in off condition. This provides open circuit load condition. The terminal voltage v_T is given to the external trigger input of the oscilloscope and the terminal current i_T is given to the channel-1 of the oscilloscope. The load will vary periodically at the main frequency. An i-v characteristic will be swept on the oscilloscope every 20ms.

Once the i-v characteristic of the PV source is measured by the above mentioned methods, the other parameters, like short circuit current, open circuit voltage, peak power point, peak voltage and currents, series resistance, shunt resistance, fill factor etc. are easily determined. The specification sheet for the PV source can then be generated.

2.9 PV source emulation

The sources that are generally available in the laboratory are voltage sources and current sources. However, if one needs to develop applications based on PV sources, one needs to have sources that have i-v characteristics that emulates the PV source. The actual PV source if available can be used only during day time when insolation is sufficient. However for development work, one needs to have the source that emulates the PV source characteristics that is available for use at all times of the day. PV source emulator is therefore an important piece of equipment that is needed for development of PV based applications.

Figure 2.9.1 shows one of the simplest forms of PV source emulation. Here several light sources are used that have spectral quality close to that of white light. The light from these light sources are incident normal to the surface of a PV cell or module. The insolation can be controlled by selectively switching ON or OFF certain lights or by electronically controlling the intensity of the light sources. The lights are powered from 230V ac mains grid. For higher power, the lamp wattages will be high and the light sources will contain significant portion of the light energy in the infra red or the thermal band of wavelengths. This will significantly increase the ambient temperature in the immediate neighbourhood of the light sources. Therefore, cooling fans need to be mounted to provide forced cooling for heat removal. This method of PV source emulation is simple in the sense that PV module itself is being used. However, for higher wattage applications, the input power requirement will be large as the PV module efficiencies are low (around 20%). This can lead to significantly large input power

2.9. PV SOURCE EMULATION

requirement and also real estate requirement to accommodate the panels. Therefore, this method is not suitable as a laboratory PV emulator especially at high powers.

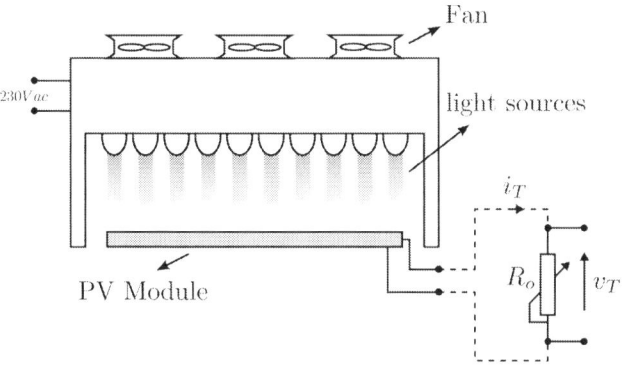

Figure 2.9.1: Insolation based emulation

The figure 2.9.2 shows a very simple circuit. It consists of a battery source connected to a resistive load. There is a resistance R_s in series which provides the necessary dynamic drop such that the i-v characteristics emulates the PV source. Alongside is given the i-v characteristic of the battery-R_s based emulator. When R_o is short circuited, the terminal voltage is zero. The operating point is on the y-axis or the i-axis where the short circuit current is $\frac{V_{dc}}{R_s}$. As the load resistor R_o is increased, the terminal voltage increases and the terminal current decreases. The terminal current is given as $i_T = \frac{V_{dc}}{(R_s+R_o)}$. As R_o increases to open circuit condition, the terminal current becomes zero and the operating point is on the x-axis or the v-axis where the open circuit voltage is V_{dc}. The power curve versus the terminal voltage is also plotted. It can be seen that the maximum power point is at a current that is half of short circuit current and voltage that is at half of open circuit voltage. Different values of R_s will result in different values of short circuit current. Thus R_s value control is equivalent to insolation variation. Another parameter for the PV source emulator is the fill factor. The fill factor is given as

$$FF = \frac{V_m I_m}{V_{oc} I_{sc}} = \frac{V_{dc}/2 \cdot I_{sc}/2}{V_{dc} \cdot I_{sc}} = 0.25 \qquad (2.9.1)$$

The fill factor for this very simple PV source emulator is only 0.25 which is poor in comparision with most commercial PV modules which have fill factors around 0.75. The efficiency is poor, but none-the-less this scheme is a practical PV source emulator that is popular for its simplicity.

Figure 2.9.3 gives the circuit diagram of resistance based PV source emulator with power drawn from the ac grid instead of from a battery. The schematic consists of an auto-transformer followed by a full bridge diode rectifier and terminated with a capacitance filter as shown. A voltage V_{dc} is developed across the capacitor. R_s is the series resistance which provides the emulator characteristics. Adjusting the auto-transformer secondary voltage will control the value of V_{dc} which in turn will control

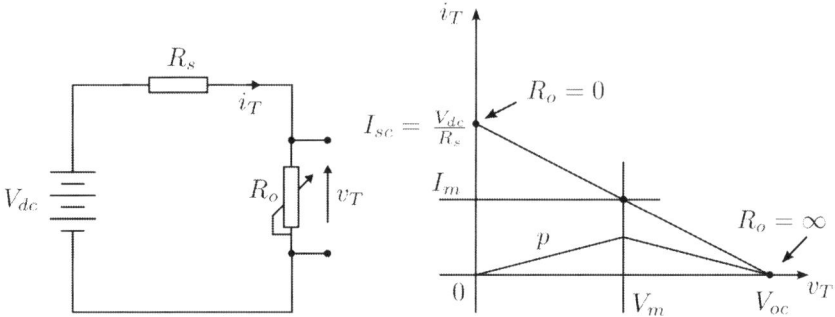

Figure 2.9.2: i-v characteristics of voltage source with R load

both the short circuit current and open circuit voltage values of the i-v characteristic. Varying only R_s is equivalent to insolation control.

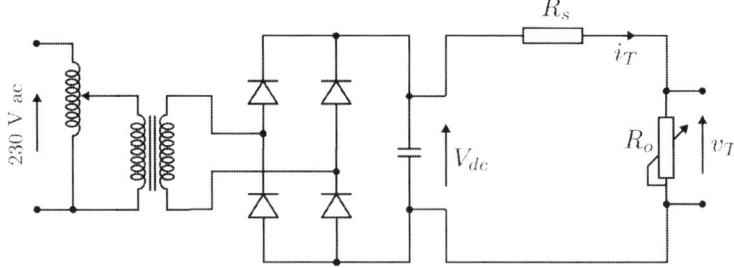

Figure 2.9.3: PV emulator circuit with R load

The PV emulator method with a resistor R_s in series is a lossy method. It has low efficiency. This can be a series disadvantage especially for high power emulation. The thermal dissipation will be large and will need heat removal mechanisms. An improved version of the PV emulator is shown in figure 2.9.4. It is based on switch mode power conversion. The system consists of a DC-DC converter which draws power from a DC source and supplies to a resistive load as shown in the figure. The switches in the DC-DC converter are controlled by a pulse width modulator (PWM) unit. The output voltage v_T is sensed and passed through a look-up table block which takes the terminal voltage as input and gives an output which is the expected terminal current. The look-up table block will be an i-v characteristic which can be even dynamically tuned. The output of the look-up table block is the reference current that should flow through the load. A current controller is used by sensing the terminal current and comparing with the reference current obtained from the look-up table block.

The error between the sensed current and the reference current is given to a current controller. The current controller output will make the PWM unit to provide pulses to the DC-DC converter such that a terminal current of i_T^* flows through the load terminals. This will tend to make the error at the input of the current controller zero. This is a very efficient method of building a PV emulator. The look-up table block plays

2.10. QUESTIONS

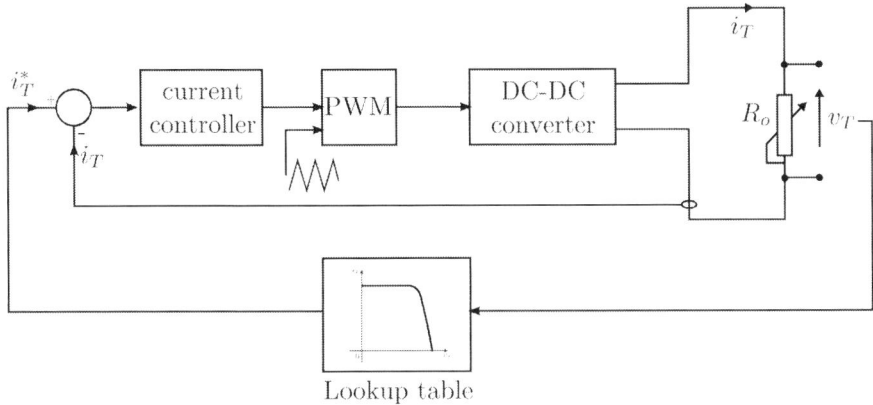

Figure 2.9.4: PV emulator using switched mode power converter

a key role in deciding the shape of the i-v characteristic that the PV source emulator will finally provide to the load. The look-up table can be programmed to shape the i-v curve as per requirement. In order to represent different insolation values, the short circuit current value of the look-up table can be appropriately changed. Likewise, the open circuit voltage values can also be varied.

2.10 Questions

1. Few photovoltaic modules are connected in series-parallel configuration. Three PV modules are connected in series string and two such strings are connected in parallel. Each module has an I-V characteristic having the x-axis intercept as 31V and the y-axis intercept as 4.2A. The short circuit current for the series-parallel system is
   a) 4.2A
   b) 8.4A
   c) 12.6A
   d) 25.2A

2. For the photovoltaic module configuration described in previous question, the open circuit voltage for the series-parallel system is
   a) 93V
   b) 31V
   c) 62V
   d) 186V

3. Two dissimilar PV cells are connected in parallel with no protection diodes. The open-circuit voltage and short-circuit current of one of the cells is 10V and 2A respectively. And the open-circuit voltage and short-circuit current of the other cell is 12V and 3A respectively. If V_{oc} and I_{sc} represent the open-circuit voltage

and short-circuit current of parallel cell combination, then
a) V_{oc} ¡ 10V
b) V_{oc} ¿12V
c) V_{oc} = 10V
d) 10V¡V_{oc}¡12V

4. Two dissimilar PV cells are connected in parallel with no protection diodes. The open-circuit voltage and short-circuit current of one of the cells is 10V and 2A respectively. And the open-circuit voltage and short-circuit current of the other cell is 12V and 3A respectively. If V_{oc} and I_{sc} represent the open-circuit voltage and short-circuit current of parallel cell combination, then
a) I_{sc} ¡ 2A
b) I_{sc} ¿ 3A
c) I_{sc} = 2A
d) I_{sc} = 3A
e) I_{sc} = 5A
f) 2A¡I_{sc}¡3A

5. Two dissimilar PV cells are connected in parallel with no protection diodes. Let V_{oc1} and I_{sc1} be the open circuit voltage and short circuit current of one cell and V_{oc2} (> V_{oc1}) and I_{sc2} (> I_{sc1}) be the open circuit voltage and short circuit current of the other cell. If V_{oc} and I_{sc} represent the open circuit voltage and the short circuit current of the parallel cell combination, then which of the following is true?
a) $V_{oc} < V_{oc1}$ and $I_{sc} = I_{sc1} + I_{sc2}$
b) $V_{oc} = V_{oc1} + V_{oc2}$ and $I_{sc} = I_{sc1} + I_{sc2}$
c) $V_{oc} > V_{oc2}$ and $I_{sc} = I_{sc1} + I_{sc2}$
d) $V_{oc1} < V_{oc} < V_{oc2}$ and $I_{sc} = I_{sc1} + I_{sc2}$

6. A photovoltaic panel has 14 modules connected in series with each module having an open-circuit voltage of 20V and short-circuit current of 3A. What should be the voltage axis intercept of the I-V characteristics of the series system?
a) 20V
b) 140V
c) 280V
d) 80V

7. A photovoltaic array has 10 PV modules connected in series forming a string and there are 5 such strings connected in parallel. Each PV module consists of 60 solar cells connected in series. For each solar cell, voltage and current corresponding to the maximum power point are 0.6V and 5.6A, respectively, at 800W/m^2 and 25°C. What is the maximum power that can be delivered by the PV array?

8. A battery having a voltage of 48V is in series with a 10 ohm resistance. Across the terminals of this battery-resistance series system, a load R_L is connected. What is the value of R_L at maximum power transfer operating point and what is the maximum power is transferred to R_L.

2.10. QUESTIONS

9. A PV module has 20 cells connected in series. There are no protection diodes. Each cell has a V_{oc} of 1V and I_{sc} of 2A. One cell is completely shaded. Across the shaded cell, a bypass diode with forward junction voltage of 0.7V is connected. The V_{oc} of the PV series module is ⋯⋯.

10. Two PV modules are connected in series as shown in the figure. PV cell1 has a short circuit current and open circuit voltage of 9A and 36V respectively. The other module PV cell2 has a short circuit current and open circuit voltage of 6A and 35V respectively. The i-v characteristics of each of the PV modules are as shown. What is the load current beyond which one of the PV modules acts as a power sink instead of a power source?

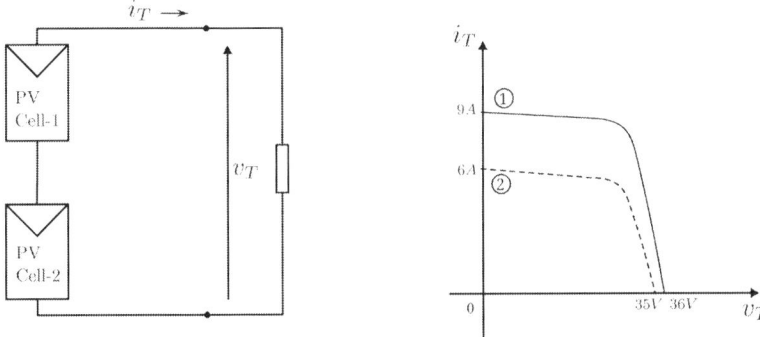

Figure 2.10.1: PV module configuration and i-v characteristics

11. In the circuit given in the figure of previous question, modify the circuit in order to prevent any PV module from acting as a power sink.

12. Consider two identical PV modules connected in parallel. The voltage and current corresponding to maximum power point operation of a module are 28V and 3.2A respectively. What is the value of load resistance connected to the parallel configuration operating at maximum power point?

13. Consider two non-identical PV modules connected in parallel. The i-v characteristics of the two PV modules are given in figure below. The terminal voltage of the parallel configuration is 25V. What is the load resistance connected across the parallel configuration?

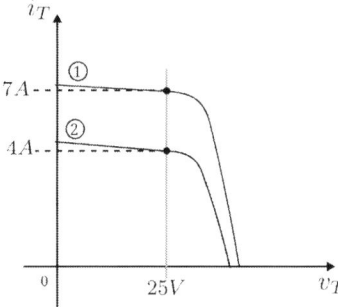

Figure 2.10.2: i-v characteristics of two non-identical PV modules

14. A battery of 48V is to be connected directly to a PV module in order to charge the battery. The solar cell used in the module has 0.6V and 5.6A as the voltage and current respectively at maximum power operating point at a given insolation and temperature. The open circuit voltage of the cell is 0.7V and the short circuit current is 6A. What is the minimum number of solar cells that needs to be connected in series to establish a potential such that the battery will be charged?

15. A PV module has 20 cells connected in series. The open circuit voltage and short circuit current of the module are 20V and 2A respectively. Each cell has negligible series resistance and shunt resistance of 10 ohms. One of the cells is completely shaded. Plot the i-v characteristics of the module.

Chapter 3

Sun Power

3.1 Introduction

A complete solar photovoltaic system contains several parts as visualised in figure 3.1.1. The photovoltaic modules or the PV source, power converter that interfaces the PV source to energy storage system and the loads. There are several energy storage mechanisms like chemical storage as in batteries, pumped hydro, compressed air storage, flywheel storage etc. Designing power interfaces between the PV source and the energy storage system is an important part of the overall system. Another important aspect is the design of power interfaces between the PV source and the load. The load is the application for which the PV source is supposed to supply power. These can be one among several applications like lighting, pumping, refrigeration, grid interaction etc.

Each one of the subsystem parts of the photovoltaic system is important and plays a significant part. It is essential to understand them and the interactions amongst them in order to design these subsystems. At the outset, even before one designs the various subsystems of the photovoltaic systems, accurate estimate of the input energy to the PV system is needed. Therefore, energy from the Sun at any given time of the day, any given time of the year should be available in order to design PV system for a specific application.

The primary objective of energy estimation from Sun is to determine the size of the PV array that would be required for a specific application at a specified place. PV sources need a significant amount of real estate to collect the solar energy. Real estate can be quite expensive and therefore serious thought on this aspect is needed during design of PV systems. Energy from the Sun has dependency on the local geography, incident radiation from the Sun, atmospheric condition, water vapor content, latitude, time of the day, and time of the year. All these parameters affect the effective energy that is available at the output of the PV source.

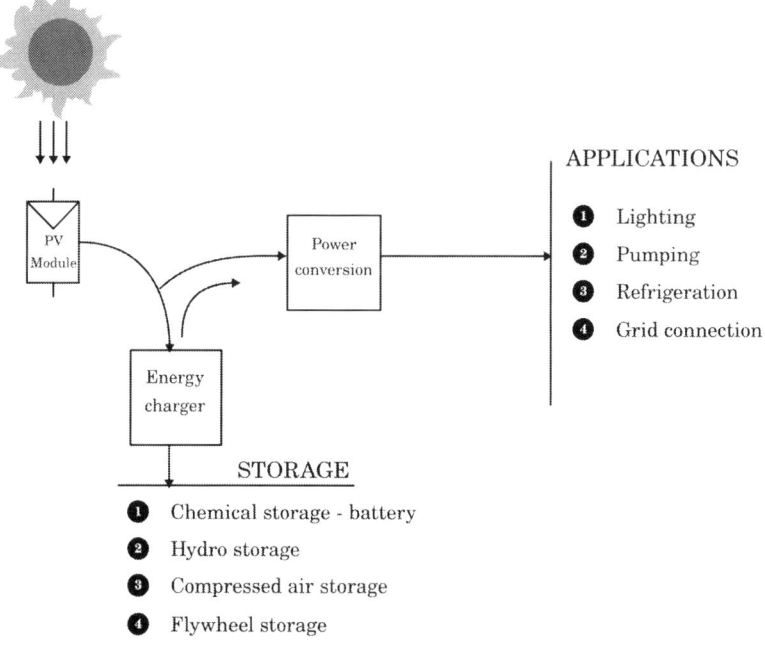

Figure 3.1.1: Typical PV system

3.2 Solar radiation terms

The incident solar energy varies with the wavelength of the incident electromagnetic wave. The term *spectral irradiance*,P_s is expressed as kW/m^2/unit wavelength. The term *insolation* is the integration of the the spectral irradiance with respect to wavelength.

Insolation is the amount of incident solar power on a unit surface area, commonly expressed in units of kW/m^2. A solar insolation level of 1kW/m^2 is often called *standard insolation* or *one peak sun*. Solar insolation in this book is denoted by ' L '. The solar insolation on a 1m^2 target surface that is placed at the earth's outer atmosphere and oriented normal to the sun's rays is called the *solar constant* and its value is about 1.37kW/m^2. The value of the solar constant varies slightly with the time of the year as the distance of the target surface from the sun varies due to an elliptic orbit followed by earth while revolving around the sun. Due to atmospheric effects, the peak insolation at the earth's surface is attenuated. On a clear day at noon, the peak solar insolation incident on a terrestrial target surface oriented normal to the sun rays is about 1kW/m^2.

The terms insolation and *irradiance* are often used interchangeably in several literature. However, the following distinction is made throughout this book. Irradiance is an amount of solar energy received on a unit surface expressed in units of kWh/m^2. Solar irradiance is essentially the solar insolation (power) integrated with respect to time. When solar irradiance data is represented on an average daily basis, the value is often called *peak sun hours* (PSH) and can be thought of as the number of equivalent

3.3. ESTIMATION OF DAILY IRRADIANCE

hours/day that solar insolation is at peak sun of $1KW/m^2$. Solar irradiance is denoted by 'H'.

3.3 Estimation of daily irradiance

The sun has a radius of 695500km as compared to 6378km of earth. The projected area of the sun is around 10000 times more than that of the earth. Therefore, one can consider that the rays from the sun are parallel lines. The cross sectional area presented to the radiation from the sun is always πr^2 at every instant of time where r is the radius of the earth. If L be the insolation incident at the outer atmosphere per unit surface area, then the yearly energy received from the sun is given as

$$E_y = L.\pi.r^2.24.365 = 27.5e3.L.r^2 KWh \quad (3.3.1)$$

The insolation is $1.37 kW/m^2$ at the earth's outer atmosphere and r is the radius of the earth and is equal to 6378000m. Using these values in equation 3.3.1, the annual energy received from the sun is 1.5214×10^{18} kWh/year.

3.4 Irradiance and insolation

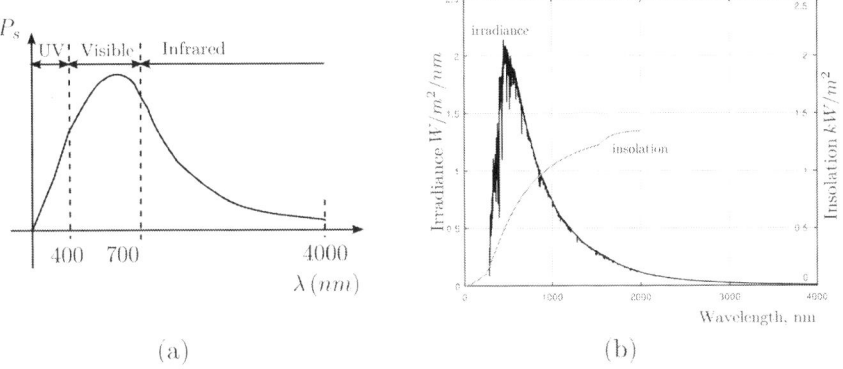

Figure 3.4.1: Solar spectrum (a) Solar irradiance and wavelengths (b) Actual measured solar irradiance and insolation

The solar spectrum will be similar to that shown in figure 3.4.1(a). The x-axis is wavelength λ in nm and the y-axis is spectral irradiance P_s. It is the amount of kW that is available in the solar radiation for each spectral unit (or wavelength unit) that is incident on a $1 m^2$ surface area. The spectral irradiance has the units of $kW/m^2/\mu m$ if the wavelength is measured in μm or $kW/m^2/nm$ if the wavelength is measured in nm.

The figure distinguishes three important regions viz. (i) the ultra-violet or UV region, (ii) the visible light region and (iii) the infrared region. The UV region gives the power associated with the high frequency or small wavelength part of the solar spectrum. The wavelengths smaller than 400nm are designated as UV part of the solar spectrum. This region amounts to around 10% of the total solar power in the incident solar radiation. The part of the spectrum associated with the visible light is between 400nm and 700nm. About 40% of the incident solar power is present in visible part of the spectrum. The remaining 50% of the solar power is in the longer wavelength zone or the infrared region. The thermal power or heat power is associated with these longer wavelengths. In the case of thermal collection, the collectors need to be sensitive in the infrared region. The PV source is sensitive to visible light and most of the power is drawn from the visible light region of the incident solar radiation.

The accumulated power available in all wavelengths together is called insolation which is given as

$$L = \int P_s . d\lambda \qquad (3.4.1)$$

Referring to figure 3.4.1, graphically this means that the area under spectral irradiance curve will give the insolation of the incident solar radiation. The insolation L is expressed in kW/m^2. If the solar radiation is measured just outside the earth's atmosphere, then it is called extra-terrestrial radiation. The spectral irradiance integrates over all wavelengths to an insolation value equal to the solar constant of about 1.37kW/m^2. Due to the elliptic nature of the earth's orbit around the sun, the extra-terrestrial insolation varies from 1.33kW/m^2 to 1.41kW/m^2. The solar constant is an annual average.

The actual measured solar radiation spectrum is shown in figure 3.4.1(b). Superimposed on this is also the insolation profile which is the integrated value of the spectral irradiance. The insolation L accumulates to solar constant value for the extra-terrestrial radiation. If the solar radiation is measured on the earth's surface, then the atmosphere will attenuate the extra-terrestrial radiation and the spectral irradiance integrates over all wavelengths to an insolation value about 1kW/m^2 or peak sun insolation.

3.5 Incident energy

In the previous section, the term insolation and spectral irradiance of the Sun is discussed for a given solar radiation spectrum. The solar radiation spectrum remains unchanged. However, with respect to a point on earth, the insolation is not constant. During night, there is no solar radiation and the insolation is therefore zero. At noon, the solar radiation is maximum and consequently the insolation is at its maximum.

Figure 3.5.1(a) shows a typical insolation profile over a day at an equatorial region. The insolation is plotted as a function of time from 0.00 hours to 24.00hours (12 midnight). From 0.00 hours to 6.00am there is no significant insolation. The insolation reaches a peak at around 12 noon and there after starts to fall again. Beyond 6.00pm the insolation will again reduce to zero as there would be no sunlight by this time. The area under the insolation curve gives the energy received in kWh/m^2 from the incident solar radiation over a day. This is denoted by 'H'. On top of the insolation curve is superimposed a rectangular profile having the same area as that under the insolation profile. The rectangular profile has an amplitude of standard insolation i.e.

3.5. INCIDENT ENERGY

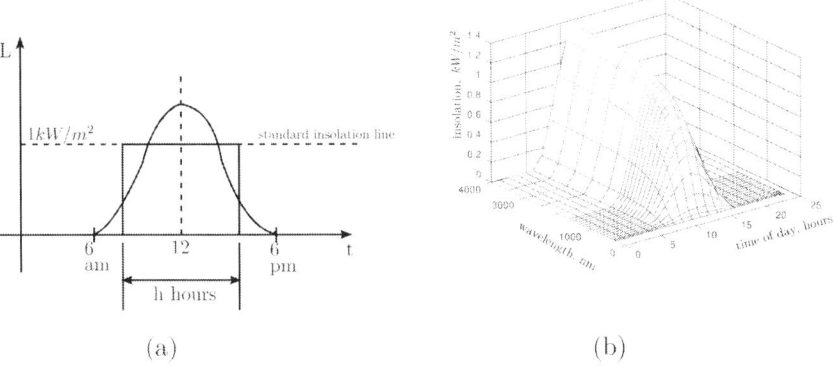

Figure 3.5.1: Daily insolation profile (a) variation of insolation over a day (b) insolation variation with wavelength and time of day

$1kW/m^2$ and a width of h hours. The energy under the rectangular profile is given as $h\,kWh/m^2$ which is equal to the energy under the actual insolation profile. h is also the equivalent peak sun hours, at the specific point of interest on earth over a day. This concept will be called upon later during the sizing of PV source for a given application.

The insolation profile from the solar spectrum as a function of wavelength and the insolation profile at a place on earth as a function of time of day are combined and visualised in figure 3.5.1(b). The z-axis is the insolation L, the x-axis is the time of day and the y-axis is the wavelength. The insolation profile for every wavelength varies with time similar to that shown in figure 3.5.1(a). The insolation profile as a function of wavelength at any given time of day varies in a manner as shown in figure 3.5.1(b).

The insolation profile at a place depends on the latitude of the place, solar collector orientation, the time of the day (hour angle), time of the year (declination), locale geography and atmospheric conditions (clearness index). The primary objective of this chapter is to estimate the insolation profile over the day and over the year in order to size the PV sources. The latitude of a place on earth is given in term of the angle that a line from the centre of the earth to the place subtends on the equatorial plane. In other words, latitude is the angular distance of a place on earth from the equatorial plane. This is denoted by ϕ. The latitude angle ϕ is positive for places north of the equatorial plane and negative for places south of the equatorial plane. It stands to reason that any place on the equator also lies on the equatorial plane and the latitude angle is zero.

The PV source is a flat plate collector. The PV modules have a flat surface. At the equator they are place horizontally on the surface. However, at other latitudes, the PV collector is placed with a tilt to optimise the amount of incident solar radiation with normal incidence. This will improve the capture efficiency. The angle of tilt of the flat plate PV collector from the horizontal of the place is denoted by β. The time of the day is denoted by the hour angle ω. At a place on the equator, the hour angle at sunrise is 0 and at sunset it is π.

Another important parameter is the declination δ. It is the angle that the line from the centre of the earth to the centre of the sun makes with the equatorial plane. This occurs due to the elliptic nature of the earth's orbit around the sun. The atmospheric conditions are difficult to predict at a place and time. This is a statistical phenomenon and is encompassed in a parameter called the clearness index k_T for the place. The locale geography also can introduce secondary reflections from the ground called the albedo effect. This is also not deterministic, but is weighted by an empirical parameter called the tilt factor r_D.

The interplay of all these parameters will eventually determine the insolation at the specific place of interest on the surface of the earth. The following sections will bring out the relationships among these parameters in determining the insolation at a given place.

3.6 Declination

The plane that is formed by the equatorial latitude is called the equatorial plane. If this plane is extended infinitely in all directions, then such a plane is called the celestial equator. Consider a line joining the centre of the earth to the centre of the sun. The angle between the earth-sun line and the celestial equator is called the *declination*. The declination is not a fixed angle. It is dynamically changing as the earth orbits around the sun. It has a periodicity of one year. The earth revolves around the sun in an elliptic orbit. This is visualised in figure 3.6.1 with respect to the equatorial axis. Figure 3.6.1 also shows the visualisation of the term declination through an earth centric viewpoint. June 21 is summer solstice wherein the earth-sun line intersects the earth at lattitude $23.5^o N$(tropic of cancer). The declination $\delta = 23.5^o$ on this day. Winter solstice occurs on 21st December when the earth-sun line passes through the southern hemisphere. The earth-sun line intersects the earth at latitude $23.5^o S$(tropic of capricorn) and the declination is $\delta = -23.5^o$. March 21st and September 21st are the two equinox days wherein the earth-sun line is along the equatorial plane and the declination $\delta = 0^o$. The declination continuously varies between $\delta = 23.5^o$ and $\delta = -23.5^o$ limits, as the earth revolves around the sun. The variation of declination takes an almost sinusoidal profile with a periodicity of one year. An empirical relation for the declination as a function of time of day N is given as

$$\delta = 23.45 \cdot sin(2\pi \frac{(N-80)}{365}) \quad (3.6.1)$$

where declination δ is given in degrees and N=1 on January 1st and 365 on December 31st.

3.7 Solar geometry

Spherical coordinate system is used to represent the sun, insolation vector, hour angle, declination, lattitude and longitude with respect to the centre of the earth. This is also referred to as the earth centric spherical coordinate system. The figure 3.7.1 shows a visualisation of the earth centric coordinate system. There are three orthogonal axes.

3.7. SOLAR GEOMETRY

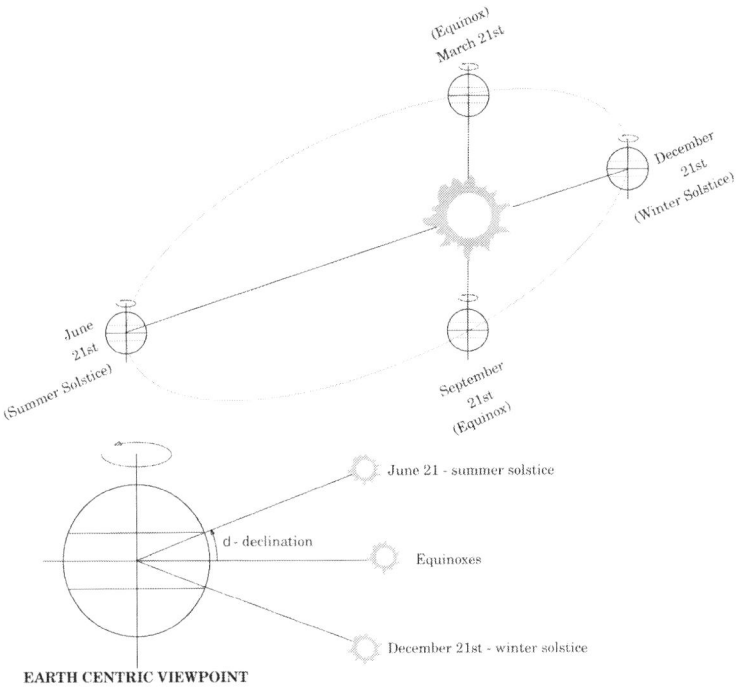

Figure 3.6.1: Declination and Earth centric viewpoint

The m-axis is in the equatorial plane and is called the meridional axis as it is in the direction of longitude of the place. The e-axis is called the east of meridional axis. It lies on the equatorial plane and points exactly in the eastern direction with respect to the longitude along m-axis. The p-axis or the polar axis is orthogonal to the equatorial plane and is shown pointing vertically up along the north pole. The earth-sun line is the insolation line or insolation vector that is incident on the earth.

The angular distance of the insolation line from the equatorial plane is the declination δ. The angle measured from the meridional axis (m-axis) to the projection of the insolation line onto the equatorial plane is called the hour angle ω. On the meridian corresponding to the m-axis, consider an arbitrary point X on the surface of the earth. The angular distance of the line joining the centre of the earth and X, with respect to the equatorial plane is the latitude ϕ.

There are two sets of periodic motions and related effects viz.

1. diurnal changes or daily changes due to earth rotating on its own polar axis. This causes the variation in hour angle ω. At sunrise the projection of the insolation line on the equatorial plane will be along the e-axis and therefore $\omega = 90^o$. At noon, the insolation line will be in the meridional plane and therefore its projection on the equatorial plane will be along m-axis and therefore $\omega = 0^o$. At sunset, the projection of the insolation line onto the equatorial plane will be along the negative e-axis and hence $\omega = -90^o$

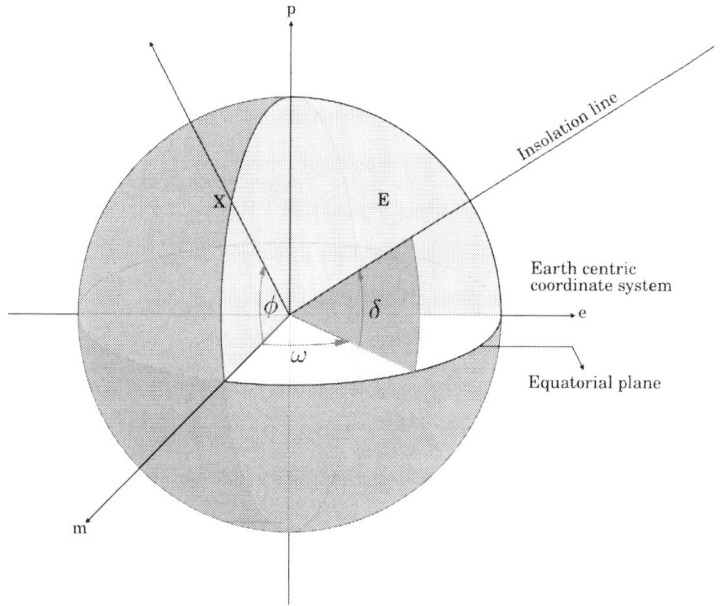

Figure 3.7.1: Spherical coordinate system with respect to center of earth

2. the yearly changes due to earth revolving around the sun. This causes changes in the declination δ which swings from $\delta = 23.5^o$ on summer solstice to $\delta = -23.5^o$ on winter solstice.

The objective is to estimate the energy that is available at an arbitrary point X on the earth considering these variabilities. The figure 3.7.2 also shows the locale centric coordinate system relative to the earth centric coordinate system. At the arbitrary place X that is selected at a point on the meridian which is along the meridional axis, a plane is placed tangential to the point X. This plane is called the *horizon plane* at the place. There are 3 axes in the locale centric coordinate system. The S-axis is on the horizon plane and also in the celestial meridional plane and points towards the celestial equatorial plane as shown. The E-axis is orthogonal to the S-axis and it lies on the horizon plane. It points in the direction east of the S-axis. The Z-axis is orthogonal to the horizon plane and points directly to the zenith at the place. The Z-axis line when extended, joins the centre of the earth.

The insolation incident at point X and the insolation line to the centre of the earth may be considered as parallel lines. The angular distance of S-axis to the projection of insolation line incident at point X on to the horizon plane is called the azimuth angle γ_s. The angular distance of the insolation line with respect to the zenith or Z-axis is called the zenith angle θ_z.

The distance of the sun from the earth can be considered to the very large with respect to the radius of the earth. Let the assumption be imposed that the radius of the earth is negligible compared to the distance of the sun from the earth and that

3.8. INSOLATION ON HORIZONTAL FLAT PLATE COLLECTOR

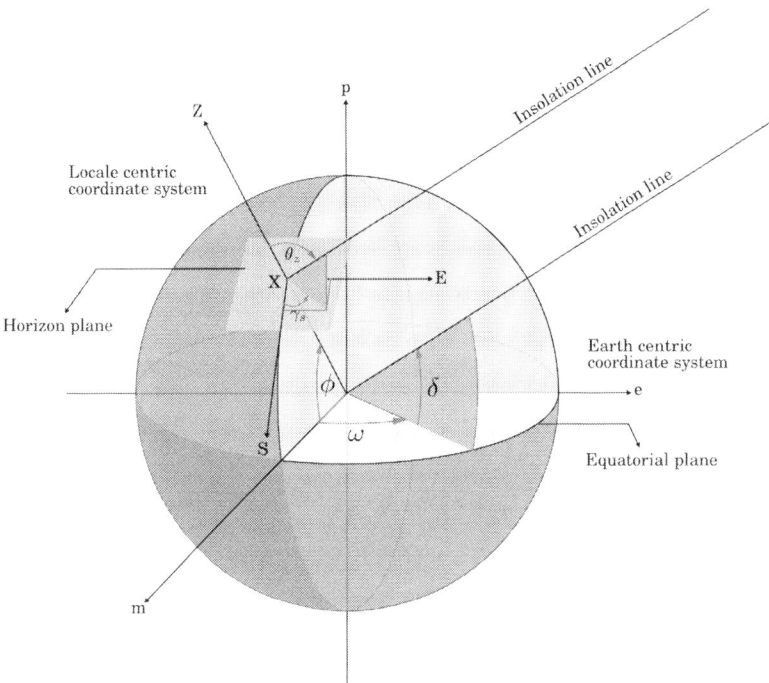

Figure 3.7.2: Locale centric coordinates relative to earth centric coordinates

parallel rays of radiation from the sun is incident on the earth. Then the locale centric coordinate system and the earth centric coordinate system can be merged together. The locale coordinate system is pushed along the Z-axis up to the centre of the earth. Consider the view wherein in the observer is standing on the e-axis or E-axis and facing towards centre of the earth. The coordinate system perspective is as shown in figure 3.7.3. The e-axis and the E-axis are now points as shown. All other axes are as shown in figure 3.7.3.

3.8 Insolation on horizontal flat plate collector

Consider a flat plate collector placed horizontally at place X on earth. Referring to figures 3.7.2 and 3.7.3, the flat plat collector will be in the horizon plane. The objective is the find the insolation incident on this flat plate collector at normal incidence. The normal incidence will be along the Z-axis or the zenith axis which is orthogonal to the horizontal flat plate collector.

Let L be the insolation incident at point X on the surface of the earth. The insolation L is along the insolation line. With respect to the locale centric coordinates, the incident insolation can be resolved along the three orthogonal axes as L_S along the S-axis, L_E along the E-axis and L_Z along the Z-axis. The incident insolation L is the vector sum of L_S, L_E and L_Z. The insolation along Z-axis is normal to the horizon-

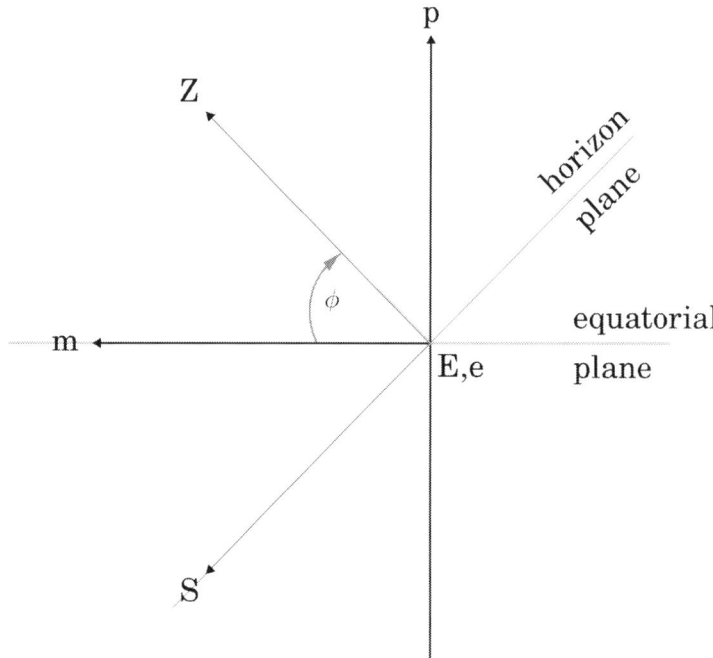

Figure 3.7.3: Merged locale and earth coordinates as viewed from east

tal flat plate collector. L_Z therefore is the component of the incident insolation L that needs to be considered for energy estimation. The resolved components of the incident insolation are

$$L_S = L \cdot sin\theta_z \cdot cos\gamma_s$$
$$L_E = L \cdot sin\theta_z \cdot sin\gamma_s$$
$$L_Z = L \cdot cos\theta_z \qquad (3.8.1)$$

The radius of the earth can be considered to be insignificant compared to the distance of earth from the sun. Therefore, the locale centric coordinate system can be transposed and merged with the earth centric coordinate system. The insolation L can be resolved into three orthogonal components in the earth centric coordinate system as well. They are L_m along the m-axis, L_e along the e-axis and L_p along the p-axis. The incident insolation L is also the vector sum of L_m, L_e and L_p. The resolved components of the incident insolation along the earth centric coordinate axes are

$$L_m = L \cdot cos\delta \cdot cos\omega$$
$$L_e = L \cdot cos\delta \cdot sin\omega$$
$$L_p = L \cdot sin\delta \qquad (3.8.2)$$

3.9. DAILY ENERGY ON A HORIZONTAL FLAT PLATE

Referring to figure 3.7.3, L_Z is along the Z-axis and L_S is along the S-axis. They are composed from two orthogonal components L_m and L_p. Therefore

$$L_Z = L_m \cdot \cos\phi + L_p \cdot \sin\phi \tag{3.8.3}$$
$$L_S = L_m \cdot \cos(90 - \phi) - L_p \cdot \cos\phi$$
$$= L_m \cdot \sin\phi - L_p \cdot \cos\phi \tag{3.8.4}$$

Substituting for L_Z, L_S, L_m and L_p from equations 3.8.1 and 3.8.2 in to equation 3.8.3, one obtains

$$L \cdot \cos\theta_z = L \cdot \cos\delta \cdot \cos\omega \cdot \cos\phi + L \cdot \sin\delta \cdot \sin\phi$$
$$\cos\theta_z = \cos\delta \cdot \cos\omega \cdot \cos\phi + \sin\delta \cdot \sin\phi \tag{3.8.5}$$

and

$$L \cdot \sin\theta_z \cdot \cos\gamma_s = L \cdot \cos\delta \cdot \cos\omega \cdot \sin\phi - L \cdot \sin\delta \cdot \cos\phi$$
$$\cos\gamma_s = \frac{\cos\delta \cdot \cos\omega \cdot \sin\phi - \sin\delta \cdot \cos\phi}{\sin\theta_z} \tag{3.8.6}$$

Equation 3.8.5 gives the cosine of the zenith angle in terms of latitude angle, declination and hour angle. Likewise, equation 3.8.6 gives the cosine of the azimuth angle in terms of latitude angle declination and hour angle. On a horizontal flat plate at a place, the insolation at normal incidence on the top surface of the plate is given by $L \cdot \cos\theta_z$. This estimate of insolation is used for all horizontal collectors. If a vertical plate is placed at the place, then the insolation at normal incidence to the south facing surface of the plate is given by $L \cdot \sin\theta_z \cdot \cos\gamma_s$. This is useful for estimating the insolation on collectors placed on south facing vertical walls and window panes at places in northern latitudes beyond the tropics.

3.9 Daily energy on a horizontal flat plate

The insolation incident normally on a horizontal flat plate at any given latitude on the surface of the earth is given by $L \cdot \cos\theta_z$ as given in equation 3.8.1. It is of interest to estimate the daily energy incident on the horizontal flat plate at that given place. The accumulated effect of the insolation over the day will give the daily energy received at the place. Let H_o be the irradiance or daily energy received at the place in kWh/m^2 under conditions where the atmospheric effects are not considered. Then,

$$H_o = \int_{day} L_z . d\omega \tag{3.9.1}$$

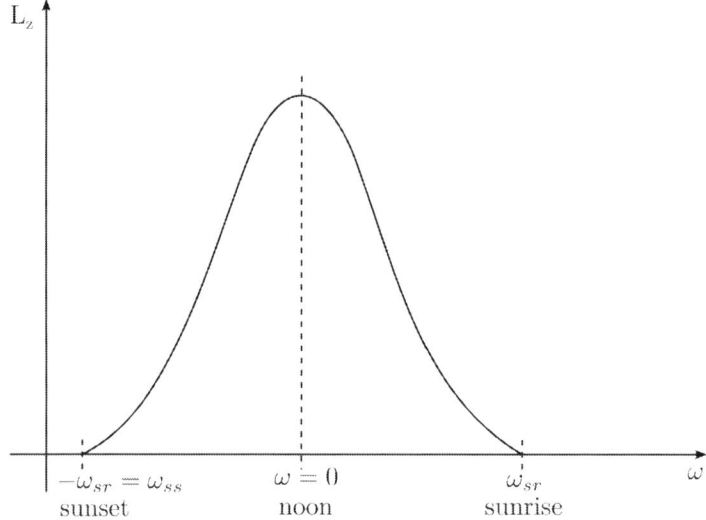

Figure 3.9.1: Insolation during daytime from sun rise to sun set

Figure 3.9.1 shows the insolation incident normally at a place as a function of the hour angle ω. Referring to figure 3.7.2, it can be noted that the hour angle $\omega = 0$ at noon when the sun is along the zenith for the place on a given meridian. The hour angle $\omega = \omega_{sr}$ at sun rise and at sun set the hour angle $\omega = \omega_{ss} = -\omega_{sr}$. The area under the insolation curve from sun set to sun rise is the daily energy received at the place. Therefore, equation 3.9.1 can be written as

$$H_o = \int_{\omega_{ss}}^{\omega_{sr}} L_z . d\omega$$

$$= 2 \cdot \int_{0}^{\omega_{sr}} L_z . d\omega \qquad (3.9.2)$$

Substituting from equation 3.8.5 into equation 3.9.2, one obtains

$$H_o = 2 \cdot \int_{0}^{\omega_{sr}} (L \cdot \cos\delta \cdot \cos\omega \cdot \cos\phi + L \cdot \sin\delta \cdot \sin\phi) . d\omega$$

$$= 2L \cdot \left[\cos\delta \cdot \cos\phi \cdot \sin\omega \big|_{0}^{\omega_{sr}} + \sin\delta \cdot \sin\phi \cdot \omega \big|_{0}^{\omega_{sr}} \right]$$

$$= 2L \cdot \left[\cos\delta \cdot \cos\phi \cdot \sin\omega_{sr} + \sin\delta \cdot \sin\phi \cdot \omega_{sr} \right] \; kW \, radians/m^2/day$$

$$= \frac{24}{\pi} \cdot L \cdot \left[\cos\delta \cdot \cos\phi \cdot \sin\omega_{sr} + \sin\delta \cdot \sin\phi \cdot \omega_{sr} \right] \; kWh/m^2/day \qquad (3.9.3)$$

Referring to equation 3.9.3, in order to estimate the incident energy on a horizontal flat plate collector at a given place, the time of year (δ), latitude of the place (ϕ),

3.9. DAILY ENERGY ON A HORIZONTAL FLAT PLATE

hour angle at sunrise (ω_{sr}) and the insolation L at the place should be known. The declination δ can be estimated using equation 3.6.1 and the latitude of the place of interest is an input specification. The insolation L at the place and the sunrise hour angle ω_{sr} need to be estimated.

3.9.1 Insolation on earth surface

The insolation L is given by an empirical relation as follows

$$L = L_{SC} \left\{ 1 + 0.033 \cdot cos \left(\frac{360 \cdot N}{365} \right) \right\} \qquad (3.9.4)$$

where

$N=$ day number or day of year. It is 1 on 1st January and 365 on 31st December.

It should be noted that $\left(\frac{360 \cdot N}{365} \right)$ is in degrees. It can be expressed in radian as $\left[\left(\frac{360 \cdot N}{365} \right) \cdot \frac{\pi}{180} \right]$.

$L_{SC} = 1.37 \, kW/m^2$. This is the mean solar constant.

3.9.2 Sunrise and sunset angles

Refer to figure 3.7.2. In the locale coordinate system, the incident insolation angle varies with the time of day. The horizon plane can be considered to be extended towards infinity in all directions. It then becomes the celestial horizon plane. If the insolation line is below the celestial horizon plane, then it is night time for the place of interest, and if the insolation line is above the horizon plane, it is day time. When the insolation line lies along the horizon plane and towards east of the S-axis, then it is sunrise at the place. When the insolation line lies along the horizon plane and toward the west of the S-axis, then it is sunset at the place.

At sunrise or sunset, the insolation line is along the horizon plane and therefore the zenith angle $\theta_z = 90^o$ at this time. The projection of the insolation line on the horizon plane is the insolation line itself. At sunrise, the hour angle measured anti-clockwise from the S-axis upto the insolation line is called the sunrise angle ω_{sr}. And at sunset, the hour angle measured clockwise from the S-axis up to the insolation line is called the sunset angle ω_{ss}. Applying these constraints to equation 3.8.5, one obtains

$$0 = cos\delta \cdot cos\omega_{sr} \cdot cos\phi + L \cdot sin\delta \cdot sin\phi$$
$$\omega_{sr} = cos^{-1}(-tan\delta \cdot tan\phi)$$
$$\omega_{sr} = -\omega_{ss} \qquad (3.9.5)$$

Equations 3.9.3, 3.9.4 and 3.9.5 together will provide an estimate of the daily energy incident on a horizontal flat plate collector at a given place on the surface of the earth.

3.9.3 Ex.3.1

What is the daily energy incident on a horizontal flat collector placed at a place on the equator on 21st of March?

The latitude being the equator, $\phi = 0$.

21st March is an equinox day. This implies that the day number $N = 80$ and the declination $\delta = 0$.

It is required that the daily energy incident at the place H_o be estimated.

$$H_o = \frac{24}{\pi} \cdot L \cdot [\cos\delta \cdot \cos\phi \cdot \sin\omega_{sr} + \sin\delta \cdot \sin\phi \cdot \omega_{sr}]$$

$$L = L_{SC}\left\{1 + 0.033 \cdot \cos\left(\frac{360 \cdot N}{365}\right)\right\}$$

Substituting $L_{SC} = 1.37$ and $N = 80$, in the above equation, one obtains $L = 1.3787\,kW/m^2$.

$$\omega_{sr} = \cos^{-1}(-\tan\delta \cdot \tan\phi)$$
$$= \cos^{-1}(-\tan 0 \cdot \tan 0) = \frac{\pi}{2}$$
$$H_o = \frac{24}{\pi} \cdot 1.3787 \cdot [1 + 0] = 10.53\,kWh/m^2/day$$

3.9.4 Ex.3.2

What is the day length for

1. $\phi = 0$
2. $\phi > 0$ during summer
3. $\phi > 0$ during winter

Day length or the length of the day is the period between sunrise and sunset. Therefore,

Day length $= 2 \cdot \omega_{sr} = 2 \cdot \frac{12}{\pi} \cdot \omega_{sr}$ in hours

Day length $= \frac{24}{\pi} \cdot \cos^{-1}(-\tan\phi \cdot \tan\delta)$ in hours

Case-1: Substituting $\phi = 0$ in the day length equation above, one obtains,

Day length $= \frac{24}{\pi} \cdot \frac{\pi}{2} = 12$ hours. At a place on the equator, irrespective of the time of year, the day length is 12 hours.

Case-2: During summer, the declination $\delta > 0$. Therefore, $\cos\omega_{sr} = -\tan\phi \cdot \tan\delta$ will be a negative value as $\phi > 0$ and $\delta > 0$. The cosine is negative in second quadrant and therefore this implies that $\omega_{sr} > \frac{\pi}{2}$. From the day length equation, one can conclude that the day length is greater than 12 hours during summer for latitudes north of the equator. The days are longer in summer for places north of the equator.

3.10. ENERGY ON A TILTED FLAT PLATE

Case-3: During winter, the declination $\delta < 0$. Therefore, $\cos\omega_{sr} = -\tan\phi \cdot \tan\delta$ will be a positive value as $\phi > 0$ and $\delta < 0$. This implies that $\omega_{sr} < \frac{\pi}{2}$. From the day length equation, one can conclude that the day length is lesser than 12 hours during winter for latitudes north of the equator. The days are shorter in winter for places north of the equator.

3.10 Energy on a tilted flat plate

The previous section discussed in detail the estimation of the daily incident energy on a horizontal flat plate collector placed at a given point on the surface of the earth. Consider the situation on June 21st i.e. on summer solstice. The declination is at positive maximum value of 23.5^o. As days progress further through the year, the sun begins to move towards the equator when on 21st September, the sun centre is on the celestial equatorial plane. For a place on the northern or southern hemisphere, it makes sense to track the sun in order to capture as much of the insolation as possible. Therefore, if one needs to maximise the amount of daily incident energy on a flat plate collector, horizontal placement of the collector will not be the best. The flat plate collector needs to be tilted. This section discusses the daily incident energy on a tilted flat plate collector placed at a point on the surface of the earth.

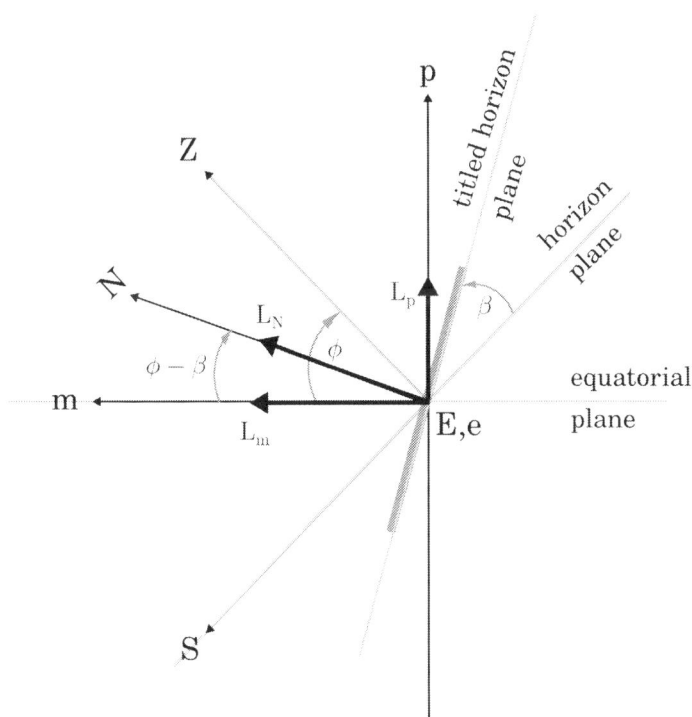

Figure 3.10.1: Tilted collector placement in merged locale and earth coordinates

In the case of a horizontal flat plate collector, the collector is placed on the horizon plane. Consider a tilt of angle β such that the collector is facing south as shown in figure 3.10.1. Due to the tilt, the new horizon for the collector is along the plane of the tilted collector. This is the new tilted horizon for the collector. The insolation normal to the tilted horizon is along L_N vector as indicated in figure 3.10.1. The objective is to estimate L_N and then consequently find the irradiance or daily incident energy on the tilted collector H_{ot}. Referring to figure 3.10.1,

$$L_N = L_m \cdot \cos(\phi - \beta) + L_p \cdot \sin(\phi - \beta)$$
$$= \cos\delta \cdot \cos\omega \cdot \cos(\phi - \beta) + \sin\delta \cdot \sin(\phi - \beta) \qquad (3.10.1)$$

where L_m and L_p are as given in equation 3.8.2.

The daily incident energy on a tilted flat plate collector is given as

$$H_{ot} = \int_{\omega_{sst}}^{\omega_{srt}} L_N . d\omega$$
$$= 2 \cdot \int_0^{\omega_{srt}} L_N . d\omega \qquad (3.10.2)$$

where ω_{srt} is the sunrise hour angle for the tilted horizon and ω_{sst} is the sunset hour angle for the tilted horizon. Substituting for L_N from equation 3.10.1 into equation 3.10.2, one obtains,

$$H_{ot} = 2 \cdot \int_0^{\omega_{srt}} (L \cdot \cos\delta \cdot \cos\omega \cdot \cos(\phi - \beta) + L \cdot \sin\delta \cdot \sin(\phi - \beta)) . d\omega$$
$$= 2L \cdot [\cos\delta \cdot \cos(\phi - \beta) \cdot \sin\omega_{srt} + \sin\delta \cdot \sin(\phi - \beta) \cdot \omega_{srt}] \; kW\,radians/m^2/day$$
$$= \frac{24}{\pi} \cdot L \cdot [\cos\delta \cdot \cos(\phi - \beta) \cdot \sin\omega_{srt} + \sin\delta \cdot \sin(\phi - \beta) \cdot \omega_{srt}] \; kWh/m^2/day$$
$$(3.10.3)$$

In equation 3.10.3, L is estimated according to equation 3.9.4. The sunrise hour angle with respect to the tilted horizon is given as

$$\omega_{srt} = \min(\omega_{sr}, \omega_{sr\beta}) \qquad (3.10.4)$$

where

$$\omega_{sr} = \cos^{-1}(-\tan\phi \cdot \tan\delta)$$
$$\omega_{sr\beta} = \cos^{-1}(-\tan(\phi - \beta) \cdot \tan\delta)$$

The sun should be above the horizon plane of the place for daylight to be incident at the place. For effective insolation to be incident on the tilted collector, the sun should also be above the tilted horizon. Therefore it is necessary that sun should be above both the horizon planes for any effective L_N to be incident on the tilted collector. ω_{sr} is the hour angle at sunrise with respect to the horizon plane of the place. $\omega_{sr\beta}$ is the sunrise hour angle with respect to the tilted horizon plane. In order to satisfy the condition that sun should be above both the horizon planes on any given day of the year for any latitude, the minimum of ω_{sr} and $\omega_{sr\beta}$ would be the effective sunrise angle at the place.

3.11 Irradiance simulation in Octave

Octave script for estimation of Irradiance H_o and H_{ot}

```
# Description : The following script calculates the incident
# energy on a horizontal/tilted surface given the latitude
# and day number.
# The input variables are
# N = Day number, N=1 for January 1st and N=365 for December 31st
# Q = latitude of the place in degrees (should be converted
#     to radians)
# B = the tilt angle for the flat plate collector
# ---------------------------------------------------------------
#Locality : Lattitude=12 deg, 58 minutes North;(Bangalore, India)
Q=12.97; #lattitude expressed in deg.
Q=Q*pi/180; #lattitude is now expressed in radians
B=Q; #tilt angle in radians

#Constants
Lsc = 1.37; #kW/m2 - mean solar constant
#Calculation of insolation and energy on all days of the year
for N=1:365,
    #Calculation of declination
    t = 2*pi*(N-80)/365;
    d = 23.45*sin(t)*(pi/180); #declination expressed in radians
    #Calculation of extra-terrestrial insolation scale factor
    k = 1 + 0.033*cos(2*pi*N/365);

    #Hour angle estimation
    wsr  = acos(-1*tan(Q)*tan(d));
    wsrB = acos(-1*tan(Q-B)*tan(d));
    wsrt= min(wsr,wsrB);

    Ho(N) = (24*k*Lsc/pi)*(cos(Q)*cos(d)*sin(wsr) + ...
```

```
                                          wsr*sin(Q)*sin(d));
  Hot(N) = (24*k*Lsc/pi)*(cos(Q-B)*cos(d)*sin(wsrt) +...
                                         wsrt*sin(Q-B)*sin(d));
  days(N) = N;
endfor #of day number loop

#Plot results
axis([0 365 0 15]); hold on;
plot(days,Ho,days,Hot);grid, xlabel('Day number, N'),...
            ylabel('kWh/m2/day'), title('Ho and Hot');
hold off;
# ----------------------------------------------------------------
```

The estimation of the irradiance on a flat horizontal collector and flat tilted collector at a place on the surface of earth is given in the form of Octave script above. For every day of the year, the irradiance values H_o and H_{ot} are calculated and stored in array variables. The daily incident energy on flat plate collector placed horizontally and at a tilt angle are plotted as a function of the time of year. This is shown in figure 3.11.1.

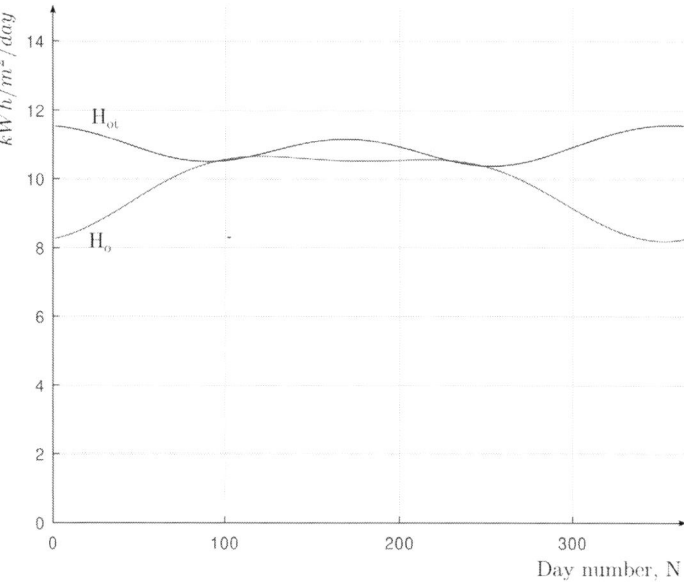

Figure 3.11.1: Octave plot of Irradiance on horizontal and tilted collector over the year

Figure 3.11.1 shows the plot profiles of H_o and H_{ot} as a function of the time of year. The day number N is plotted on the x-axis and H_o and H_{ot} on the y-axis. The latitude of a place on the surface of the earth is chosen as 12 deg, 58 minutes North. This is the latitude of Bangalore, India. The daily incident energy on a horizontal flat plate collector varies from around just about 8.2 $kWh/m^2/day$ to about 10.5 $kWh/m^2/day$.

When the flat plate collector is tilted at an angle same as that of the latitude and facing towards south, the irradiance on the tilted collector improves significantly. It varies between 10.5 $kWh/m^2/day$ and 11.5 $kWh/m^2/day$.

3.12 Large tilt generalisation

Consider a situation wherein the tilt is large. The merged locale and earth coordinates shown in figure 3.10.1 is modified to indicate a large tilt angle and depicted in figure 3.12.1. When the tilt angle is made equal to the latitude angle i.e. $\beta = \phi$, then the tilted horizon plane would be aligned along the polar axis or p-axis. The normal to the tilted horizon plane would be aligned along the meridional axis m-axis. If the tilt is larger than this i.e. $\beta > \phi$, then the profile as seen from the east is as shown in figure 3.12.1.

The insolation normal to the tilted horizon is along L_N vector. The objective remains the same and that is to estimate L_N and then consequently find the irradiance or daily incident energy on the tilted collector H_{ot}.

Referring to figure 3.12.1,

$$L_N = L_m \cdot cos(\phi - \beta) - L_p \cdot sin(\phi - \beta) \tag{3.12.1}$$

where L_m and L_p are as given in equation 3.8.2. The normal incident insolation L_N has two components, viz. (i) $L_m \cdot cos(\phi - \beta)$ is the meridional component which is the positive part that is the insolation incident normal to the tilted flat plate *front face*, and (ii) $-L_p \cdot sin(\phi - \beta)$ is the polar component which is the negative part that is the insolation incident normal to the tilted flat plate *back face*.

Equation 3.12.1 indicates that L_N is the equivalent algebraic sum of the two components. However, in reality, the polar or the negative component of the insolation does not diminish the intensity of the meridional or the positive component. It means that the two opposing components fall on opposite faces of the flat plate collector. The intervening flat plate collector decouples the two opposing components. If one wishes to estimate the insolation falling on only the front face of the flat plate collector, then

$$L_{Nf} = L_m \cdot cos(\phi - \beta) \tag{3.12.2}$$

The insolation falling on only the back face of the flat plate collector is given as

$$L_{Nb} = L_p \cdot sin(\phi - \beta) \tag{3.12.3}$$

Similar reasoning can be made when the tilt angle is negative. At places located in lower latitudes, one may wish to track the sun over the year as the sun travels up to Tropic of Cancer (23.45^oN). The flat plate collector needs to be tilted northwards in order to track the sun. This implies that the tilt angle $\beta < 0$. This also means that beyond a particular negative tilt angle, the normal to the tilted horizon can fall in the 1st quadrant of figure 3.12.1. Under such a condition, the polar component is the

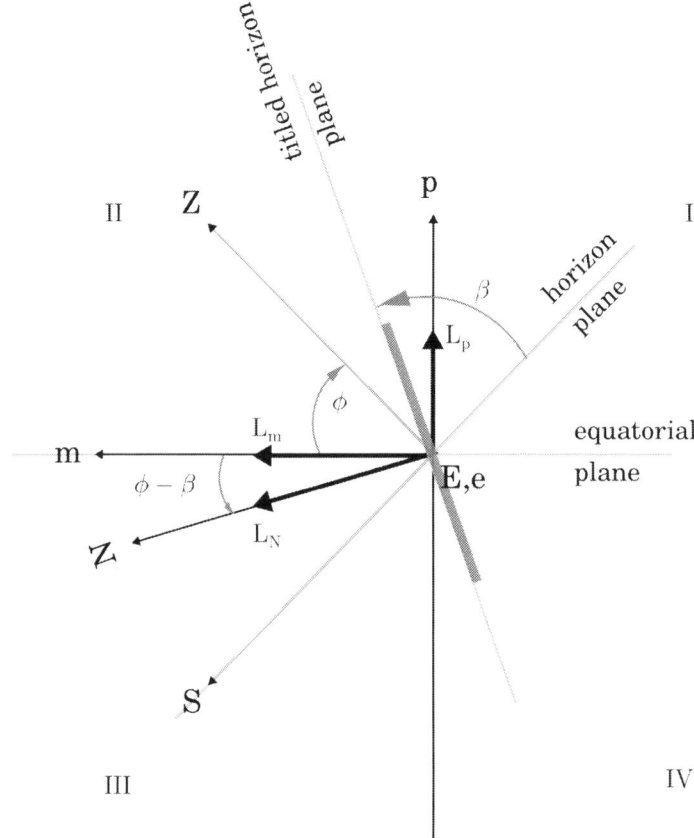

Figure 3.12.1: Collector placement in merged locale and earth coordinates with large tilt angle

positive component and is incident on the front face of the collector and the meridional component is the negative component and is incident on the back face of the collector. It should be noted that this is not a common occurrence. The traversal of the normal to the tilted horizon moving into the 1st quadrant will happen for collectors that are placed very near the north pole. For estimating the insolation falling on only the front face of the flat plate collector one needs to consider only the polar component, $L_p \cdot \sin(\phi - \beta)$. Therefore, the generalised normal incident insolation L_N for estimating insolation incident on the front face of the flat plate collector can be written as

$$L_N = c_m \cdot L_m \cdot \cos(\phi - \beta) + c_p \cdot L_p \cdot \sin(\phi - \beta) \qquad (3.12.4)$$

where c_m and c_p are conditional parameters and

3.12. LARGE TILT GENERALISATION

$$c_m = 1 \ for \ \phi - \beta \leq 90^o$$
$$= 0 \ otherwise$$

and

$$c_p = 1 \ for \ \phi - \beta \geq 0^o$$
$$= 0 \ otherwise$$

Equation 3.12.4 is the generalised insolation equation. When $0 \leq (\phi - \beta) \leq 90^o$, the L_N vector will lie in the 2nd quadrant wherein both the polar and meridional components are incident on the front face of the collector. Therefore, both c_m and c_p are both unity. When the tilt angle is positive and larger than the latitude angle, then the L_N vector will lie in the 3rd quadrant. Here $c_m = 1$ and $c_p = 0$ as the meridional component is incident on the front face of the collector and the polar component is incident on the back face of the collector. When the tilt angle is negative and $\phi - \beta \geq 0^o$, then the L_N vector will lie in the 1st quadrant. Here $c_m = 0$ and $c_p = 1$ as the meridional component is incident on the back face of the collector and the polar component is incident on the front face of the collector.

The daily incident energy on a tilted flat plate collector is given as

$$H_{ot} = \int_{\omega_{sst}}^{\omega_{srt}} L_N . d\omega$$

$$= 2 \cdot \int_{0}^{\omega_{srt}} L_N . d\omega \qquad (3.12.5)$$

where ω_{srt} is the sunrise hour angle for the tilted horizon and ω_{sst} is the sunset hour angle for the tilted horizon. Substituting for L_N from equation 3.12.4 into equation 3.12.5 and using equation 3.8.2 for L_m and L_p, one obtains,

$$H_{ot} = 2 \cdot \int_{0}^{\omega_{srt}} (c_m \cdot L \cdot cos\delta \cdot cos\omega \cdot cos(\phi - \beta) + c_p \cdot L \cdot sin\delta \cdot sin(\phi - \beta)) . d\omega$$
$$= 2L \cdot [c_m \cdot cos\delta \cdot cos(\phi - \beta) \cdot sin\omega_{srt} + c_p \cdot sin\delta \cdot sin(\phi - \beta) \cdot \omega_{srt}] kW \cdot rad/m^2/day$$
$$= \frac{24}{\pi} \cdot L \cdot [c_m \cdot cos\delta \cdot cos(\phi - \beta) \cdot sin\omega_{srt} + c_p \cdot sin\delta \cdot sin(\phi - \beta) \cdot \omega_{srt}] kWh/m^2/day$$
$$(3.12.6)$$

In equation 3.12.6, L is estimated according to equation 3.9.4. The sunrise hour angle with respect to the tilted horizon is given as

$$\omega_{srt} = min\left(\omega_{sr}, \omega_{sr\beta}\right) \qquad (3.12.7)$$

where

$$\omega_{sr} = cos^{-1}\left(-tan\phi \cdot tan\delta\right)$$
$$\omega_{sr\beta} = cos^{-1}\left(-tan(\phi - \beta) \cdot tan\delta \cdot \frac{c_p}{c_m}\right)$$

When $0 \leq (\phi - \beta) \leq 90°$, the L_N vector will lie in the 2nd quadrant and equation 3.12.6 reduces to equation 3.10.3. The sun should be above the horizon plane of the place for daylight to be incident at the place. For effective insolation to be incident on the tilted collector, the sun should also be above the tilted horizon. Therefore it is necessary that sun should be above both the horizon planes for any effective L_N to be incident on the tilted collector. ω_{sr} is the hour angle at sunrise with respect to the horizon plane of the place. $\omega_{sr\beta}$ is the sunrise hour angle with respect to the tilted horizon plane. In order to satisfy the condition that sun should be above both the horizon planes on any given day of the year for any latitude, the minimum of ω_{sr} and $\omega_{sr\beta}$ would be the effective sunrise angle at the place.

3.13 Optimum fixed tilt for collectors

If the collector should continuously tracks the sun, two trackers are needed, one for tracking the sun daily in the east-west direction and another for tracking the sun yearly for changes in declination. This will be in the north-south direction. Tracking systems will significantly increase the incident energy on the collector. However, the tracking system will increase the cost and maintenance. In most case, especially roof top systems that feed energy into the grid, the solar panels are kept tilted at the angle of the latitude. This gives almost uniform daily incident energy on the collector throughout the year with least maintenance.

The following Octave script gives further insight into the selection of the optimum tilt angle for a place. The estimation of irradiance at a place is slightly modified to include tilt angle as a parameter. The H_{ot} at a place is calculated for several tilt angles over the year.

Octave script for optimum tilt angle

```
# Description : The following script calculates the incident
# energy on a horizontal/tilted surface given the latitude
# and day number using tilt angle as a parameter
# The input variables are
# N = Day number, N=1 for January 1st and N=365 for December 31st
```

3.13. OPTIMUM FIXED TILT FOR COLLECTORS 61

```
# Q = latitude of the place in degrees (should be converted
#     to radians)
# B = the tilt angle for the flat plate collector
# ---------------------------------------------------------------
clc
clear
#Locality: Lattitude=12 deg, 58 minutes North;(Bangalore, India)
Q=12.97; #expressed in deg.
Q=Q*pi/180; #lattitude is now expressed in radians
B=Q; #tilt angle in radians
#constants
Lsc = 1.37; #kW/m2 - mean solar constant
disp(['Hot-MIN',"\t\t",'H-RIPPLE',"\t",'LATITUDE,deg',"\t",...
                                            'TILT ANGLE,deg']);
Bmax=90*pi/180;tilt=[];meanHot=[];
for B=-pi/2:Bmax/30:Bmax
   #Calculation of insolation and energy on all days of the year
   for N=1:365,
      #Calculation of declination
      t = 2*pi*(N-80)/365;
      d = 23.45*sin(t)*(pi/180); #declination expressed in radians
      #Calculation of extra-terrestrial insolation scale factor
      k = 1 + 0.033*cos(2*pi*N/365);
      cm = Q-B<=pi/2;
      cp = Q-B>=0;

      wsr = acos(-1*tan(Q)*tan(d));
      wsrB = acos(-1*tan(Q-B)*tan(d)*cp/cm);
      wsrt= min(wsr,wsrB);

      Ho(N) = (24*k*Lsc/pi)*(cos(Q)*cos(d)*sin(wsr) + ...
                                         wsr*sin(Q)*sin(d));
      Hot(N) = (24*k*Lsc/pi)*(cm*cos(Q-B)*cos(d)*sin(wsrt) + ...
                                      cp*wsrt*sin(Q-B)*sin(d));
      days(N) = N;
   endfor #of day number loop
   #Show results
   disp([num2str(min(Hot)),"\t\t",num2str(max(Hot)-min(Hot)),...
           "\t\t",num2str(Q*180/pi),"\t\t",num2str(B*180/pi)]);
   plot(days,Ho,days,Hot,days,mean(Hot)*ones(size(Ho)));grid,...
              xlabel('Day number, N'), ylabel('kWh/m2/day'),...
              title('Ho and Hot');
   tilt=[tilt B*180/pi];
   meanHot=[meanHot mean(Hot)];
   pause
```

```
endfor #of tilt angle loop
plot(tilt,meanHot);grid, xlabel('Tilt angle, deg'), ...
   ylabel('mean Hot, kWh/m2/day'), ...
   title('Finding optimum tilt angle');
# ---------------------------------------------------------------
```

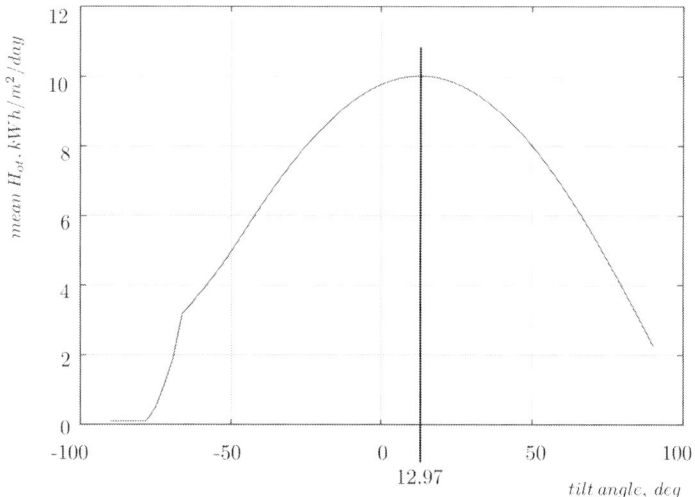

Figure 3.13.1: Octave plot of tilt angle versus annual mean irradiance for the tilt

The tilt angle versus the annual mean irradiance on a tilted surface is plotted as shown in figure 3.13.1. The figure shows that the optimum tilt angle for the place having latitude as 12.97^oN is also 12.97^o. The collector is tilted such that it faces the equator. As a consequence the collector is facing south wards. It can be verified that for any given place the optimum tilt angle is same as the latitude angle of the place.

3.14 Atmospheric effects

The previous sections discussed estimation of incident energy on a flat collector placed at a point on the earth surface, without considering the attenuation effects of atmosphere. With out the effects of atmospheric attenuation, the irradiance at the surface of the earth will be same as the extra-terrestrial irradiance. However, atmosphere has a tendency to absorb the incident radiation. The ozone layer in the atmosphere absorbs a significant amount of the high frequency (small wavelength) portions of the visible radiation. Apart from ozone layer, clouds too absorb and scatter the incident radiation. This will reduce the amount of radiation that will eventually reach the surface of the earth. The water vapour content in the vertical column of airmass just above the place of interest is also another component of the atmosphere that will attenuate the radiation received at the surface of the earth. Aerosols, other gases and dust present in the

3.14. ATMOSPHERIC EFFECTS

vertical column of airmass also will add to the attenuation of radiation. H_o is the incident energy on the surface of the earth without considering atmospheric attenuation. If atmospheric attenuation is considered then the incident energy on the horizontal flat plate collector is denoted by H_a. The incident energy on a tilted flat plate collector considering atmospheric effects is denoted by H_{at}.

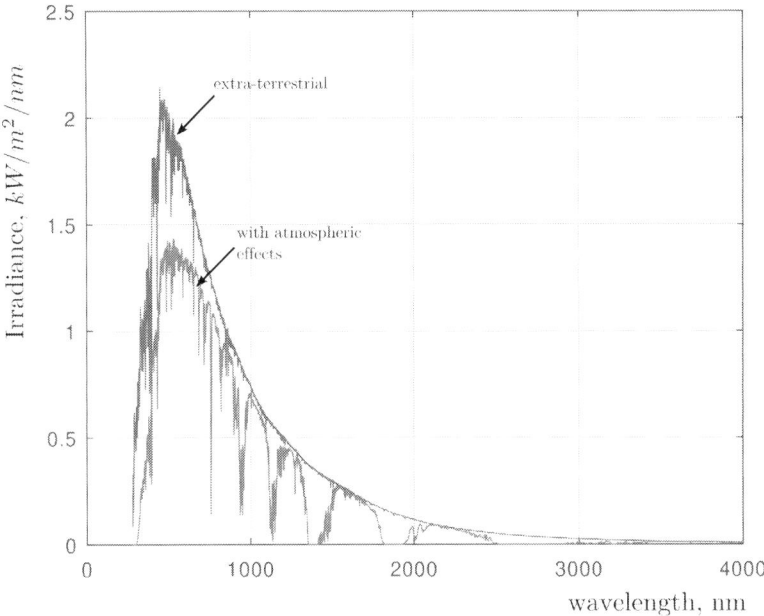

Figure 3.14.1: Solar spectrum with atmospheric attenuation

Figure 3.14.1 shows the solar spectra with and without atmospheric effects. The curve with higher amplitude represents the spectral irradiance without atmospheric effects. It is the same as the extra-terrestrial radiation spectrum as given before in figure 3.4.1(b). The extra-terrestrial spectrum will become attenuated after passing through the earth's atmosphere. The solar spectrum of the radiation received at the surface of the earth is shown superimposed in figure 3.14.1. The attenuation of the ultraviolet components or high frequency components are due to ozone layer absorption. Clouds provide significant attenuation in the visible frequency part of the spectrum. The figure also shows notches in the spectrum caused by significant attenuation due to water vapour content present in the atmosphere. This is predominant at lower visible and infrared frequencies.

It can be seen that there are 5 factors that significantly contribute to the incident irradiation on the surface of the earth. They are

1. Location : It is seen that the latitude of the place has a significant bearing on the amount of incident solar radiation. Apart from latitude, the height of the place above mean sea level (m.s.l) also matters. The airmass in the vertical column above the place is directly related to the height of the place above mean sea level.

The airmass column has a significant attenuation effect on the extra-terrestrial irradiance before it reaches the surface of the earth.

2. Time of day : Another location parameter is the longitude or the meridian of the place. The daily change in the insolation from sunrise to sunset at a place directly relates to the amount of irradiance that is received in a day.

3. Time of year : The declination relates to the yearly effects. The variation in the incident energy over the year is encompassed in the declination term.

4. Collector tilt : The previous section discussed the effects of collector plate tilt. The optimum fixed tilt angle needed for the collector to capture the maximum incident energy is also discussed.

5. Local climate : While points (1) to (4) are deterministic, the local climate has a measure of uncertainty about it. This is a statistical phenomenon. If local climate is measured over a period of several years, then a statistical correction factor can be used for the local climate. This statistical correction factor is called the clearness index k_T. It is the ratio of the incident energy at a place with atmospheric effects to the incident energy at the place without atmospheric effects. Therefore, the clearness index can be written as
$K_T = \frac{H_a}{H_o}$.

3.14.1 Airmass coefficient

The airmass coefficient gives a measure of the optical path length through which the spectral irradiance or the insolation vector passes through the atmosphere before it reaches the flat plate collector placed on the surface of the earth at a given locality. The path length of the insolation vector is dependent on the time of the day, time of the year and the geography of the place. Based on the optical path length, the factor by which the extra-terrestrial radiation undergoes attenuation will vary.

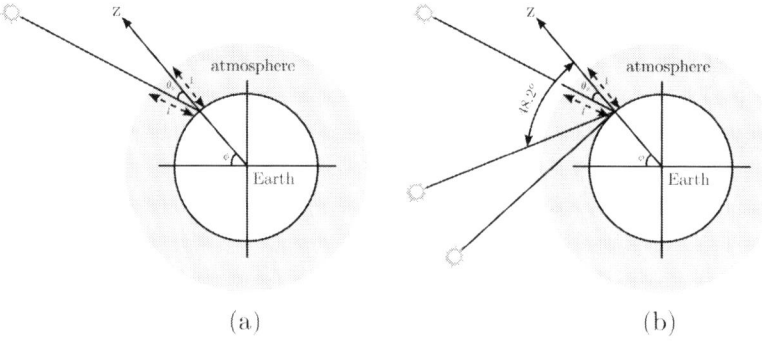

Figure 3.14.2: Airmass (a) Illustration of optical path length (b) optical path lengths at varying zenith angles

3.14. ATMOSPHERIC EFFECTS

The airmass coefficient defines the optical path length of the insolation vector through the atmosphere. This coefficient will not provide any information on the climatic condition at the location. This however uses the geometrical properties like the zenith angle to provide a measure of the absorption or attenuation that may happen. It is a benchmark parameter which is used to compare various panels under standard conditions or standard air masses. It is also used as a benchmark for comparison of spectral irradiance data of various measurements done by different groups with respect to standard air masses.

Consider the illustration of optical path length as shown in figure 3.14.2(a). The zenith axis is considered as the reference. If the insolation vector from the sun is along the zenith axis, then the optical path length through the airmass is the shortest. This airmass is called AM1. It is one of the standard air masses. If the insolation vector is incident at the zenith angle θ_z with respect to zenith axis, then the optical path length is longer. The airmass coefficient is expressed as AMl where 'l' is a number that relates to the optical path length. In figure 3.14.2(a), the path length along the zenith axis is considered as unity. The path length along any other path is 'l'. For small zenith angles,

$$l = \frac{1}{\cos\theta_z} = \sec\theta_z \qquad (3.14.1)$$

Table 3.1: Airmass coefficient at different zenith angles

Airmass coefficient	Zenith angle, θ_z	Remarks
AM0	0	No atmosphere
AM1	0	With atmosphere
AM1.1	25^0	
AM1.5	48.2^0	Standard airmass
AM2	60^0	

Figure 3.14.2(b) gives a visualisation of the optical path lengths for different zenith angles. Using equation 3.14.1, the airmass coefficients are tabulated in table 3.1. The airmass coefficient for the optical path along the zenith axis with no atmosphere is AM0 and AM1 for that with atmosphere and along the zenith axis. AM1.1 is the airmass coefficient for an optical path along a line that is at an angle of 25^0 with respect to the zenith axis. AM1.5 is considered as standard airmass which is used as a benchmark for comparisons. PV panel datasheets also provide parameters with respect to AM1.5. This is the airmass coefficient at a zenith angle of 48.2^0. AM2 is the airmass coefficient at a zenith angle of 60^0.

The airmass coefficient discussed till now assumes that the place on the surface of the earth is at sea level. However, the place may be quite above sea level too. In that case the airmass coefficient needs a correction factor to account for the reduction in the path length due to the height of the place above mean sea level. Let l^* be the optical path length at a place which is above mean sea level. As usual l would mean the optical path length of the place at the same location but at sea level. The pressure at the place gives a measure of the path length correction. Let P^* be the pressure at the place which is at a height about sea level and P_o is the pressure at the same coordinates which is at sea level. Then the airmass coefficient of the place which is at a height about sea level is given as AMl^*, where

$$l^* = l \cdot \frac{P^*}{P_o} \quad (3.14.2)$$

The pressure at a place above sea level will be lesser that that at sea level. Therefore, the correction factor for l in equation 3.14.2 will be less than 1. This implies that for places at heights above sea level, the optical path length will be lesser than the place at the location at sea level. The corrected airmass coefficient can address places that are situated above mean sea level.

3.14.2 Clearness Index

The clearness index K_T is the ratio of daily energy incident on a horizontal flat plate with atmosphere to the daily energy incident on the horizontal flat plate without atmosphere. It is expressed as

$$K_T = \frac{H_a}{H_o} \quad (3.14.3)$$

The clearness index is a measure of the attenuation due to presence of atmosphere. The attenuation of radiation due to atmosphere depends on the place, water vapour content in the vertical column, weather condition prevalent during the day and during the year. There is a measure of uncertainty in the weather conditions at a place which makes it difficult to estimate the energy incident at a place. All these climatic uncertainties are included into the clearness index variable. The value of H_o that is the incident irradiance without atmospheric effect can be estimated using equation 3.9.3. The value of H_a, the irradiance at a place with atmospheric effects is measurable using a pyranometer. Pyranometer is an instrument that is used to measure insolation at a place. Using this measured value of the irradiance at a place, one can calculate the actual clearness index using equation 3.14.3. This measurement needs to be done over several days and years to obtain a large enough data sample in order to obtain the clearness index at a place that is a statistical average of the weather condition at the place.

One may wish to estimate H_a at a place where actual measured data is unavailable. In such a case, an estimate of H_a needs to be made. For this, an estimate of the clearness index at the place is necessary. It is impossible to obtain the true clearness index for the place. There will exist an error due to the presence of various uncertainties. The estimate of the clearness index K_{Te} will differ from the true clearness index of the place K_T by an error value e as expressed below.

3.14. ATMOSPHERIC EFFECTS

$$K_T = K_{Te} + e$$

The atmospheric conditions at a place are dependent on latitude angle, tilt, time of day, time of year and other uncertain parameters like water vapour content in the vertical column, geographical position and clouds. The clearness index should be modeled to fit the actual clearness index using only measurable variables. A curve fit model can be used for this purpose. One can use regression fit with an appropriate curve profile. The atmospheric conditions have a yearly periodicity at any given place. A fourier fit with a periodicity of one year is a realistic fit for modeling atmospheric uncertainties. The clearness index estimate equation can be of the following form as given in equation 3.14.4,

$$K_{Te} = A_1 + (A_2 \cdot \sin\tau + A_3 \cdot \cos\tau) + (A_4 \cdot \sin2\tau + A_5 \cdot \cos2\tau) + \cdots \quad (3.14.4)$$

$$A_i = a_{i1} + a_{i2} \cdot x + a_{i3} \cdot x^2 + \cdots$$

$$x = f(\phi, \omega, \beta, \cdots)$$

where the variable x is a function of latitude ϕ, hour angle ω, tilt angle β and other measurable parameters.

Equation 3.14.4 is a generic model that can be used to fit with weather data at a given region. It is required that the curve fit coefficients be estimated in order to define the curve fit model. This process of estimating the curve fit coefficients or parameters a_{ij} will be illustrated using a 1st harmonic model for the purpose of understanding the fourier fit model. As a demonstration, the clearness index estimate equation can be written as

$$K_{Te} = A_1 + (A_2 \cdot \sin\tau + A_3 \cdot \cos\tau)$$

where

$$\tau = \frac{2\pi N}{365}$$

with N being the day number in the year, and

$$A_i = a_{i1} + a_{i2} \cdot x$$

$$x = \phi$$

Therefore,

$$K_{Te} = (a_{11} + a_{12} \cdot x) + (a_{21} + a_{22} \cdot x)\sin\tau + (a_{31} + a_{32} \cdot x)\cos\tau \qquad (3.14.5)$$

In order to model the clearness index, one needs to estimate the parameters a_{ij} of the estimate equation 3.14.5. K_T is the actual clearness index at a place that is measured using an instrument like a pyranometer. K_{Te} is the estimate of the clearness index at the place based on an estimate model such as given in equation 3.14.5. The error in the estimate is given as,

$$e = K_T - K_{Te}$$

There are several places where the clearness index is measured and the errors are placed into an error vector. In order to develop the estimate model for clearness index, it is required to obtain the coefficient parameters a_{ij} under the constraint that the sum of the square of the errors is minimised. This would be the constraint for obtaining the optimal values of a_{ij}. Thus, the constraint cost function for minimisation is

$$J = \sum_i e_i^2 = \sum_i (K_T - K_{Te})_i^2 \qquad (3.14.6)$$

In order to minimise J, the partial derivatives with respect to each of the coefficient parameter $\frac{\partial J}{\partial a_{ij}}$ should be set to zero. Therefore, set $\frac{\partial J}{\partial a_{11}}$, $\frac{\partial J}{\partial a_{12}}$, $\frac{\partial J}{\partial a_{21}}$, $\frac{\partial J}{\partial a_{22}}$, $\frac{\partial J}{\partial a_{31}}$, $\frac{\partial J}{\partial a_{32}}$ to 0 individually. Thus,

$$\frac{\partial J}{\partial a_{11}} = -2\sum_i (K_T - K_{Te})_i = 0$$

$$\frac{\partial J}{\partial a_{12}} = -2\sum_i (K_T - K_{Te})_i \cdot x_i = 0$$

$$\frac{\partial J}{\partial a_{21}} = -2\sum_i (K_T - K_{Te})_i \cdot \sin\tau_i = 0$$

$$\frac{\partial J}{\partial a_{22}} = -2\sum_i (K_T - K_{Te})_i \cdot x_i \sin\tau_i = 0$$

$$\frac{\partial J}{\partial a_{31}} = -2\sum_i (K_T - K_{Te})_i \cdot \cos\tau_i = 0$$

3.14. ATMOSPHERIC EFFECTS

$$\frac{\partial J}{\partial a_{32}} = -2\sum_i (K_T - K_{Te})_i \cdot x_i \cos\tau_i = 0$$

Substituting for K_{Te} from the estimate equation 3.14.5 and re-arranging, one obtains

$$a_{11}\sum_i 1 + a_{12}\sum_i x_i + a_{21}\sum_i \sin\tau_i + \ldots$$

$$\ldots a_{22}\sum_i x_i \sin\tau_i + a_{31}\sum_i \cos\tau_i + a_{32}\sum_i x_i \cos\tau_i = \sum_i K_{T_i}$$

$$a_{11}\sum_i x_i + a_{12}\sum_i x_i^2 + a_{21}\sum_i x_i \sin\tau_i + a_{22}\sum_i x_i^2 \sin\tau_i \cdots$$

$$\cdots + a_{31}\sum_i x_i \cos\tau_i + a_{32}\sum_i x_i^2 \cos\tau_i = \sum_i x_i K_{T_i}$$

$$\vdots$$

$$\begin{bmatrix} \sum_i 1 & \sum_i x_i & \sum_i \sin\tau_i & \sum_i x_i \sin\tau_i & \sum_i \cos\tau_i & \sum_i x_i \cos\tau_i \\ \sum_i x_i & \sum_i x_i^2 & \sum_i x_i \sin\tau_i & \sum_i x_i^2 \sin\tau_i & \sum_i x_i \cos\tau_i & \sum_i x_i^2 \cos\tau_i \\ \vdots & \vdots & \vdots & \vdots & \vdots & \vdots \\ \vdots & \vdots & \vdots & \vdots & \vdots & \vdots \\ \vdots & \vdots & \vdots & \vdots & \vdots & \vdots \\ \vdots & \vdots & \vdots & \vdots & \vdots & \vdots \end{bmatrix} \begin{bmatrix} a_{11} \\ a_{12} \\ a_{21} \\ a_{22} \\ a_{31} \\ a_{32} \end{bmatrix} \cdots$$

$$\cdots = \begin{bmatrix} \sum_i K_{T_i} \\ \sum_i x_i K_{T_i} \\ \sum_i \sin\tau_i K_{T_i} \\ \sum_i x_i \sin\tau_i K_{T_i} \\ \sum_i \cos\tau_i K_{T_i} \\ \sum_i x_i \cos\tau_i K_{T_i} \end{bmatrix} \quad (3.14.7)$$

In the above estimation matrix, a_{ij} are the parameters that need to be evaluated. Equation 3.14.7 is of form $Ax = b$ with a trivial solution of $A^{-1}b$. One may use environments such as Octave or MATLAB to solve for the parameters a_{ij} from equation 3.14.7. The above method can be adopted even for K_{Te} estimation equation that contains several harmonic orders. Using the above approach the model for India is developed.

The clearness index estimate equation is taken as

$$K_{Te} = A_1 + A_2 sin\tau + A_3 sin2\tau + A_4 sin3\tau \\ + A_5 cos\tau + A_6 cos2\tau + A_7 cos3\tau \quad (3.14.8)$$

where

$$\tau = 2\pi \frac{(N-80)}{365}$$

and

$$A_i = a_{i1} + a_{i2}x + a_{i3}x^2 + a_{i4}w + a_{i5}w^2$$

$$x = (\phi - 35)$$

where w is the water vapour content in the vertical column above the place. The water vapour content is estimated by a similar fourier curve fit method and this estimate is used in the clearness index estimate equation. The water vapour estimate model is given as

$$w = G_1 + G_2 sin\tau + G_3 sin2\tau + G_4 sin3\tau \\ + G_5 cos\tau + G_6 cos2\tau + G_7 cos3\tau \quad (3.14.9)$$

$$G_i = g_{i1} + g_{i2}x + g_{i3}x^2$$

The estimate of the water vapour content is plugged into the estimate coefficients A_i to estimate the clearness index as per equation 3.14.8. The Octave script for estimation of water vapour content for India is given in Appendix D.1 and this estimate is used in estimating the clearness index. The model for clearness index and the octave script to estimate it for India is given in Appendix D.2.

3.14.3 Tilt factor

In discussion till now, the solar flat plate collector is placed horizontally at a place. However, apart from places on the equator, horizontal placement of the solar flat plate collector is not the optimal orientation. Generally the flat plate collector is tilted with respect to the horizon plane. The angle of tilt with respect to the horizon plane is called the tilt angle, β. H_o is the daily energy that is incident on a horizontally placed flat plate collector. It the same plate were tilted at an angle β with respect to the horizon plane, then the amount of energy incident on such a tilted plate is represented by H_{ot}. Tilt factor R_D is defined as the ratio of amount of energy incident on a tilted flat plate to that incident on a horizontally placed flat plate. It is given as

$$R_D = \frac{H_{ot}}{H_o} \tag{3.14.10}$$

Equations 3.9.3 and 3.10.1 are applied in the tilt factor equation 3.14.10 to obtain the relationship given below

$$\begin{aligned} R_D &= \frac{cos(\phi-\beta) \cdot cos\delta \cdot sin\omega_{srt} + \omega_{srt} \cdot sin(\phi-\beta) \cdot sin\delta}{cos\phi \cdot cos\delta \cdot sin\omega_{sr} + \omega_{sr} \cdot sin\phi \cdot sin\delta} \\ &= \frac{cos(\phi-\beta)}{cos\phi} \cdot \left(\frac{sin\omega_{srt} + \omega_{srt} \cdot tan(\phi-\beta) \cdot tan\delta}{sin\omega_{sr} + \omega_{sr} \cdot tan\phi \cdot tan\delta} \right) \\ &= \frac{cos(\phi-\beta)}{cos\phi} \cdot \left(\frac{sin\omega_{srt} - \omega_{srt} \cdot cos\omega_{srt}}{sin\omega_{sr} - \omega_{sr} \cdot cos\omega_{sr}} \right) \end{aligned} \tag{3.14.11}$$

where ω_{srt} is as defined in equation 3.10.4.

3.15 Energy on tilted surface with atmospheric effects

The energy incident on a tilted surface considering atmospheric effects will provide estimates that are closest to reality. One can categorise the incident energy on a tilted surface as (i) direct radiation, (ii) ground reflected (albedo) radiation and (iii) diffuse radiation. All these different categories of insolation will produce the cumulative energy H_{at} that is incident on a tilted surface considering atmospheric and geographic conditions. This can be visualised with the help of figure 3.15.1.

Figure 3.15.1 shows typical placement of a tilted solar flat plate collector on the ground. The terrain is undulating and also shows a nearby elevated ground or hill. The flat plate is mounted on struts and placed at an angle β with respect to the horizon. On this flat plate collector is incident the three types of radiation viz. the direct radiation, ground reflected radiation and the diffuse radiation. As can be observed from figure 3.15.1, an extra-terrestrial insolation vector enters the atmospheric boundary of the planet and is incident on the flat plate collector. This is the direct radiation incident on the collector. Apart from the direct incident radiation, there are subsidiary radiations like the albedo effect. It can be visualised from the figure 3.15.1 wherein the extra-terrestrial radiation is incident on the slope of the hill and then is reflected from it. The reflected radiation is incident on the flat plate collector. This is called the ground reflected radiation or the albedo effect. A third effect is the scattering of the extra terrestrial insolation due to clouds. Some amount of the scattered radiations also called

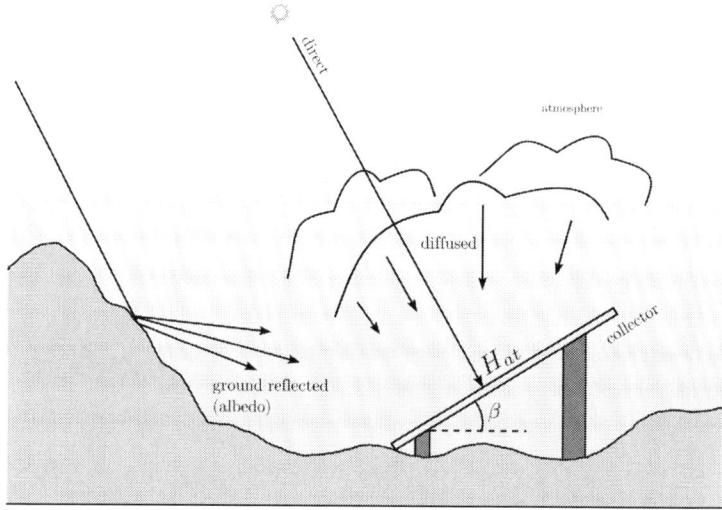

Figure 3.15.1: Insolation on tilted flat plate collector

diffuse radiations will be incident on the flat plate collector. The cumulative effect of all these three types of incident radiation will give the total incident energy on the flat plate collector, H_{at}.

H_{at} is actually a composition of energy from direct radiation plus energy from diffused radiation plus energy from ground reflected radiation or the Albedo. All the three effects put together will be the energy available on the tilted flat plate collector. This is a practical and real situation, and it is not trivial to estimate this. There are several uncertainties unlike the cases that involved estimation of H_o and H_{ot} where the effect of atmosphere and its uncertainties were not present. However, the estimates were idealistic and as a result deterministic relations were obtained.

There is considerable uncertainty in modeling the variations in atmosphere, cloud, water vapour and other gaseous content in the vertical column above the place where the collector is placed. The topography of a place too varies from place to place having a bearing on the albedo. The uncertainty in the locality and atmospheric conditions at a place is modeled in the clearness index K_T of a given place. Based on statistical weather data at the place, the clearness index is estimated as discussed. The energy incident on a horizontal flat plate collector considering the atmospheric conditions, H_a is given as

$$H_a = K_{Te} \cdot H_o$$

where K_{Te} is the estimated clearness index and H_o is the energy incident on a horizontal flat plate without atmospheric effects.

Consider H_d as only the diffuse component of the radiation that is incident on the flat plate collector if it is placed horizontally. Then $(H_a - H_d)$ would represent the direct component of the radiation incident on a horizontal flat plate collector. Using

3.15. ENERGY ON TILTED SURFACE WITH ATMOSPHERIC EFFECTS

the tilt factor obtained in equation 3.14.11, $(H_a - H_d) \cdot R_D$ will represent the direct component of the radiation incident on a flat plate collector that is placed with a tilt angle β. The diffuse component of the radiation incident on a tilted surface is given by an empirical relation $H_d \frac{(1+\cos\beta)}{2}$. It can be seen that for a horizontal placement of the flat plate collector, the tilt angle β becomes 0 and the diffuse component becomes just H_d. The albedo component or the ground reflected component of the radiation that is incident on the tilted flat plate collector is also expressed by an empirical relation given as $H_a \cdot \rho \cdot \frac{(1-\cos\beta)}{2}$. It can be observed that when the collector is placed horizontal, then the tilt angle β is zero and there is no ground reflected component. This is an expected conclusion as the albedo effect will be minimal when the collector is horizontal. A new variable ρ called the reflection coefficient is introduced in the albedo component. It is a weighting factor for the albedo component. The reflection coefficient is dependent on the terrain of the place. If the terrain is plain, then there is not much possibility of ground reflected radiation incident on the collector and therefore, the reflection coefficient is low (around 0.1). If the neighbourhood of the place has hills, snow capped mountains and undulations, then the ground reflection component incident on the collector is higher. The reflection coefficient varies between 0.1 and 0.7 and is chosen based on the terrain of the place. Combining all the above effects, H_{at} the energy incident on a tilted flat plate collector considering atmospheric effects too is given by

$$\begin{aligned} H_{at} &= (H_a R_D - H_d R_D) + \left(H_d \frac{(1+\cos\beta)}{2}\right) + \left(H_a \cdot \rho \cdot \frac{(1-\cos\beta)}{2}\right) \\ &= H_a \left(1 - \frac{H_d}{H_a}\right) R_D + H_a \left(\frac{H_d}{H_a}\right) \frac{(1+\cos\beta)}{2} + H_a \cdot \rho \cdot \frac{(1-\cos\beta)}{2} \\ &= H_a \cdot \gamma_T \end{aligned} \qquad (3.15.1)$$

where γ_T is the modified tilt factor considering atmospheric effects too. It is the ratio of the energy incident on a tilted flat plate collector with atmospheric effect to the energy incident on a horizontal flat plate collector with atmospheric effects. The ratio $\frac{H_d}{H_a}$ has a strong dependence on the clearness index. It is given by the following relation

$$\frac{H_d}{H_a} = 1 - 1.13 K_T$$
$$\rho = 0.1 \, to \, 0.7$$

H_{at} estimate will give a close approximation to the incident energy on a flat plat collector at a given place. This estimate will later be used to determine the size of the photovoltaic panels for a specific application.

3.16 Energy script in Octave

There are several steps that has been discussed in the previous sections that leads to ultimately the estimation of the incident energy on a tilted flat plate collector taking into consideration the atmospheric effects. The steps are as follows,

Determine the location of the place, ϕ

Determine the water vapour content estimate in the vertical column at the given place as per equation 3.14.9

Use the water vapour content estimate for estimating the clearness index model for the place as per equation 3.14.8

Determine the tilt angle, β

Estimate H_o

Estimate H_{ot}

Estimate H_a

Estimate R_D

Estimate γ_T which includes estimation of clearness index too

Estimate H_{at}

Based on the above steps an octave script can be written to estimate H_{at} on a given day of the year and every day of the year. A typical octave script is given below where the plot of H_{at} over the year can be visualised.

```
# This script calculates the incident energy on a
# horizontal/tilted surface given the latitude and
# day number with atmosphere
# INPUTS
#       N = Day number, N=1 for January 1st & N=365
#           for December 31st
#           (February 29th not considered).
#       Q = latitude of the place in degrees
#           (convert to radians)
#       B = Tilt angle in degrees (convert to radians)
clc
clear
#INPUTS
Q=12.97; #expressed in deg.
x=(Q-35); #Fourier curve fit variable
Q=Q*pi/180; #lattitude is now expressed in radians
B=Q; #tilt angle in radians as found optimal
```

3.16. ENERGY SCRIPT IN OCTAVE

```
    #without atmosphere

#constants
Lsc = 1.37; #kW/m2 - mean solar constant
rho = 0.2; #reflection co-efficient
# water vapour estimate equation co-efficients
G=[ 1.6204   -0.2291   -0.0068;
    1.6857   -0.0073   -0.0020;
   -1.2423   -0.0810   -0.0010;
    0.5626    0.0708    0.0019;
   -1.2140    0.0064    0.0014;
   -0.0990   -0.0133   -0.0004;
    0.4972    0.0605    0.0015];

# clearness index estimate equation co-efficients
A=[ 0.5563    0.0089    0.0002    0.0743   -0.0089;
   -0.2350    0.0119    0.0004    0.1473   -0.0237;
   -0.1011   -0.0091   -0.0004    0.1029   -0.0201;
    0.0136    0.0041    0.0002   -0.0071    0.0010;
    0.1300   -0.0133   -0.0003   -0.0848    0.0098;
   -0.0600    0.0048    0.0002    0.0733   -0.0132;
    0.0970    0.0058    0.0002   -0.0282    0.0010];


# Calculation of insolation and energy on all days of the year
for N=1:365,
    #Calculation of declination
    t = 2*pi*(N-80)/365;
    d = 23.45*sin(t)*(pi/180); #declination expressed
                                #in radians

    #Calculation of extra-terrestrial insolation scale factor
    k = 1 + 0.033*cos(2*pi*N/365);
    cm = Q-B<=pi/2;
    cp = Q-B>=0;
    wsr = acos(-1*tan(Q)*tan(d));
    wsrB = acos(-1*tan(Q-B)*tan(d)*cp/cm);
    wsrt= min(wsr,wsrB);

    Ho(N) = (24*k*Lsc/pi)*(cos(Q)*cos(d)*sin(wsr)...
            + wsr*sin(Q)*sin(d));
    Hot(N) = (24*k*Lsc/pi)*(cm*cos(Q-B)*cos(d)*sin(wsrt) ...
            + cp*wsrt*sin(Q-B)*sin(d));
    days(N) = N;
```

```
    #Introduce atmospheric effects
    #estimate water vapour content
      XX=[1;x;x*x];
      G1=G(1,1:3)*XX;
      G2=G(2,1:3)*XX;
      G3=G(3,1:3)*XX;
      G4=G(4,1:3)*XX;
      G5=G(5,1:3)*XX;
      G6=G(6,1:3)*XX;
      G7=G(7,1:3)*XX;
      W(N)=G1+G2*sin(t)+G3*sin(2*t)+G4*sin(3*t)...
           +G5*cos(t)+G6*cos(2*t)+G7*cos(3*t);
    #estimate clearness index
      YY=[1;x;x*x;W(N);W(N)*W(N)];
      A1=A(1,1:5)*YY;
      A2=A(2,1:5)*YY;
      A3=A(3,1:5)*YY;
      A4=A(4,1:5)*YY;
      A5=A(5,1:5)*YY;
      A6=A(6,1:5)*YY;
      A7=A(7,1:5)*YY;
      Kte(N)=A1+A2*sin(t)+A3*sin(2*t)+A4*sin(3*t)...
             +A5*cos(t)+A6*cos(2*t)+A7*cos(3*t);
      Rd(N)= Hot(N)/Ho(N);%tilt factor
      Kd(N)= 1-1.13*Kte(N); % Hd/Ha diffuse radiation factor
      rt(N)= ((1-Kd(N))*Rd(N)) + (Kd(N)*(1+cos(B))/2) ...
             + (rho*(1-cos(B))/2);  % overall tilt factor
endfor #of day number loop
Hat_direct=Rd.*Kte.*Ho;
Hat = rt.*Kte.*Ho; #tilted surface with atmospheric effects
#Show results
plot(days,Ho,days,Hat);grid, xlabel('Day number, N'),
ylabel('kWh/m2/day'), title('Ho and Hat versus Day of year');
```

3.17 Irradiance on vertically placed collectors

In the temperate regions of the planet north of the tropic of cancer and also south of the tropic of capricorn towards the poles, the component of the insolation vector that gets resolved along the horizon plane is significant in the respective summer days. One can place the flat plate solar PV collectors on the vertical walls of the building and windows of the building that are facing towards the equator or facing south for regions in the northern hemisphere. For such applications, it is of interest to estimate the amount of solar energy that is incident on a south facing vertical surface. The normal insolation vector L_N will be along the south of the horizon plane of the place. This will be along L_S direction which is in the third quadrant.

The insolation incident on the vertical surface at a place can be estimated using equation 3.12.6, which is the generalized insolation equation for large tilts. By setting the tilt angle as $\beta = 90°$, the parameters $c_m = 1$ and $c_p = 0$. The daily energy incident on a vertical flat plate collector H_{vo} is given as

$$H_{vo} = 2 \cdot \int_0^{\omega_{srt}} (L \cdot \cos\delta \cdot \cos\omega \cdot \cos(\phi - 90°)) . d\omega$$

$$= 2L \cdot [\cos\delta \cdot \cos(\phi - 90°) \cdot \sin\omega_{srt}] \; kW\,radians/m^2/day$$

$$= \frac{24}{\pi} \cdot L \cdot [\cos\delta \cdot \sin\phi \cdot \sin\omega_{srt}] \; kWh/m^2/day \qquad (3.17.1)$$

L is estimated according to equation 3.9.4. The sunrise hour angle with respect to the tilted horizon is given as

$$\omega_{srt} = \min(\omega_{sr}, \omega_{sr\beta}) \qquad (3.17.2)$$

where

$$\omega_{sr} = \cos^{-1}(-\tan\phi \cdot \tan\delta)$$
$$\omega_{sr\beta} = \cos^{-1}(0)$$

The clearness index is a statistical scaling factor that is dependent on the local climatic conditions. It is same for both the horizontal and vertical collectors. So also is the overall tilt factor. Therefore, H_{vat} which is the energy incident normally on a south facing vertical flat plate collector including atmospheric effects of the locality is given as

$$H_{vat} = H_{vo} \cdot K_{Te} \cdot \gamma_T \qquad (3.17.3)$$

3.18 Questions

1. The output of a PV cell is 2 W at an insolation of $1kW/m^2$. The area of a cell is $150 \times 150 mm^2$. If the efficiency of a cell is 15 %, then the angle of incidence of the insolation vector with respect to the normal to solar cell is
   a) 53.65 deg
   b) 36.34 c) 42.28
   d) 47.72

2. The sensitivity of the solar cell to the spectral irradiance of sun light is maximum at

a) wavelengths less than 400nm
b) wavelengths between 400nm and 700nm
c) wavelength between 700nm and 2000nm
d) wavelength between 2000nm and 5000nm

3. The value of solar constant is approximately:
   a) $1\ kW/m^2$
   b) $1.37\ kW/m^2$
   c) $1\ kWh/m^2$
   d) $1.37\ kWh/m^2$

4. The extra terrestrial insolation on March 27 is
   a) 1.3747
   b) 1.3341
   c) 1.3837
   d) 1.3447

5. At sunset and sunrise, the zenith angle is
   a) 0 deg
   b) 90 deg
   c) 180 deg
   d) 45 deg

6. If the solar declination is 23.5 deg and latitude angle is 45 deg , then the sunrise hour angle is
   a) 45 deg
   b) −45 deg
   c) 90 deg
   d) 115 deg

7. In which location (latitude) it is possible to have a day without sunrise:
   a) 23.45
   b) −66.5
   c) −23.45
   d) 0

8. The Albedo is the term used for
   a) The direct component of incident energy
   b) The diffused component of incident energy
   c) The ground reflected component of incident energy
   d) Algebraic sum of direct, diffused and ground reflected components

9. If the clearness index at a location is 0.1, then the ratio of energy incident on flat plate collector without atmosphere to that with atmosphere will be
   a) 10
   b) 0.1
   c) 0.5
   d) 2

3.18. QUESTIONS

10. At a given point on earth located on the 30 deg. latitude line, if the solar declination is −20 deg and the hour angle is 45 deg, then calculate the zenith angle.

11. On what day approximately the sun will be overhead (noon) at a place located on the latitude 8.57 ?

12. On which day will zenith angle be equal to latitude of the place ?

13. What is the solar declination angle on August 30th?

14. At a place on the equator, what is the hour angle at 11.00am?

15. At some latitude and longitude, the airmass index at sea level is 1.5. For a place at the same latitude and longitude, but above sea level where the ambient pressure is 0.8 atmospheres, the airmass index is ....

16. Consider the daily energy incident on the last day of the year, at a specific location on the 10 deg. latitude line. What is the tilt factor without considering atmospheric effects, if the collector is tilted at an angle that is same as the latitude?

17. A solar collector is placed, at a specific location on the tropic of cancer. What is the sun-rise hour angle on summer solstice if the collector is tilted at an angle that is same as the lattitude?

18. A vertical pole of length 4 m is placed on a particular point on earth. If the length of the shadow of this pole is 2 m, then what is the airmass index?

19. For a given place an angle of 20° is formed between the insolation line passing through the place and the equatorial plane of the earth-centric coordinate system.

    (a) What is the maximum possible insolation that is achievable if atmospheric absorption is not considered?

    (b) On which day does the maximum insolation occur?

    (c) What is the zenith angle on the day when the insolation is maximum?

20. What is the length of the day (in hours) on September 21st for a given latitude angle of 45°?

21. At a place on the 45^oN lattitude, how many hours are there in the longest and shortest day, respectively?

22. What is the incident energy in $kWh/m^2/day$ on May 20th at a place lying on 30^oN lattitude? Plot the incident energy versus day of the year.

23. What is the incident energy in $kWh/m^2/day$ on May 20th at a place lying on 30^oN lattitude if the flat plate collector is tilted southwards at lattitude angle? Plot the incident energy versus day of year.

24. What is the incident energy in $kWh/m^2/day$ on May 20th at a place lying on $30°N$ lattitude if the flat plate collector is mounted on the south facing vertical wall? Plot the incident energy versus day of year.

25. A PV panel acts as a horizontal flat plate collector that is placed at a place that lies on the equator. The electrical output of the PV panel gives a measure of the incident energy on the panel. On the summer solstice day, the measured incident energy is $5 kWh/m^2$. What is the clearness index at the place on summer solstice?

26. Consider the daily energy incident on the last day of the year at a specific location on the tropic of cancer. Let the collector be tilted at an angle that is the same as the latitude of the place. Neglect atmospheric effects.

    (a) Find the hour angle for the tilted collector in rad/s
    (b) Find the incident energy per day at the place on a horizontal surface.
    (c) Find the incident energy per day on the tilted collector surface
    (d) What is the tilt factor?

Chapter 4

Photovoltaic Sizing

For any given application it is essential to arrive at the proper size of the PV collector system. The solar PV collection system needs quite a significant amount of real estate on account of the conversion efficiency being low i.e. less than 20% in most types of solar PV technologies. Therefore it is essential to match the energy required for the application and the energy that is obtainable from the PV collector system. This design process is also loosely termed as sizing of photovoltaics. This chapter will discuss and demonstrate the sizing process. There are several applications for which PV is used. They can be broadly classified as applications that require batteries and those that do not require batteries. Solar PV sizing for both these categories will be considered in the sections to follow.

4.1 Sizing for applications without battery

One of the popular applications without using batteries, is solar PV power being pumped into the grid. The roof top solar PV modules are interconnected as strings and connected to the grid through a grid-tied inverter which pumps the PV power into the grid. Another application which does not use batteries, is water pumping. The solar PV power is used to operate a water pump which lifts water from an underground storage (sump) to an overhead reservoir or overhead tank. The pumping of water is performed only when solar energy is available thereby avoiding use of batteries. Pumping water for irrigation wherein vast expanses of agricultural lands need to be watered, is also an application that falls into the classification of non-battery applications. Applications like refrigeration, peltier cooling and AC can also be considered as non-battery based applications in some cases. In order to understand the sizing of solar PV modules for non-battery applications, the rooftop PV power being pumped into the grid is being considered to gain insights into PV sizing. The same principles can be used for sizing PV for other non-battery applications as well.

Consider the system in figure 4.1.1. Solar PV modules are fitted on the roof of a building as shown. There are two electrical interfaces to the building, viz. (i) an incomer wherein power is tapped from the grid and supplied to the loads with in the

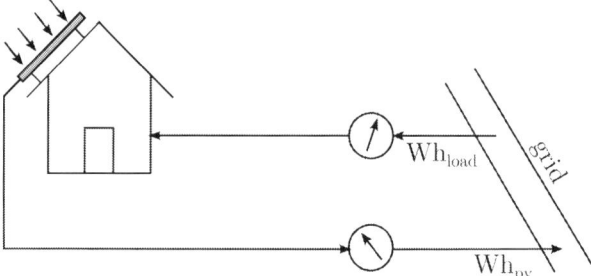

Figure 4.1.1: Grid tied system

building, and (ii) an outcomer wherein the PV power generated from the roof top PV modules are injected into the grid. The grid is represented by 2 parallel lines that may represent a 3-phase system or a 1-phase system.

The power that is drawn from the grid to feed the electrical loads within the building is metered and the daily energy consumed by the building in watt-hours is called Wh_{load}. Likewise, the power that is fed into the grid from the solar power generated by the rooftop solar PV modules is also metered and the daily energy injected into the grid in watt-hours is called Wh_{pv}. If the solar rooftop PV modules are absent then the electricity energy billing would be for Wh_{load} amount of energy. However, with the solar PV generation, the net amount of energy that is consumed by the building as a whole is $(Wh_{load} - Wh_{pv})$. The electricity energy bill will necessarily be only for the difference amount. If $Wh_{load} = Wh_{pv}$, then the building consumes net zero energy. It would be a zero energy building. If $Wh_{load} < Wh_{pv}$, then the building is a generator and needs to be paid by the power distribution company for the net energy generated. If $Wh_{load} > Wh_{pv}$, then the building is a load or power sink and needs to pay the power distribution company for the net energy consumed. Such a type of metering arrangement is called net metering.

Many power distribution companies set different tariffs for power drawn from the grid and that for power injected into the grid. In order to encourage renewable energy generation and usage, power distribution companies set a higher tariff for power that is injected into the grid from green energy sources compared to the tariff for power drawn from the grid. If C_{load} is the cost per kWh for the energy drawn from the grid and C_{pv} is the cost per kWh for the green energy that is injected into the grid, C_{pv} is generally greater than C_{load}. However, some distribution companies may not give differential tariffs, and in such cases C_{pv} will be same as C_{load}. The net cost payable to the distribution company will be as given by equation 4.1.1.

$$NetCost = (C_{load} \cdot Wh_{load} - C_{pv} \cdot Wh_{pv}) \qquad (4.1.1)$$

For zero net cost, the amount of energy generated by the solar PV system should be

4.1. SIZING FOR APPLICATIONS WITHOUT BATTERY

$$Wh_{pv} = Wh_{load} \cdot \left(\frac{C_{load}}{C_{pv}}\right)$$

It is evident that when the tariff for energy consumed from the grid and the tariff for the energy injected into the grid are same, i.e. $C_{load} = C_{pv}$, then in order to achieve zero net cost, the amount of energy generated by the solar PV system should be equal to the amount of energy consumed from the grid by the building loads.

4.1.1 Estimate H_{at}

The first step towards sizing the solar PV collectors is to estimate the daily incident energy at the place. Based on the discussions in the previous chapter, the daily incident energy on a tilted flat plate collector considering effects of atmosphere is estimated. One can visualise the variation of H_{at} at the place over the year. Figure 4.1.2 provides a sample visualisation of H_{at} over the entire year. On some day of the year, the incident energy is a minimum, H_{atmin}. This is the minimum daily energy available at the place considering all days of the year. The design must be performed for this worst case value of incident energy, H_{atmin} so that the performance can be maintained on all days of the year. For a place like Bangalore, the Octave script discussed in previous chapter can used to obtain H_{atmin}. It is found to be $4.58 kWh/m^2$.

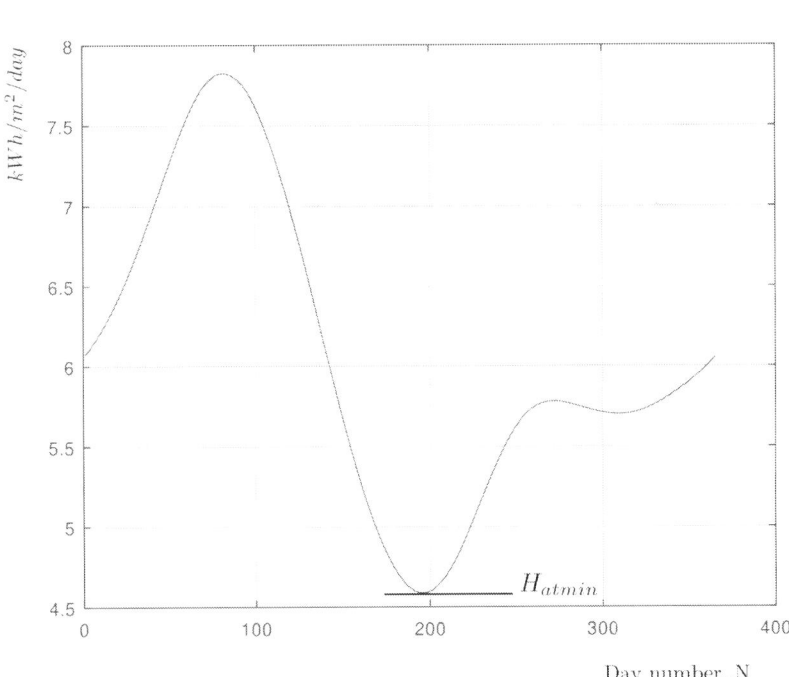

Figure 4.1.2: H_{at} estimation

4.1.2 Estimate peak sun insolation hours

The insolation at a place is not a constant but varies with time. From sunrise to sunset it varies with a profile as shown in figure 4.1.3a. The area under this insolation curve over the whole day gives the daily incident energy at the place, H_{at}. On the day of the year corresponding to minimum peak insolation, the energy obtained on that day would be H_{atmin}. Consider a rectangular profile having constant insolation of $1kW/m^2$. This insolation of $1kW/m^2$ is called peak sun which is also the standard insolation. Under constant peak sun insolation, for the same equivalent energy H_{atmin}, a rectangular insolation profile is shown in figure 4.1.3b.

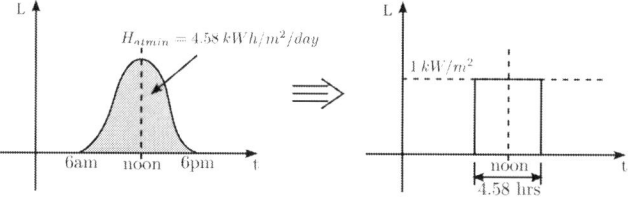

Figure 4.1.3: Actual and equivalent rectangle insolation profile

The width of the rectangle is peak sun hours t_{ps} that gives H_{atmin} amount of daily energy at peak sun. Thus,

$$H_{atmin} = 1kW/m^2 \cdot t_{ps}$$
$$t_{ps} = \frac{H_{atmin}}{1kW/m^2}$$

The peak sun hours t_{ps} is numerically equal to H_{atmin}. Therefore, at Bengaluru, the t_{ps} is equal to 4.58 hours.

4.1.3 Estimate the intrinsic area

Referring to equation 1.5.3, it can be re-written as follows

$$P_o = \eta \cdot L \cdot A$$

where P_o is the output power of PV modules for an insolation of L incident on area A of the PV modules. From figure 4.1.3b, the incident insolation L is $1kW/m^2$. It is constant and available for a period of t_{ps} such that the area of the rectangular profile gives the incident energy H_{atmin} for the place. As the peak sun hours t_{ps} is numerically equal to H_{atmin}, the output power equation can be written as

$$Wh_{pv} = \eta \cdot 1 \cdot A \cdot t_{ps}$$
$$= \eta \cdot A \cdot H_{atmin}$$

4.1. SIZING FOR APPLICATIONS WITHOUT BATTERY

where Wh_{pv} is the energy delivered by the set of PV modules in a day. The area of the PV modules can be written as

$$A = \frac{Wh_{pv}}{\eta \cdot H_{atmin}} \quad (4.1.2)$$

The area A estimated from equation 4.1.2 is called the intrinsic area of the PV modules. It is the actual area of the PV modules that is exposed to the incident insolation.

4.1.4 Estimate the estate area

The power output from the solar panels is directly proportional to the intrinsic area of the solar panels. This is the actual flat plate collector area A. However, the real estate area required would be higher than the intrinsic area A. There are several unmodelled areas that needs to be included to arrive at an estimate of the required real estate inorder to deliver Wh_{pv} amount of energy from the solar PV panels every day. The actual real estate area A_{estate} should include

> The intrinsic area A of the solar PV panels.
>
> The solar PV modules have aluminum frames around each module. The frames thickness are around 20mm wide and they appear as a margin on all sides of the solar PV module. The amount of area corresponding to the module frames should also be factored into the required real estate area.
>
> The individual solar PV modules in the array should be accessible for cleaning and repairs. This means that a passage should be allowed for a person to reach and work on a specific PV modules during maintenance operations. This is an unmodelled amount of area that is important and needs to be accounted for.
>
> The PV modules from the array need to be connected to a distribution panel. The distribution panel will then direct the power to grid tied inverters which are connected to the grid through an AC distribution panel. All these require space.

An empirical rule of thumb is to allow 30% of the intrinsic area as the extra space required for the unmodelled areas mentioned above. The is reflected in the estimate of the estate area as given in equation 4.1.3.

$$A_{estate} = 1.3A \quad (4.1.3)$$

Example : Consider a household in Bangalore which consumes 200 units of energy on an average per month. Calculate the area required for a roof top PV solar system setup for at least net zero energy.

Given

$\phi = 12.97^o$ for Bangalore
Consumption = 200 units per month

Calculate Wh_{load}

$1\,unit = 1\,kWh$

Therefore, the daily consumption Wh_{load} is obtained as
Assuming 30 days in a month,
$Wh_{load} = \frac{200\,kWh/month}{30} = 6.67 kWh$

Determine H_{atmin}

The daily consumption Wh_{load} should be fully compensated by the PV generation system even on the least insolation day so that net zero energy is achieved. Therefore, $Wh_{pv} = 6.67 kWh$ even on a day where the daily incident energy is least, i.e. H_{atmin}.
Using the insolation model for India and the Octave script discussed in the previous chapter,
$H_{atmin} = 4.58 kWh/m^2/day$ for Bangalore.

Determine real estate required

Consider efficiency of a PV module as 18%. The exact value of the efficiency can be obtained from the datasheet of the PV modules that will be purchased.
$A = \frac{Wh_{pv}}{\eta \cdot H_{atmin}} = \frac{6.67}{0.18 \cdot 4.58} = 8.1 m^2$
$A_{estate} = 1.3 A = 1.3 \cdot 8.1 = 10.53 m^2$

4.2 Roof top PV example

Consider the picture given in figure 4.2.1. 100 PV modules are mounted on part of the roof of the Electronic Systems Engineering department building at Indian Institute of Science, Bangalore. Each PV module has the dimension of $1.64m \times 0.992m$. The datasheet for the PV module gives the terminal voltage at peak power as 30.1V and the current at peak power as 8.32A. Calculate the peak power that the solar PV array system is capable of delivering.

The intrinsic area of the PV array is given as

$$A = 1.64m \times 0.992m \times 100\,modules \simeq 163 m^2$$

$$Wh_{pv} = \eta \cdot A \cdot H_{atmin}$$
$$= \frac{V_{mp} I_{mp}}{A_{module} 1000 W/m^2} \cdot A \cdot H_{atmin}$$

Considering the value of H_{atmin} at Bangalore as $4.58 kWh/m^2$, one obtains

Figure 4.2.1: Roof top example

$$Wh_{pv} = \frac{30.1 \times 8.32}{1.63 \times 1000} \times 163 \times 4.58$$
$$= 114.7 kWh/day$$

Peak watts delivered by the array is $\frac{Wh_{pv}}{H_{atmin}} = \frac{114.7}{4.58} = 25kW/day$

4.3 Parking lot PV example

Consider the picture given in figure 4.3.1. This is an example of a PV system on a parking lot. This can be used to provide power for charging electric vehicles. During times when EVs are not being charged, the power is pumped into the grid through grid connected inverter. There are 90 PV modules mounted on the parking lot at Department of Electronic Systems Engineering at Indian Institute of Science, Bangalore. Each PV module has the dimension of $1.985m \times 1m$. The datasheet for the PV module gives the terminal voltage at peak power as 40V and the current at peak power as 9.51A. Calculate the peak power that the solar PV array system is capable of delivering.

The figure 4.3.1 shows the parking lot that is located alongside the road at the road level. The department building is at a slight elevation that is reachable through a flight of 12 steps. In order to have better utilisation of the area underneath the solar panels, the solar panels are mounted on raised pedestal structure as shown in figure 4.3.2. The figure 4.3.2(b) gives the view from the front of the parking lot. The figure 4.3.2(a) gives the details of the pedestal mount structure for the solar panels. The panels are mounted on the south facing side at a tilt angle ϕ which is same as the latitude of the place. For this specific installation at Bangalore, the tilt angle provided is 12.97^o. The northside of the panel mount also has a slope which is needed to act as a wind breaker.

Figure 4.3.1: The top view of the parking lot of department of Electronic Systems Engineering building at Indian Institute of Science, Bangalore

The north side is covered with 3mm mild steel sheets. In order to drain away the water during rains, a 100mm wide gutter is provided as shown. The actual built structure is seen in figure 4.3.2(b).

The intrinsic area of the PV array is given as

$$A = 1.985m \times 1m \times 90\,modules = 178.65m^2$$

$$Wh_{pv} = \eta \cdot A \cdot H_{atmin}$$
$$= \frac{V_{mp}I_{mp}}{A_{module}1000W/m^2} \cdot A \cdot H_{atmin}$$

Considering the value of H_{atmin} at Bangalore as $4.58 kWh/m^2$, one obtains

$$Wh_{pv} = \frac{40 \times 9.51}{1.985 \times 1000} \times 178.65 \times 4.58$$
$$= 156.8 kWh/day$$

Peak watts delivered by the array is $\frac{Wh_{pv}}{H_{atmin}} = \frac{156.8}{4.58} = 34.23 kW/day$

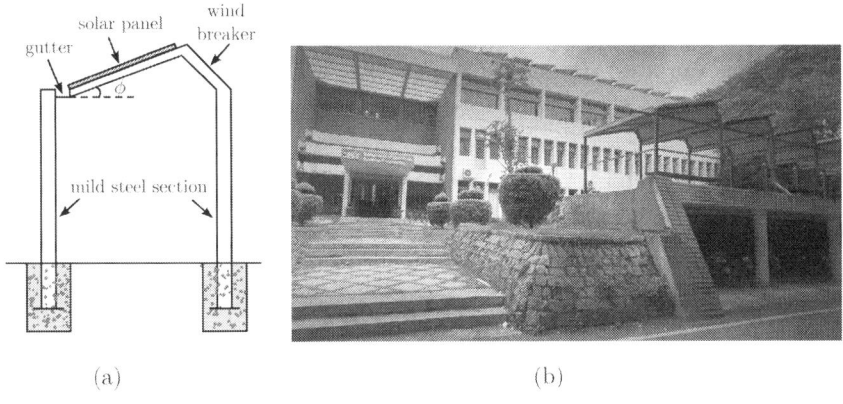

Figure 4.3.2: Parking lot example (a) structure of the PV panel mounts (b) actual photo of the parking lot PV system at department of Electronic System Engineering, Indian Institute of Science, Bangalore

4.4 1MVA grid connected PV system example

Here is an example where a PV array field is setup for a 1MVA generation and it feeds the generated power into the grid at 11kV through a distribution transformer. Figure 4.4.1 shows the picture of the 1MVA array installation at Indian Institute of Science, Challakere campus. There are 2500 PV modules each rated for 400W peak. The power from the PV modules are routed through a set of inverters that provide power at 400V. A 400V:11kV distribution transformer interfaces the inverters on the low tension (LT) side to the 11kV high tension (HT) grid.

Figure 4.4.2 gives the single line diagram of the grid connection 1MVA PV system. The PV modules are grouped into a set of 9 arrays. Each array has 14 strings in parallel with each string containing 20 PV modules. Each PV module is rated at 400Wp. Therefore, each PV array generates $14 \times 20 \times 400W = 112kW$. There are 9 PV array groups each capable of generating $112kW$. Thus a total of $112 \times 9 = 1008kW$ can potentially be generated from this PV array field. The metal frame of the PV modules are connected to the physical earth. This will maintain the PV module frames at zero or earth potential and make them safe for human maintenance and interaction.

Each string has 20 PV modules. Each PV module is rated for 40V at peak power. Thus the string voltage is 800V. There are 14 strings connected in parallel and each string provides 10A at peak power. Therefore, each PV array has a terminal voltage of 800V and terminal current of 140A at peak power operating point. The power from the PV array is processed through an inverter that is rated to handle 112kW. There are nine such inverters. Three inverters are grouped together and routed through an AC distribution box (ACDB). There are three such ACDBs as shown in the figure 4.4.2. If the currents are injected into the 400V LT side of the transformer at unity power factor, then each inverter drives a line current I_L of

Figure 4.4.1: The 1 MVA solar field at Indian Institute of Science, Challakere campus

$$I_L = \frac{P_{inv}}{\sqrt{3}V_L} = \frac{112000}{\sqrt{3} \times 400} = 161.66A$$

Each inverter output is connected to an ACDB. At the incomer of the ACDB, a 200A three phase MCCB with neutral pole is used. It is specified for a breaking current of 25kA. It also has thermal and magnetic based overload and short circuit release. The ACDB has three incomers, one each from an inverter. They are connected to an internal bus. The ACDB out going cable is given to a low tension (LT) panel as shown. There are three incomers for the LT panel, one each from ACDB. Each LT panel incomer services three inverters. Thus, a 630A rated three phase MCCB is used. This also has a 25kA breaking current. Thermal and magnetic based overload and short circuit release is also included. The LT panel is connected to the 400V:11kV distribution transformer which is in turn connected to the HT grid through a HT panel. A 4 pole 2500A MDO air circuit breaker (ACB) is connected at the out going cable of the LT panel as shown.

The intrinsic area of the PV array field is given as

$$A = 2m \times 1m \times 2500\,modules = 5000m^2$$

4.4. 1MVA GRID CONNECTED PV SYSTEM EXAMPLE

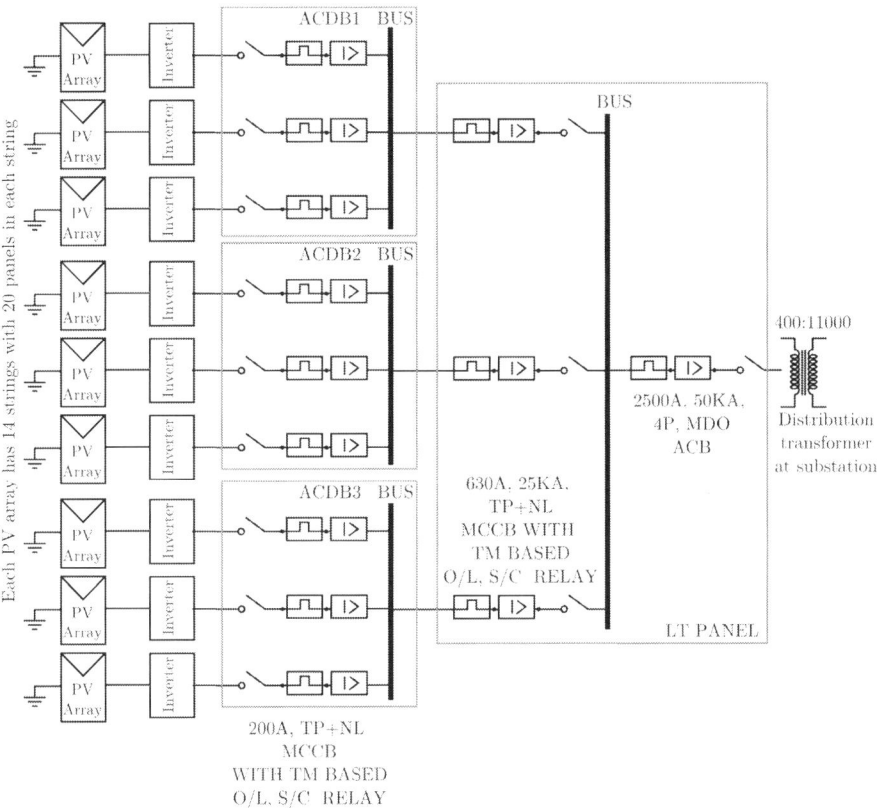

Figure 4.4.2: Single line diagram of a 1 MW plant at Indian Institute of Science, Challakere campus

$$Wh_{pv} = \eta \cdot A \cdot H_{atmin}$$
$$= \frac{V_{mp}I_{mp}}{A_{module}1000W/m^2} \cdot A \cdot H_{atmin}$$

One can use the Octave script discussed in the previous chapter for obtaining the value of H_{at} at Challakere. The lattitude of Challakere is 14.31^oN. The value of H_{atmin} at Challakere is found to be $4.51kWh/m^2$, then

$$Wh_{pv} = \frac{40 \times 10}{2 \times 1000} \times 5000 \times 4.51$$
$$= 4510kWh/day$$

Peak watts delivered by the array is $\frac{Wh_{pv}}{H_{atmin}} = \frac{4510}{4.51} = 1MW/day$. The real estate area required will be 30% more than A and is given as $1.3 \times 5000 = 6500m^2$ which is about 1.6 acres.

4.5 Cleaning and maintenance

Over a period of time, dust settles on the PV modules and partially blocks the insolation. As a result the output power from the PV module can be significantly lower compared to a clean PV module. Therefore, it is imperative that a systematic cleaning and maintenance process be adopted in order to maximise the generated power from the PV modules.

The simplest method to keep the PV panels dust free is to employ persons to perform daily water wash of the PV panels. Alternately, one may install a pipe system where a water nozzle is placed in front of each PV module. Water is sourced from a sump or reservoir and pumped through the nozzle to provide a water wash to the PV module. This water wash cleaning can be done once a day. This can be automated by means of a timer mechanism. The timer can be preset to a predetermined time of day when the pump will be switched on and the PV modules get sprayed from the nozzle.

4.6 Questions

1. Consider a roof top PV system wherein the PV is interfaced to the grid through a grid connected inverter. This is capable of pumping power into the grid at grid voltage. The building consumes 300kWh of energy per day. The energy cost that needs to be paid to the power distribution company is at Rs. 5/kWh. The energy cost that is paid to the consumer when PV power is injected into the grid is at Rs. 6/kWh. What should be the energy rating of the PV system in order to have zero net cost?

2. In a roof top PV system, the PV modules are designed to achieve zero net cost for power distribution company energy cost of Rs. 4/kWh for consumption and Rs. 6/kWh for generation. The daily energy consumption is at 200kWh. Later on the power distribution company revised the generation cost to be same as the cost for consumption i.e. Rs. 4/kWh. What is the net cost for the consumer?

3. At a place located on the 10^oN latitude, the incident energy on a flat plate collector on a given day is $5kWh/m^2/day$. What is the time period for an equivalent standard insolation incident on the collector in the time period on the given day?

4. The area of a PV module is $1.8m^2$. The module efficiency is 18%. What is the peak wattage of the PV module?

5. For a place on the 10^oN latitude, the minimum daily incident energy in a year is $4kWh/m^2/day$. A set of PV modules have 20% efficiency are used to inject power into the grid. An energy of $100kWh/day$ is required to be injected into the grid. What should be the intrinsic area of the set of PV modules?

4.6. QUESTIONS

6. At a given place, the minimum daily incident energy in a year is $4kWh/m^2/day$. What should be the energy rating of the PV modules if the module efficiency is 20% and has an intrinsic area of $10m^2$?

7. A set of PV modules are injecting unity power factor power into a three phase 400V grid. A grid interactive inverter is used for this purpose. The PV panels are cummulatively rated at 100kW peak. What should be the current rating of the circuit breaker connected between the inverter and the grid?

8. A set of PV modules are connected to the 400V grid through a grid connected inverter. The place has a daily incident irradiance of $6kWh/m^2$. The PV modules are located in a real estate of $30m^2$. The PV panels have a conversion efficiency of 18%. If the grid connected inverter is 95% efficient, what is the daily energy injected into the grid?

9. For the problem in the previous question, what is the rms line current that is injected into the grid?

10. Consider a system wherein a setup of PV modules are supplying power to a 400V grid through a grid connected inverter. The daily incident irradiance of the place is $4kWh/m^2$. The PV modules inject 20kWh of energy into the grid daily. The PV module intrinsic area is $25m^2$. The terminal voltage of the PV module system is 380V at peak power operating point. What is the current delivered by the PV system at peak power operating point?

Chapter 5

Energy Storage

Consider a system wherein a PV source supplies a load through a power converter. Several applications have no energy storage devices. A typical application is PV power injection into the grid through grid-tied inverters. There are many applications where the load may demand a steady, stable power. However, the power from the PV source is very fluctuating as it is dependent on insolation, time of day, time of year, and weather conditions. One needs to use an energy buffer like a battery in such a case as shown in figure 5.0.1. The PV source charges up the energy buffer (eg. battery) through a charger and stores the energy. The battery supplements the PV source and releases steady power to the load.

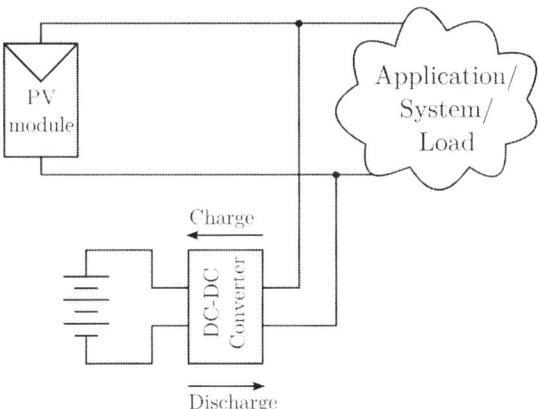

Figure 5.0.1: Typical PV system

The energy buffer may not store the energy in the electrical domain. Battery is an example of an energy buffer wherein the energy is stored in the chemical domain. Likewise, flywheel is an energy buffer wherein the energy is stored in the mechanical domain. The energy buffer is generally interfaced with the PV system through a power interface which also will act as an energy converter to convert the energy from electrical domain to the energy domain of the buffer and vice-versa. In this chapter, energy storage solutions will be discussed specifically with respect to renewable energy applications. The following sections will discuss the batteries, flywheel, pumped hydro and compressed air storage mechanisms.

5.1 Batteries

Batteries are of two main types: one is primary cell and the other is known as secondary cell. The cells used for torch lights, clocks, remote controls, and watches are a few example applications where primary cells are used. Few primary cell examples are zinc manganese oxide also called the dry cell, zinc silver oxide, lithium thyionylchloride. The most important feature is that primary cells can be discharged only once.

On the other hand, secondary cells can be charged and discharged several times. Car batteries, two-wheeler batteries, UPS batteries are all examples of secondary cells. Lead oxide is more commonly called the lead-acid battery, which has diluted sulphuric acid as the electrolyte. The lead acid battery is one of the more popular batteries and is used in automobiles and home UPS. Nickel metal hydride batteries, lithium-ion, lithium polymer batteries are other secondary cells. Lithium-ion and lithium polymer batteries have become very popular due to their high power and energy densities. They are used in almost all electronic gadgets like cell phones, tablets, laptops, cameras, video cam, and electric vehicles.

The secondary batteries are capable of undergoing 100 to 20,000 charge-discharge cycles. This is one of the major advantages of secondary cells. For photovoltaic based systems, most of the applications will use only secondary cells. The following sections will discuss the characteristics, features and important parameters of the secondary cells that will be useful for selection of batteries.

5.1.1 Capacity

One of the important parameters needed in selection of a battery is its *capacity*. Consider a battery such that the voltage across the battery terminals is v_b. Let a current i_b flow into the positive terminal of the battery. The battery is said to be *charging up*. If the current is flowing out of the positive terminal of the battery, then it is said to be *discharging*. When i_b flows into the battery through the positive terminal, power flows into the battery and accumulates over a period of time as energy. A battery can store only upto a certain amount of energy that it is designed for. The upper limit of energy that a battery can hold or store is called the capacity of the battery. It is the maximum energy that the battery can store.

$$E_{in} = \int v_b \cdot i_b \cdot dt \qquad (5.1.1)$$

In a specific amount of time, if i_b flows into the battery through the positive terminal, then E_{in} amount of energy is fed into the battery. However not all the input energy E_{in} is converted to chemical form for storage. The actual stored energy is given as

5.1. BATTERIES

$$E_{store} = \eta_{ec} \cdot E_{in} \quad (5.1.2)$$

wherein η_{ec} is the efficiency of the energy conversion from electric domain to chemical domain. The energy is stored in the chemical domain in the case of batteries. The maximum value of E_{store} is the capacity of the battery in energy units. However, commercially the capacity is generally expressed in terms of charge units. The nominal voltage across the battery V_b is considered constant dc and the charge into the battery C_{in} and charge stored in the battery C_{store} are given as

$$C_{in} = \frac{E_{in}}{V_b} = \int i_b \cdot dt \quad (5.1.3)$$

$$C_{store} = \eta_{ec} \cdot \frac{E_{in}}{V_b} \quad (5.1.4)$$

The energy capacity as given by E_{store} will give a measure of the true capacity of the battery. If the voltage across the battery is considered constant value, then the charge capacity C_{store} will also give a measure of the capacity of the battery. In practice, the battery voltage v_b is never constant and therefore the energy capacity E_{store} is a truer measure of capacity of the battery. E_{store} uses the units of watt-hours (Wh) and C_{store} uses the units of ampere-hours (Ah).

Consider the visualisation shown in figure 5.1.1. When the battery is fully charged, the battery has capacity C_{store}. One cannot discharge the battery completely. There must exist some residual charge or minimum watt-hours in the battery. The analogy of a water tank can be effectively used here to understand the various terms. In case of the water tank, when the water tank is full of water, one can say that the tank is filled to capacity. Likewise even in the case of the battery, when it is fully charged, it is storing maximum energy. Under full charge condition, the open circuit voltage across the terminals will be maximum. For example, in the case of a lead acid battery, a 12 volt lead acid battery has 14.3V as voltage under full charge condition. Consider level x as shown. The amount of charge available at level x is the minimum watt-hour required for the battery to function without deterioration. At level x the voltage across the terminals of the battery is lesser than that at full charge. In the case of 12V nominal lead acid battery, the voltage at level x will be around 10.8V.

Suppose the battery is discharged to the level x, then the depth to which it has discharged is called depth of discharge (DoD). From 100% C_{store} to x% C_{store}, (100 - x)% is the depth from full charge condition. It is called the depth of discharge and normally expressed as percentage and it is a very important specification parameter that can be seen in battery specifications and it has significant influence on life of the battery. For batteries of different chemistries, the DoD will vary. For start-lighting-ignition (SLI) batteries used in automobiles, the DoD is 20%. This means that only 20% of C_{store} is available for use by the load or application. In home UPS applications, the batteries usually have a DoD of 80%. These are called deep discharge batteries where in 80% of C_{store} can be utilised.

Another term that one commonly encounters in battery specification is state of charge (SoC). It is the amount of charge present in the battery. It is the complement of DoD. The state of charge is given as

$$SoC = \left(1 - \frac{DoD}{100}\right) \quad (5.1.5)$$

One charge of the battery followed by a discharge of the battery to the load is called a charge-discharge cycle. The life of the battery is expressed in terms of charge-discharge cycles. Typical lead-acid batteries have a life of 1000 charge-discharge cycles. It should also be noted that the number of charge-discharge cycles is a function of the DoD that is used for the battery.

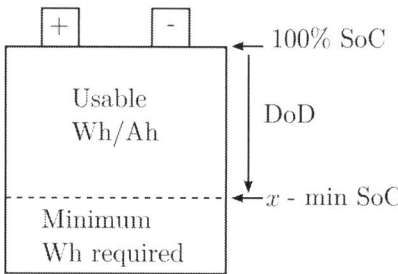

Figure 5.1.1: Battery capacity and DoD

5.1.2 C-rate

C-rate is another important parameter that is used in selection of battery. It is the nominal rate at which the capacity of the battery must be either replenished or depleted. C-rate is related to the capacity of the battery. It normally refers to the discharge current of the battery. The charging current is generally set same as the discharging C-rate parameter of the battery. The C-rate of the batteries are specified as C10, C5, 2C etc.

Let C be the ampere hour capacity of the battery and i_b be the instantaneous current out of the battery. The average current out of the battery is denoted as I_b. Consider a battery having C-rate specification as Cn where n can be 0.5, 1, 2, 5, 10 etc. In the C-rate nomenclature, n represents the time period or hours for which the charge or discharge current flows in order to accumulate or remove C ampere hours of capacity.

Consider a simple example of a battery having capacity 20 Ah and C-rate specified as C10. This means that when an average current of $\frac{C}{10} = \frac{20}{10} = 2A$ is allowed to be discharged from the battery, the battery will deliver power for at least 10 hours.

There are several ways the C-rate is represented in the datasheets and literature. However, one of the more popular way of nomenclature is to have a numeric value to the left side or right side adjacent to the letter 'C'. For example, C10 indicates that the 10 divides the numeric value of the capacity C to estimate the nominal average

5.1. BATTERIES

discharge current. Therefore, the nominal average discharge current for batteries with C-rate as C10 is $\frac{C}{10}$ amps. A C5 battery will imply a nominal average discharge current of $\frac{C}{5}$ amps. A battery with a C-rate of C0.5 means that the nominal average discharge current is $\frac{C}{0.5} = 2C$ amps. If there is a fractional number on the right of C adjacent to it, then one can use its inverse value and C appended to it. Thus, C0.5 and 2C are equivalent. A C-rate of 5C means that the discharge current is 5 times the numeric value of C. It can also be written equivalently as C0.2.

Consider the figure 5.1.2(a) wherein a battery is connected to a load resistance R_L which is a variable resistor like a rheostat. The battery has an internal resistance R_b. A current of i_d is being discharged from the battery and flowing through the load. The voltage across the terminals of the battery is v_b. The figure 5.1.2(b) shows the graph having time as x-axis and voltage as the y-axis. This gives a visualisation of the voltage profile of v_b as a constant discharge current is drawn from it.

The elapsed time of discharge and the discharge current value will provide the estimate of ampere-hours that is discharged from the battery. Ideally one would expect the battery voltage to be constant with time as a current i_d is drawn from it. This is shown as the dashed line in figure 5.1.2(b). The discharge current flows till the entire ampere-hour charge is removed from the battery. In an ideal battery, the internal resistance R_b is considered zero. The voltage v_b is constant with discharge and becomes zero when battery is completely discharged. However, in practical batteries R_b is neither zero nor constant. As a result the voltage v_b is not constant with elapsed discharge time.

The value of R_L is adjusted to set the discharge current $i_d = \frac{C}{10}$. As time progresses, one can observe that v_b has a drooping profile as seen in figure 5.1.2(b). The battery is recharged to full capacity again in order to reset the experiment. Then reduce R_L further to double the discharge current to $\frac{C}{5}$ and repeat the experiment. The v_b profile with time can be obtained as a family of curves with various discharge currents. Observe that the voltage droop increases as the discharge current increases. This drop in the voltage is due to voltage lost in the dynamic internal resistance of the battery. The dynamic internal resistance R_b is a nonlinear resistance which is a function of several parameters relating to the chemistry of the battery. R_b is a dominant function of depth of discharge (DoD).

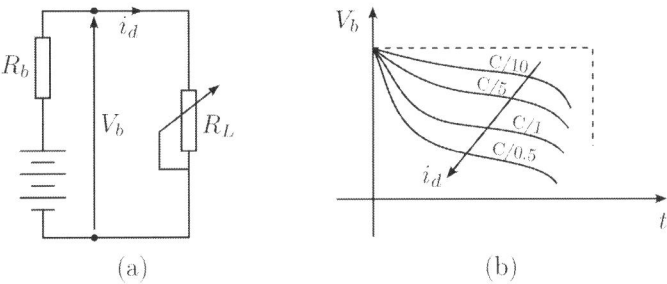

Figure 5.1.2: Internal resistance and discharge rate

When the battery is fully charged, R_b is very low. However, as the stored charge is removed from the battery, R_b tends to rise. At time closer to zero, the value of R_b

is low and the terminal voltage is closer to open circuit voltage. As time progresses, the R_b value increases with loss of charge in the battery. This results in the terminal voltage drooping more rapidly.

It is very important to note that C-rate plays a very significant role in the selection of battery. Given a specific discharge time, one can specify that the battery terminal voltage to be within a particular limit. Then by choosing an appropriate C-rate and discharge profile, appropriate batteries can be selected for a given application.

5.1.3 Efficiency

The energy is stored within the battery in a chemical form. The charge for the battery is supplied from the electrical energy domain and the discharge to the application or load is also in the electrical energy domain. While charging the battery, there is a conversion of energy from electrical to chemical domain and during discharging there is again a conversion of energy from chemical to electrical domain. These conversions and losses within the respective domains imply that part of the input energy will go off as heat. This leads to the specification of the energy efficiency of the battery which is also an important selection criteria.

The battery energy efficiency is defined as the ratio of watt-hour delivered to the load to the watt-hour drawn from the input source. Referring to figure 5.1.3, the energy efficiency can be written as

$$\eta = \frac{Wh_{load}}{Wh_{pv}} \qquad (5.1.6)$$

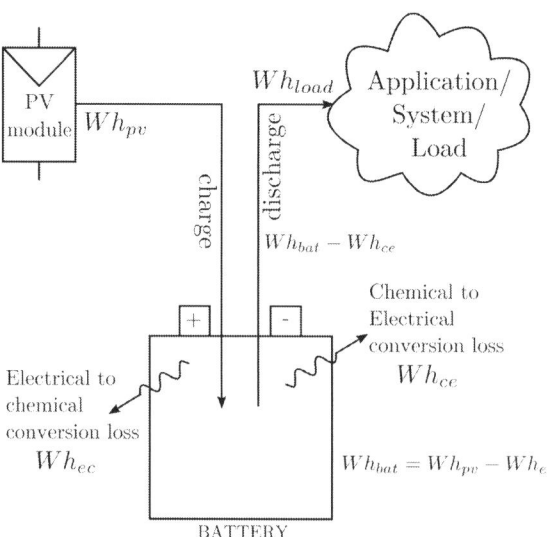

Figure 5.1.3: Energy buffering

5.1. BATTERIES

The energy that is stored in the battery is Wh_{bat}. In order to store an energy of Wh_{bat} in the battery, one must supply Wh_{pv} from the input source and an amount of energy Wh_{ec} is lost due to energy conversion from electrical to chemical domain and other heat losses. Therefore,

$$Wh_{bat} = Wh_{pv} - Wh_{ec} \qquad (5.1.7)$$

From the equation 5.1.7, the charging efficiency can be estimated as follows,

$$\eta_{ec} = \frac{Wh_{bat}}{Wh_{pv}} = 1 - \frac{Wh_{ec}}{Wh_{pv}} \qquad (5.1.8)$$

Again referring to the figure 5.1.3, the energy stored in the battery Wh_{bat} can be expressed with respect to load side energies. Therefore,

$$Wh_{load} = Wh_{bat} - Wh_{ce} \qquad (5.1.9)$$

where Wh_{ce} is the energy lost in conversion from chemical to electrical form during discharge process. Referring to equation 5.1.9, the discharge efficiency can be estimated as

$$\eta_{ce} = \frac{Wh_{load}}{Wh_{bat}} = 1 - \frac{Wh_{ce}}{Wh_{bat}} \qquad (5.1.10)$$

From equations 5.1.8 and 5.1.10, one can observe that the energy efficiency of the battery is

$$\eta = \frac{Wh_{load}}{Wh_{pv}} = \eta_{ec} \cdot \eta_{ce} \qquad (5.1.11)$$

For most batteries, the efficiency during charge and discharge are about 85%. This means that the overall battery energy efficiency is around 72% for charging and discharging taken together.

5.1.4 Energy and power densities

Another set of parameters that have a bearing on the volume and weight of the batteries is energy density and power density. The energy density is classified into two broad types viz. (i) the gravimetric energy density and (ii) the volumetric energy density. As the name indicates, the gravimetric energy density is with respect to the weight of the battery and it has the units of Wh/kg and the volumetric energy density relates to the volume of the battery and is expressed in units of Wh/m^3 or more commonly as watt-hour per litre. Gravimetric energy density is also known as specific energy in some literature.

Battery	Wh/kg	W/kg	Wh/lt	$T_{op} \,^oC$	η %	Life
Lead Acid	35	100	80	ambient	70	200
Nickel Metal Hydride	100	250+	300	ambient	70	500
Nickel Iron	25	100	30	ambient	65	10000+
Sodium Sulphur	100	150	300	350	75	2500
Lithium Polymer	265	250+	670	ambient	80	1000+
Sodium Ion	200	1000	375	ambient	92	1000+

Table 5.1: Battery comparison

The gravimetric energy density Wh/kg has much significance on mobile platforms like electric vehicles. The weight of the batteries for a given amount of capacity will determine the payload that the electric vehicle can carry. On the other hand volumetric energy density has much significance in applications where space is a constraint.

Like energy densities, battery datasheets also provide the power densities. Here again there are two types gravimetric power density and volumetric power density. As the name suggests, gravimetric power density has the unit's W/kg and volumetric power density is expressed in W/m^3 or $W/litre$. Gravimetric power density is also known as specific power. Here again the W/m^3 is having a direct bearing on the size of the battery. It indicates the volume of the battery for a given instantaneous power from the battery. This provides an estimate of the instantaneous discharge current that can be drawn from the battery. The gravimetric energy density Wh/kg has a bearing on the range of the electric vehicles and the gravimetric power density W/kg has a bearing on the acceleration or the overtaking ability of an electric vehicle.

The batteries have different values for energy and power densities. Some batteries may have high energy densities and lower power densities. Some other batteries may have low energy densities but higher power densities. One must choose batteries based on the application requirement. For the example of an electric vehicle, depending on whether one needs range or acceleration as the primary requirement, the choice of batteries with higher energy density or higher power density is decided.

5.1.5 Comparison of batteries

There are a host of batteries in the market for various applications. The table 5.1 gives a comparison of few types of batteries with respect to the parameters discussed in this chapter. The parameters for comparison considered in table are specific energy or gravimetric energy density Wh/kg, the power density W/kg, the volumetric energy density Wh/lt, operating temperature T_{op}, the energy efficiency of the battery η, and the life of the battery in terms of charge-discharge cycles. The charge-discharge cycles are tabulated at 100% DoD to have a common reference benchmark for comparison.

The first three columns of the table relate to the energy and power densities of the battery. They given an impression about the compactness and heaviness of the battery. The gravimetric energy density Wh/kg indicates the heaviness of the battery for a given capacity. Apparently among the batteries provided in the table, the Lithium Polymer battery is the least heavy for a given capacity. This implies that it is a light

5.1. BATTERIES

battery. One can observe that the lead acid battery has a very low specific energy of 35 Wh/kg. This means that for a given capacity the Lithium Polymer battery is around 8 times lighter. One can also see that the volumetric energy density Wh/lt is also best for the Lithium Polymer battery. The nearest competitor is the Sodium Ion and Nickel Metal Hydride batteries.

The power density W/kg gives a measure of the amount of discharge current that can be drawn from the battery for a given weight. The Sodium Ion batteries seem to have very high acceleration or over-taking capability if used in an EV application. Both Lithium Polymer and Nickel Metal Hydride batteries have similar values. Lead acid battery and Nickel Iron batteries have low discharge current values. This implies that they are slow discharge batteries.

For most batteries the operating temperature T_{op} is at ambient temperature. However, it is not so for all battery chemistries. The Sodium Sulphur battery has an operating temperature of 350^oC. With respect to energy and power densities, Sodium Ion batteries compares reasonably well with respect to the Lithium Polymer batteries. The Lithium resource is not easily available. The Sodium Ion batteries are being considered as a close alternative especially due to the abundance of sodium in sea water.

The penultimate column in the table gives the energy efficiency of the batteries. It can be seen that the Sodium Ion and Lithium Polymer batteries are the more efficient ones followed closely by Sodium Sulphur battery. the Nickel Iron battery is the least efficient one. In most battery types, the energy efficiency hovers around 70%.

The last column in the table gives a measure of the life of the battery. The life of the battery is expressed in terms of the number of charge-discharge cycles that the battery is able to perform before its performance starts deteriorating. The discharge is done up to 100% DoD and then charged up to full capacity. This would form one charge-discharge cycle. One can observe here that the Nickel Iron battery is way ahead of any other battery type. The Nickel Iron batteries are said to have life times of 30 to 50 years. This kind of life is not possible by any other battery type. A rather distant second is the Sodium batteries with 2500 charge-discharge cycles.

In conclusion, one must select a battery that is most appropriate for a particular application. For example the Nickel Iron battery may show poor performance metrics with respect to most parameters but it has an unbeatable life specification. Even though it may be heavier and larger for a given capacity. For applications like solar PV energy being pumped into the grid along with storage, the Nickel Iron shows promise as real estate is any way available for collecting the solar energy. The batteries can be placed under the solar panels thereby removing the compactness and heaviness constraint. Here the life constraint become very significant and probably the only battery that can match the life of the solar PV. Though this is not yet popular, it is a battery type that is being actively researched for this application.

Lithium polymer batteries are very popular nowadays, almost all gadgets use lithium polymer batteries due to very good all round parameters. Even though it has a high initial cost, high capital cost, it is compact and has lot of advantages in terms of energy and power densities. However, Lithium is not abundantly available. The Sodium Ion batteries show much promise. It has characteristics that are close to Lithium polymer with the added advantage of being abundantly available in nature. The lead acid battery, even though it has very poor numbers with respect to most performance metrics, is one of the most popular batteries and the one which has maximum usage in automotive start-lighting-ignition (SLI) and housekeeping application. It is also used in UPS applications. Even though it is very large with low energy densities, it has low initial cost. Not only that, it is easy to maintain, repair and recycle.

The fossil fuels like charcoal, petrol, diesel have a specific energy of around 10000 to 12000 Wh/kg which is way ahead of any battery type existing. This gives us an insight on the how well nature stores energy in a very small given volume and weight.

5.1.6 Battery selection

The following steps will summarize the selection of batteries for a specific application.

Step-1: Identify and select battery chemistry for the given application. For example in the case of an application wherein the PV energy is being injected into the grid which is also supported by a battery bank, specific energy and power density may not be a very significant issue. However cost, life, availability, recyclability of the battery could be the dominant factors to consider. The lead-acid battery may be a good choice or even Nickel-Iron battery. On the other hand if the application is for a mobile platform like electric vehicles, energy and power densities may form the main factors to consider in the selection. The Lithium polymer or the Sodium ion could be potential candidates. Therefore the best battery is dictated by application consideration

Step-2: Find out the depth of discharge DoD that is allowable for the particular type of battery. This will given an indication of the usable amount of energy that can be removed from the battery. Several battery chemistries will need a base amount of energy to be present in the battery if deterioration in life is to be avoided. For example the lead-acid batteries used for automobile starting, lighting and ignition applications called the SLI batteries, have high power densities but low DoD. They have a DoD of only 20%. On the other hand, the tubular lead-acid batteries that are common in home UPS systems have a deeper DoD of 80%. Knowledge of allowable DoD will allow estimation of the capacity of the battery.

Step-3: Next step is to find out the watt-hour requirement of the load Wh_{load}. This is essential to calculate the capacity of the battery. From equation 5.1.7, the amount of energy that is required to be made available in the battery for given load energy is $\frac{Wh_{load}}{\eta_{ce}}$.

Step-4: The capacity of the battery is given as

Watt-hour capacity = $\frac{Wh_{load}}{DoD \cdot \eta_{ce}}$ and the

Amp-hour capacity = $\frac{Watt\ hour\ capacity}{Vbat_{nominal}}$

Step-5: Battery datasheets will provide nomographs of $I^n t$ versus discharge current i_d. Typical curves will be like that shown in figure 5.1.4. For very low discharge currents, n can be taken as unity and for larger discharge currents, n moves closer to 2. On the x axis is given the discharge time in hours and on the y axis is the discharge current in amps. Consider a point on the $I^n t$ curve on the extreme right portion of the discharge time side. Here the discharge current is small and the profile is linear. The value of n=1. The area of the rectangle formed by the x and y projections of a point on the curve is $i_d \times t$ which is also the capacity units. As one moves towards higher discharge currents, the curve profile takes on a square law pattern and becomes $I^2 t$ where n=2. Here the area of the rectangle formed by the x and y projections of a point on this part of the curve will provide lower amp-hours. This means that as the discharge current increases, the time for which it can be drawn continuously will be less than if the curve were linear. This further implies that the capacity that can be drawn from the battery is not the same for all discharge currents. The $I^n t$ curve locus will provides the C-rate of the battery. A point on the locus will indicate the following: the projection on the x-axis will provide the discharge time t_d for which the current i_d can be continuously discharged. The capacity is $C = i_d \times t_d$ and the rate is t_d. For example if $i_d = 5A$ is being drawn continuously up to $t_d = 10$ hours, then the C-rate is a C_{10} battery wherein 50 Ah can be removed from the battery at a continuous discharge of 5A.

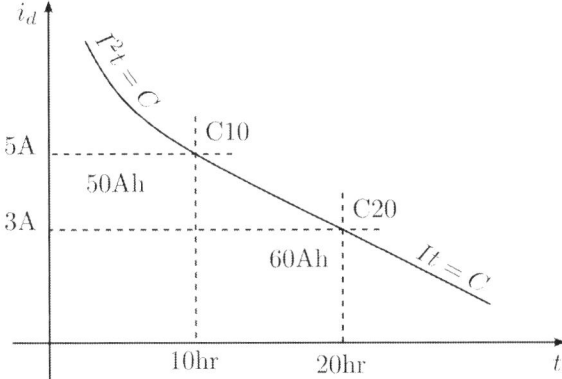

Figure 5.1.4: Capacity and discharge rate

5.2 Flywheel storage

Though battery is the most popular energy storage device, there are several other devices that are being used for different applications. One such storage device is the flywheel. The battery is a chemical storage device, where as the flywheel is a mechanical energy storage device. A typical system with flywheel energy storage is shown in figure 5.2.1. The energy domain of the PV system is electrical and that of the flywheel is mechanical. Therefore, one must have an electrical to mechanical energy domain converter to make the flywheel device compatible for PV systems.

The flywheel is connected to the shaft of a DC generator. The DC generator electrical terminals are connected to the PV source as shown in figure 5.2.1. Let ω_1 be the angular velocity of the flywheel. The the energy stored E_1 in the flywheel is the kinetic energy by virtue of the angular velocity and is given as

$$E_1 = \frac{1}{2}J\omega_1^2$$

where J is the inertia of the flywheel. If energy is either removed from the flywheel or injected into the flywheel, the angular velocity changes to ω_2. If energy is injected into the flywheel i.e. the flywheel is being charged up, then $\omega_2 > \omega_1$. If the energy is being discharged i.e. the energy is being removed from the flywheel, then $\omega_2 < \omega_1$. The energy E_2 of the flywheel at angular speed of ω_2 is

$$E_2 = \frac{1}{2}J\omega_2^2$$

The amount of energy injected into or removed from the flywheel is

$$\begin{aligned} \triangle E = E_2 - E_1 &= \frac{1}{2}J\left(\omega_2^2 - \omega_1^2\right) \\ &= J(\omega_2 - \omega_1)\frac{(\omega_2 + \omega_1)}{2} \\ &= J \cdot \triangle\omega \cdot \overline{\omega} \end{aligned} \quad (5.2.1)$$

It can be seen from equation 5.2.1 that the differential energy injected into or removed from the flywheel is proportional to the differential angular velocity. The power drawn is the rate of energy transfer. Therefore, the rate of change in the angular velocity is proportional to the amount of power flow into or out of the flywheel.

When the power is flowing from the electrical side to be stored in the flywheel, then the electromechanical machine acts as a motor and accelerates the machine shaft. This will tend to increase the angular velocity thereby increasing the amount of energy stored in the flywheel. When the energy from the flywheel is being discharged, then the electromechanical machine will act as a generator and allow power to flow from the flywheel to the electrical side and supplements the PV source power to load.

In a practical situation, the flywheel will be mounted onto the shaft of the machine which in turn will be isolated from the yoke of the machine through bearings. The bearing losses, the iron and copper losses of the electromechanical machine will lead to an energy conversion efficiency in the neighbourhood of 80%.

5.3 Pumped hydro storage

Figure 5.3.1 shows another method of energy storage/buffering wherein the electrical power is stored as potential energy of a mass of pumped up water. The system uses

5.4. COMPRESSED AIR STORAGE

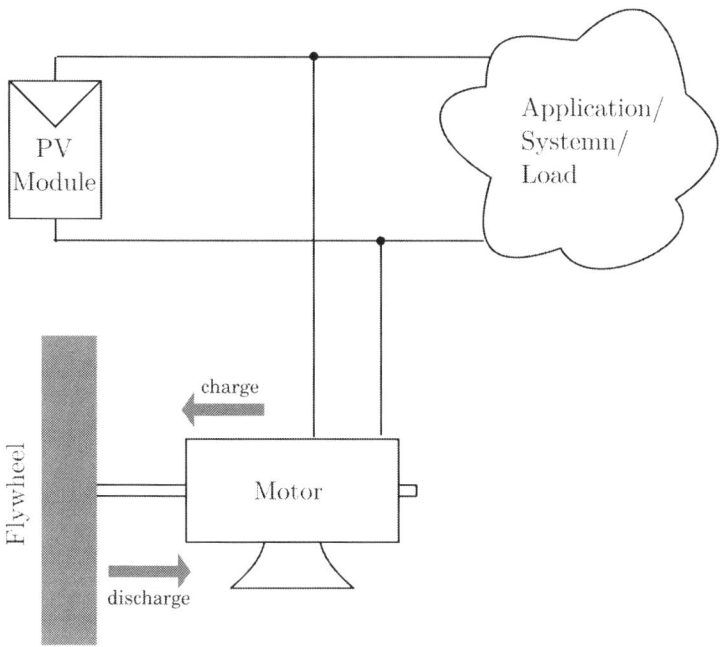

Figure 5.2.1: PV with flywheel energy storage

an electromechanical machine and a turbine which is mounted on the shaft of the machine.

During the charging mode, power flows from the PV source side and into the hydro system. The electromechanical machine is connected to the PV source through the electrical terminals on one side and to the centrifugal pump mounted on its shaft on the mechanical side. The electromechanical machine acts like a motor and rotates the centrifugal pump which provides a pressure difference across its inlet and outlet pipes. The centrifugal pump lifts the water from the sump or reservoir and fills up an overhead tank or reservoir which is at a height h from the sump. A mass m of water is lifted through a height h and stored as potential energy. This is the charge up process.

During the discharging mode, the water from the overhead reservoir is let to flow by opening a valve. The water flows from the overhead reservoir to the sump through a penstock and through a nozzle which impinges on the blades of the pump that now acts as a turbine. The turbine rotates the shaft of the electromechanical machine which now behaves as a generator and supplements the PV source supplying power to the load. Typical charging efficiencies for this type of storage system is about 85%. Similarly, the efficiency for the discharge cycle is also around 85%. For one charge-discharge cycle, the efficiency is $0.85^2 \cdot 100 = 72\%$.

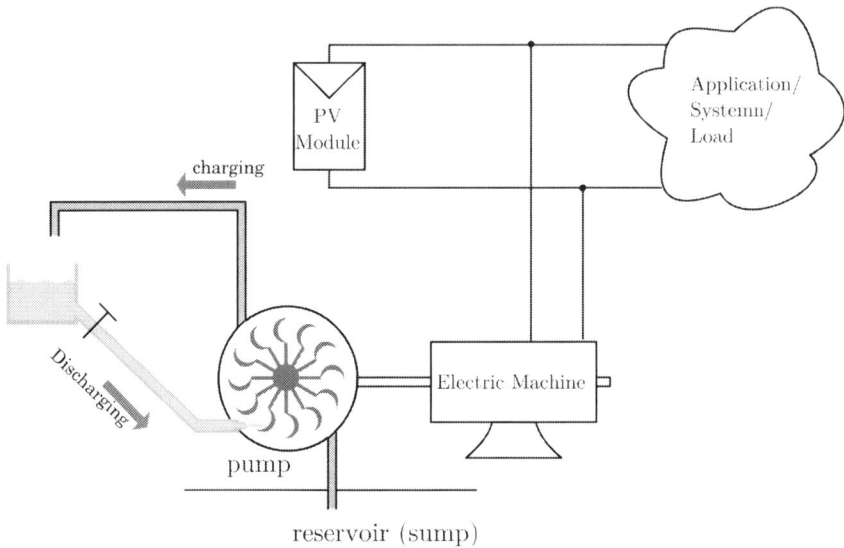

Figure 5.3.1: PV with pumped hydro energy storage

5.4 Compressed air storage

Another energy storage/buffering mechanism is compressed air storage. Here, the excess energy from the PV source is stored in compressed air by virtue of the increased pressure. Here again, the system uses a motor which converts electrical energy into mechanical energy at the shaft and the shaft of the motor is connected to an air compressor. The air compressor takes in the ambient air i.e. air at atmospheric pressure, compresses it and puts it into a container through a non-return valve.

The air pump will push the compressed air into this container and the pressure within the container keeps increasing at a constant volume and energy is stored by virtue of the air pressure. In this way one can charge up the compressor by taking the electrical energy from the PV source, pass it through the motor to convert the energy into mechanical form and then from the mechanical to hydraulic form and pushed into the container. This is the charging process.

The energy is stored by the virtue of the pressure and the discharge volume Q. The instantaneous power can be controlled by the discharge rate. $P \cdot \dot{Q}$ or $P \cdot \frac{dQ}{dt}$ will be the power that is used to charge up the compressor or discharge it. Note that during the charging mode, electromechanical machine is acting as a motor which converts electrical energy to mechanical energy. The air pump converts mechanical energy into hydraulic energy and stores the energy by virtue of the increased pressure of the air in the container.

In order to discharge the stored energy into the load, one need to open the outlet valve of the container. The discharge rate of the air exit is controlled by a nozzle. The air jet impinges on the air turbine which is used to convert the kinetic energy of the air jet into mechanical energy. The mechanical energy at the shaft of the air turbine is

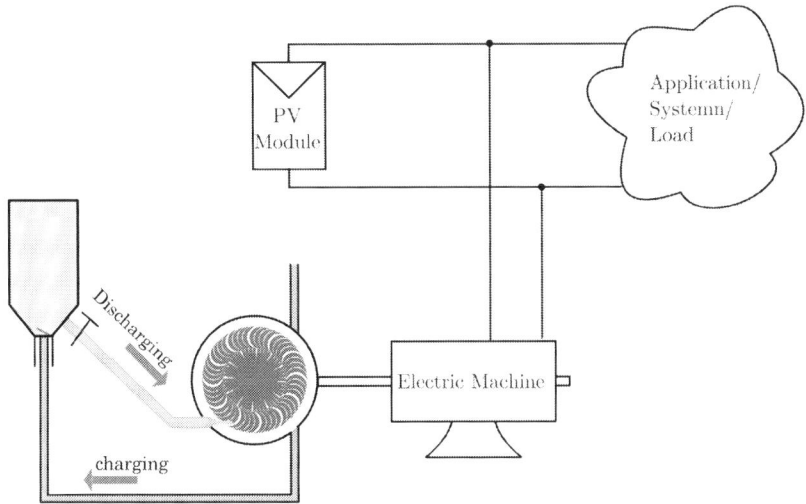

Figure 5.4.1: PV with compressed air energy storage

converted to the electrical energy through the electromechanical machine which now behaves as a generator. This is the discharging process.

Here too, the charging efficiency is around 85% and for the overall charge-discharge cycle, the efficiency is around 72%.

5.5 Questions

1. Which of the following is appropriate for SLI batteries ?
    a) They can be used for longer period
    b) They can give very large amount of power but for a very short amount of time
    c) They can give very large amount of power for a very long time
    d) They have very low power density but very high energy density

2. Which of the given type of battery cannot be re-charged?
    a) Zinc-silver oxide
    b) Nickel metal hydride
    c) Lithium-ion polymer
    d) Lead acid

3. Depth of discharge for Li-polymer batteries is
    a) 20 %
    b) 30 %
    c) ¡ 50%
    d) ¿ 50%

4. Consider a 80 Ah battery which has a nominal voltage of 12V. The state of charge for the battery is at 20%. The DoD of the battery is

a) 60 %
b) 70 %
c) 80 %
d) 20 %

5. A battery has a pre-determined charging and discharging levels. Among the following scheme, which has probably more life cycle?
   a) Battery with charging limit up to 80% and discharging limit at 40% of battery capacity
   b) Battery with charging limit up to 100% and discharging limit at 0% of battery capacity
   c) Battery with charging limit up to 80% and discharging limit at 10% of battery capacity
   d) Battery with charging limit up to 90% and discharging limit at 20% of battery capacity

6. If one were to consider the life of the battery in terms of charge-discharge cycles, the battery of choice would be
   a) Lithium Polymer
   b) Nickel Iron
   c) Lead Acid
   d) Sodium Ion

7. Which type of battery has the lowest energy density?
   a) Lithium Polymer
   b) Lead Acid
   c) Sodium Ion
   d) Nickel Metal Hydride

8. For a given weight of the batteries in an electric vehicle, the one that can give the best acceleration or overtaking ability is
   a) Lithium Polymer
   b) Sodium Ion
   c) Lead Acid
   d) Nickel Metal Hydride
   e) Nickel Iron

9. For a given weight of the batteries in an electric vehicle, the one that can give the best range per charge is
   a) Lithium Polymer
   b) Sodium Ion
   c) Lead Acid
   d) Nickel Metal Hydride
   e) Nickel Iron

10. For a particular application, the watt hour requirement is 500Wh. Select a battery with a nominal voltage of 12V and DoD of 50%. If the energy efficiency of the battery is 70%, then calculate the capacity of the battery.

11. A battery has a capacity of 40Ah. The c-rate specification of the battery is 3C. What is the maximum discharging current allowed for the battery?

12. A 20Ah battery with a voltage of 12V is connected to the terminals of a PV module for charging. The PV module has a short circuit current of 2.5A and an open circuit voltage of 14.5V. The battery has a c-rate of C10. If the internal resistance of the battery is 0.5 ohm, then what is the input resistance seen by the PV module across its terminals?

13. A PV array has an intrinsic area of 25 m^2. The daily incident energy at the place is 3.5 $kWh/m^2/day$. The PV array has an efficiency of 16% and it supplies power to an electric motor that drives a centrifugal pump to lift water from a sump to an overhead tank placed 10m above the sump. The motor has an efficiency of 88% and the pump has an efficiency of 92%. What is the expected energy stored in the overhead tank?

14. For the above question, calculate the capacity of the overhead tank if the density of water is 1000 kg/m^3.

15. In a flywheel storage system, the flywheel is rotating at 50 rad/s. The flywheel is charged up with more energy and it starts rotating at 100 rad/s. By how much is the increase in the energy stored in flywheel?

Chapter 6

Photovoltaic Application

6.1 Load Profile

Sizing solar PV panels for applications without the use of battery has been discussed. This sections will discuss the sizing of solar PV panels for applications that also include batteries. Figure 6.1.1 shows the load distribution over the entire day. The x-axis is the time axis showing the time from 0 hours to 12 midnight. The load distribution can be broadly divided into two parts, viz. (i) Day load, Wh_{day} which would occur sometime after 6 am upto about 6 pm and (ii) Night load, Wh_{night} which would occur from 0 hours to around 6 am and after 6 pm upto 12 pm midnight.

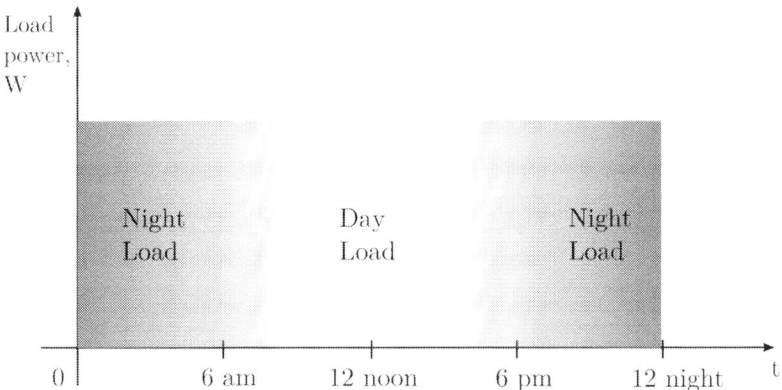

Figure 6.1.1: Day and night load distribution

During the day, major portion of the day load, Wh_{day} will be serviced by the PV panels. Some portion of the day load will be supplemented by the battery. In the time periods around noon time, the solar PV will dominate. During the night, almost all of the night load will be serviced by the battery as there is no insolation and PV panels will not generate any power during night time.

For any given application, load profiling for the entire day needs to be performed so that the daily energy requirement can be partitioned into day load and night load. The specification of the peak load current and the average load currents over the day will be useful to estimate the battery size.

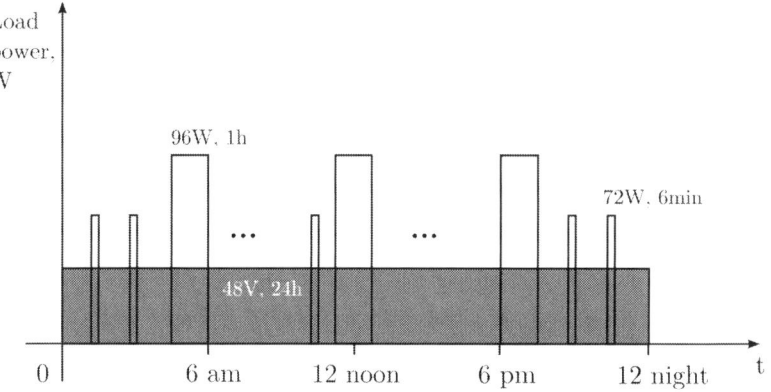

Figure 6.1.2: Example load distribution over the day

Consider an example on determining the load profile. Let there be three loads whose distribution profile over the day is shown in figure 6.1.2. The character of the three loads are,

1. Load-1 is characterised by 48 volts dc, 48 watts and it is a day and night continuous load, which occurs for full 24 hours of the day. The day light zone is demarcated as the period between 7 am and 5 pm. During this time the PV panel delivers power to the load depending on the insolation and supplements the battery. The remaining part of the 24 hours is designated as night zone wherein only the battery is participating in delivering power to the load.

2. Load-2 is a water pumping device. It is switched-on three times daily for one hour duration each. It is switched-on once before sunrise, once at noon and once after sunset. This load has an average running current of 4 A at 24 V dc. This load has 3 operational periods one before sunrise, one at noon and the last one after sunset. The one at noon is considered as day light load and the other 2 operational periods are considered as night loads.

3. Load-3 is considered as another type of load where in it draws 3 A at 24 V dc. It occurs every 2 hours and remains on for a period of 6 minutes only.

The day load can be calculated as follows

The daylight hours from 7 am to 5 pm is $h_d = 10$ hours

$$Wh_{day} = (48V \times 10h) + (24V \times 4A \times 1h) + \left(24V \times 3A \times \frac{6}{60}h \times 5\,times\right)$$
$$= 480 + 96 + 36 = 612Wh \qquad (6.1.1)$$

Similarly, the night load is calculated as

$$Wh_{night} = (48V \times 14h) + (24V \times 4A \times 1h \times 2\,times) + ...$$
$$\left(24V \times 3A \times \frac{6}{60}h \times 7\,times\right)$$
$$= 672 + 192 + 50.4 = 914.4 Wh \qquad (6.1.2)$$

Let the voltage of the battery that supplies the above loads be set at 24V. From figure 6.1.2 the peak battery current and the average battery current can be deduced which is essential to decide the battery capacity and C-rate specifications. The peak battery current I_{Bm} is calculated as follows:

$$I_{Bm} = \left(\frac{48W}{24V}\right) + \left(\frac{24V \times 4A}{24V}\right) + \left(\frac{24V \times 3A}{24V}\right)$$
$$= 2A + 4A + 3A = 9A$$

The worst case peak load current if loads 2 and 3 overlap in time is 9A. However, if the operation of the loads are so scheduled that they do not overlap, then the peak load current under such condition would be 6A.

The average load current I_B is estimated based on the duty cycle of operation and is given as

$$I_B = 2 \cdot \left(\frac{24h}{24h}\right) + 4 \cdot \left(\frac{3h}{24h}\right) + 3 \cdot \left(\frac{\left(\frac{6}{60}\right) \cdot 12}{24h}\right)$$
$$= 2A + 0.5A + 0.15A = 2.65A$$

6.2 Days of autonomy and recharge

In the case of PV based energy storage systems, there is a notion called "days of autonomy" that has a direct bearing on the determination of the battery size. Consider the diagram shown in figure 6.2.1. The x-axis is the time line in days. Every day, the solar insolation contributes to supplying the load power. The battery supplements the solar power during the daytime and entirely supports the night time load.

Consider a situation that on two contiguous days, k^{th} and $(k+1)^{th}$ days, the weather is very cloudy with negligible insolation. This would imply that the the battery should almost entirely supply the daytime load too apart from the night time load for these two days. The system is said to be in operation autonomously without the aid of solar power. The two contiguous days are said to be "days of autonomy". If there were 'n' contiguous days where solar power is negligible, then there would be "n days of autonomy". This would imply that the battery needs to be sized in such a manner that it should cater to the night time loads as well as the day time loads on these "days of autonomy". Evidently the battery size will be larger compared to the situation where there are no "days of autonomy". The "days of autonomy" is denoted by n_a.

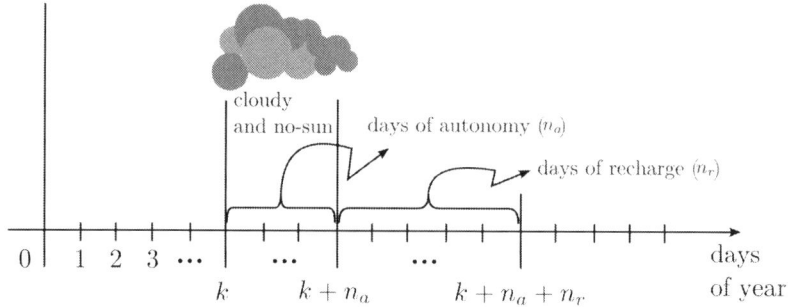

Figure 6.2.1: Days of autonomy and recharge

In the days following the "days of autonomy", the batteries need to be replenished with extra charge that it lost to support the day time loads during the days of autonomy. The number of days in which this lost charge is replenished is called the "days of recharge". This happens on contiguous days following the "days of autonomy". The "days of recharge" is denoted by n_r.

The impact of "days of autonomy" is on the battery size. More the n_a larger will be the size of the battery storage system as it needs to cater to the full day time loads too during the "days of autonomy". The impact of "days of recharge" is on the PV size. Smaller the n_r larger will be the size of the PV panels, as the PV panels will need to supply in lesser time the extra charge that the battery storage system had to give away during the "days of autonomy".

The default value of n_a will be 0. This implies 0 number of days of autonomy. Every day the PV panel or the PV source will contribute to the day time load. The number of contiguous days that the PV source does not participate in supplying power to the day time load is associated with autonomous operation. Likewise, the default value of n_r is also 0. It is the number of days needed to replenish the charge lost during the immediately preceding autonomous operation.

6.3 Battery sizing

Consider the night load to be Wh_{night}. This load has to be supplied entirely by the battery. If there are no "days of autonomy" then $n_a = 0$ and consequently the "days of recharge", $n_r = 0$. If η_B is the energy efficiency of the battery then,

$$Wh_{load} = \frac{Wh_{night}}{\eta_B}$$

If $n_a \neq 0$ then,

6.3. BATTERY SIZING

$$Wh_{load} = \frac{Wh_{night}}{\eta_B} + \frac{(Wh_{day} + Wh_{night})}{\eta_B} \cdot n_a \qquad (6.3.1)$$

The first term in equation 6.3.1 is the regular night load that the battery has to supply and the second term is the day and night load that the battery has to supply during n_a days of autonomy. Equation 6.3.1 can be simplified as

$$Wh_{load} = \frac{Wh_{night}}{\eta_B} \cdot (n_a + 1) + \frac{Wh_{day}}{\eta_B} \cdot n_a \qquad (6.3.2)$$

The depth of discharge DoD will be the effective capacity utilisation of the battery. The watt-hour capacity of the battery Wh_{bat} can be estimated as

$$Wh_{bat} = \frac{Wh_{load}}{DoD} \qquad (6.3.3)$$

If the nominal voltage of the battery is $V_{bat-nom}$, then the ampere-hour capacity of the battery is given as

$$Ah_{bat} = \frac{Wh_{load}}{DoD \cdot V_{bat-nom}} \qquad (6.3.4)$$

Example

Consider the example related to figure 6.1.2 wherein the nominal battery voltage is 24V for a deep discharge tubular lead acid battery. The depth of discharge is 80% and the efficiency of the battery is 70%. For a case where there are no days of autonomy, $n_a = n_r = 0$. Then from equation 6.3.2,

$$Wh_{load} = \frac{Wh_{night}}{\eta_B} = \frac{914.4}{0.7} = 1306Wh$$

$$Ah_{bat} = \frac{1306Wh}{0.8 \cdot 24V} = 68Ah$$

The peak discharge current for the application is 6A and the average discharge current is calculated as 2.65A. Considering only the average discharge current of 2.65A, $\left(\frac{68Ah}{2.65A}\right)$ a C25 battery is indicated as sufficient. However, the application draws a peak discharge current of 6A which implies a C10 battery. A battery with a capacity of 70Ah, C10 rate would be appropriate for this application.

6.4 PV array sizing

The PV array has to deliver the day load, Wh_{day} and the night load, Wh_{night} that will be stored by charging the battery during the day. Along with this, the PV also has to deliver the energy related to battery charge and discharge losses. During days of autonomy, there is no significant Sun power for the PV to provide either the day load or the night load. Therefore, for the designated number of days of autonomy, the PV should pre-load the battery with appropriate amount of energy corresponding to the day and night loads for these days of autonomy.

Consider the worst case scenario wherein there is no significant sun power during n_a days of autonomy. This is visualised through figure 6.4.1 wherein the x-axis is the number of days. Each segment represents a day. There are n_a days of autonomy followed by n_r days of recharge. The battery is expected to be recharged such that the energy lost during the days of autonomy is replenished back to the battery.

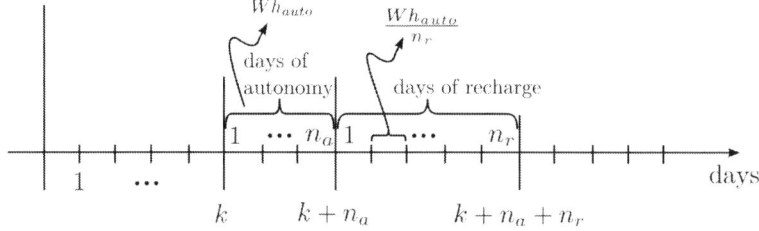

Figure 6.4.1: Autonomous days and recharge days

The total energy lost during the days of autonomy is Wh_{auto}. The entire Wh_{auto} amount energy has to be put back into the battery in n_r number of days. This implies that on each day after the days of autonomy, the PV panels have to supply $\frac{Wh_{auto}}{n_r}$ amount of extra energy so that the battery will recover and recuperate the energy lost during the days of autonomy. The the PV panels have to supply

$$Wh_{pv} = \left(Wh_{day} + \frac{Wh_{night}}{\eta_B}\right) + \frac{Wh_{auto}}{n_r} \qquad (6.4.1)$$

where

$$Wh_{auto} = \left(\frac{Wh_{day} + Wh_{night}}{\eta_B}\right) \cdot n_a \qquad (6.4.2)$$

From equation 6.4.2, it can be seen that during the days of autonomy, the extra energy that the battery needs to supply is the day load and the night load along with the battery charge and discharge losses.

Equation 6.4.1 gives the energy that is drawn from the PV source. The peak wattage of the PV array can be calculated as

6.4. PV ARRAY SIZING

$$P_m > \frac{Wh_{pv}}{H_{atmin}}$$

where H_{at} is the number of hours of standard insolation incident at the place in one day and is as given in equation 3.15.1.

From equations 4.1.2 and 4.1.3, the PV array area can be calculated as

$$A_{intrinsic} = \frac{P_m}{\eta_{pv}}$$

$$A_{estate} = 1.3 \cdot A_{intrinsic}$$

The intrinsic area of the PV array is calculated without taking the maintenance space into account. Making allowance for space to maintain the PV array, the actual estate area is A_{estate}.

For the place Bangalore located at latitude of 12.97 deg. N, the minimum daily insolation that is incident on a titled flat plate collector is around 4.58 $kW/m^2/day$. Thus,

$$H_{atmin} = 4.6 \, kW/m^2/day$$

Numerically, the H_{atmin} is equivalent to the number of hours i.e. 4.6 hours in this example, when standard insolation of $1 \, kW/m^2$ is incident. For the case of the running example in this chapter, referring to equations 6.1.1 and 6.1.2,

$$Wh_{pv} = 612 Wh + \frac{914.4 Wh}{0.7} = 1918 Wh$$

Using the estimate of incident energy H_{atmin}, the peak power requirement for the PV array can be estimated as

$$P_m > \frac{Wh_{pv}}{H_{atmin}} = \frac{1918}{4.58} = 418.8 \, W \, peak$$

If one chooses to use monocrystalline PV cells having efficiencies of about 18%, then the intrinsic area of the PV array can be estimated as

$$A_{intrinsic} = \frac{P_m}{\eta_{pv}} = \frac{0.42 \, kW}{0.18} = 2.33 \, m^2$$

The actual estate area required will be

$$A_{estate} = 1.3 \cdot A_{intrinsic} = 1.3 \cdot 2.33 = 3.03 \, m^2$$

6.5 Design toolbox in Octave

This section lists the Octave code for estimating the size of the battery and PV array for a given set of specifications. The first part of the system design Octave file is specifications section. It contains the specifications related to loads, specs related to battery and specs related to PV and insolation. This is followed by the section on sizing of battery. In order to size the PV array, the estimate of incident insolation at the place on a given day is required. The H_{atmin} is estimated like in chapter 3. The latitude and the tilt angle are provided as inputs to the H_{atmin} estimate function. This is followed the sizing of the PV array as discussed in the previous section.

```
#*******************************************************************
#
#      This is a script for the solar PV system design.
#      This will call other modules of the toolbox
#
#*******************************************************************
clc;
clear;
#*******************************************************************
#                       SPECIFICATIONS
#*******************************************************************
       #SPECS RELATED TO LOAD
WHday = 612;     #Wh day load
WHnight = 914.4;   #Wh night load
ILm = 6;     #A, peak load current
IL = 2.65;   #A, avg. load current
na = 3; #days of autonomy
nr = 2; #days of recharge after days of autonomy

#SPECS RELATED TO BATTERY
Vbatnom = 24;    #V nominal battery voltage
effB = 0.7; #battery efficiency
DoD = 0.8;  #depth of discharge

#SPECS RELATED TO PV and INSOLATION
effpv = 0.18;    #efficiency of PV cells
Q = 12.97;  #lattitude in degrees
B = 10; #tilt angle in degrees
#*******************************************************************
#                       SIZE BATTERY
#*******************************************************************
WHa = WHnight/effB;
WHb = (WHday+WHnight)*na/effB;
WHload = WHa + WHb; #battery to handle night load energy plus
```

6.5. DESIGN TOOLBOX IN OCTAVE

```
              #energy for days of autonomy
#  Battery Amp.hr. capacity
AHbat = WHload/(DoD*Vbatnom);
#  Battery C-rate
crate_m = AHbat/ILm;   #c-rate for peak load discharge
crate_avg = AHbat/IL; #c-rate for avg load discharge
#****************************************************************
#                     GET Hat_min kWH/m2/day
#****************************************************************
Hatmin=Hat_est(Q,B);
#****************************************************************
#                         SIZE PV
#****************************************************************
if (nr==0)
   nrr=1;
else
   nrr=nr;
endif
WHpa = WHday + (WHnight/effB);
WHpb = (WHday + WHnight)/effB;
WHpv = WHpa + (WHpb*na/nrr);   #daily energy requirement of
                               #PV array

# Peak watt of array
Pm = WHpv/Hatmin; #Watt peak
# Intrinsic array area
Aintrinsic = Pm/(effpv*1000); #Pm in W
    # Real estate area required for PV
Aestate = 1.3*Aintrinsic;
#****************************************************************
#                       DISPLAY RESULTS
#****************************************************************
    fprintf(1,'%s\n\n','PV SYSTEM DESIGN');
    fprintf(1,'%s\n\n','SPECIFICATIONS');

    fprintf(1,'%s\n','SPECS RELATED TO LOAD');
    fprintf(1,'%s\n',['WH day load, Wh = ',num2str(WHday)]);
    fprintf(1,'%s\n',['WH night load, Wh = ',num2str(WHnight)]);
    fprintf(1,'%s\n',['Peak load current, A = ',num2str(ILm)]);
    fprintf(1,'%s\n',['Average load current, A = ',num2str(IL)]);
    fprintf(1,'%s\n',['Days of autonomy, na = ',num2str(na)]);
    fprintf(1,'%s\n\n',['Days of recharge, nr = ',num2str(nr)]);

    fprintf(1,'%s\n','SPECS RELATED TO BATTERY');
    fprintf(1,'%s\n',['Nominal battery voltage, V = ',...
```

```
                                            num2str(Vbatnom)]);
fprintf(1,'%s\n',['Battery WH efficiency, % = ',...
                                            num2str(effB*100)]);
fprintf(1,'%s\n\n',['Depth of Discharge, % = ',...
                                            num2str(DoD*100)]);

fprintf(1,'%s\n','SPECS RELATED TO PV and INSOLATION');
fprintf(1,'%s\n',['PV cell efficiency, % = ',...
                                            num2str(effpv*100)]);
fprintf(1,'%s\n',['Lattitude of place, deg. = ',...
                                            num2str(Q)]);
fprintf(1,'%s\n\n',['Tilt angle of PV modules, deg. = ',...
                                            num2str(B)]);

fprintf(1,'%s\n','BATTERY SIZING');
fprintf(1,'%s\n',['AH capacity of battery, Ah = ',...
                                            num2str(AHbat)]);
fprintf(1,'%s\n',['C-rate w.r.t peak load ...
                    discharge = C/',num2str(crate_m)]);
fprintf(1,'%s\n',['C-rate w.r.t avg load ...
                    discharge = C/',num2str(crate_avg)]);

fprintf(1,'%s\n','PV SIZING');
fprintf(1,'%s\n',['Hat-min, kWh/m2/day = ',...
                                            num2str(Hatmin)]);
fprintf(1,'%s\n',['WH required from PV, Wh = ',...
                                            num2str(WHpv)]);
fprintf(1,'%s\n',['PV array peak watts, W-peak = ',...
                                            num2str(Pm)]);
fprintf(1,'%s\n',['Intrinsic area of PV array, m2 = ',...
                                            num2str(Aintrinsic)]);
fprintf(1,'%s\n',['Actual real estate area ...
                    required, m2 = ',num2str(Aestate)]);
#***************************************************************
```

6.6 Questions

1. Days of autonomy occurs during
   a) Rainy cloudy days
   b) Summer
   c) Winter
   d) Spring

2. The requirement for battery capacity increases as
   a) number of recharge days increases

6.6. QUESTIONS

   b) number of days of autonomy decreases
   c) number of recharge days decreases
   d) PV array size increases

3. Consider a PV application operated using tubular lead acid battery. The nominal voltage of the battery is 24V. The battery has a DoD of 80% and energy efficiency of 70%. Let the load energy requirement during sunlight hours be 560Wh and that at night be 1120Wh. The number of days of autonomy is 0.

   (a) The load energy requirement is ....
   (b) The energy that the battery is to be supplied with is ....

4. Consider the PV application described in the previous question. The Ah capacity requirement of the battery is ......

5. Consider a PV application operated using battery energy storage. The nominal voltage of the battery is 48V. The battery has a DoD of 60% and energy efficiency of 70%. Let the load energy requirement during sunlight hours be 800Wh and that at night be 1500Wh. If the number of days of autonomy is 0, then the watt-hour required from the PV is ....

6. Consider a PV application using battery energy storage. The nominal voltage of the battery is 24V. The battery has a DoD of 80% and energy efficiency of 60%. Let the load energy requirement during sunlight hours be 420Wh and that at night be 1380Wh. If the number of days of autonomy is 2 and the number of days of recharge following the days of autonomy is 5, then the watt-hour required from the PV is ......

7. Consider a load distribution profile wherein one of the loads is characterised by 48V dc requiring 5A of current. This occurs every 2 hours and remains ON for a period of 30 minutes starting at 12 midnight. Another load occurs once during the night time for a period of 4 hours. It occurs at 24V dc and requires 10A of current. Consider daytime to be between 7.00am and 5.00pm. Calculate the peak and the average load currents if a 24V battery supplies the above loads?

8. A set of loads can be classified into day time load of 600 Wh and night time load of 1800 Wh. A PV array and 24V battery system are used to supply the loads. The battery has a DoD of 70% and conversion efficiency of 75%. Calculate the capacity of the battery.

9. A PV array and 24V battery system are used to supply a set of loads. The peak current drawn by the battery is 20A and the average current drawn by the battery is 4A. The battery has a DoD of 70% and conversion efficiency of 75%. A 140Ah battery is selected for the application. What should be the c-rate of the battery?

10. A set of loads can be classified into day time load of 600 Wh and night time load of 1800 Wh. A PV array and 24V battery system are used to supply the

loads. The battery has a DoD of 80% and conversion efficiency of 70%. The PV array has an efficiency of 18%. The daily incident energy at the place is 5 $kWh/m^2/day$. The system must have uninterrupted supply to the loads for 2 days of autonomy. The batteries must be recharged within 3 days after the days of autonomy. Calculate the battery capacity and the PV size.

Chapter 7

Maximum power point tracking

7.1 Introduction

The i-v characteristics of the PV source indicates that the power versus voltage characteristic has a peak power point. If the solar PV source is operated with a load such that the operating point corresponds to the peak power point, then maximum power can be extracted from the PV source for a given insolation at any given time. However, the load on the PV source may be neither constant nor at a value conducive to extract maximum power. Therefore, a power converter is inserted between the solar PV source and the load such that the PV source is presented with an equivalent resistance across its terminals that is conducive for extracting maximum power, whatever may be the value of the load. Such a power converter that manages the operating point in such a manner as to extract maximum power from the PV source at all times is called maximum power point tracking (MPPT).

Figure 7.1.1(a) shows a block schematic of a typical MPPT system. The PV source is connected to the load resistance R_o through a DC-DC converter as shown. Figure 7.1.1(b) shows the i-v characteristic and the p-v characteristic of the PV source. For a terminal resistance R_T across the terminals of the PV source, the operating point is such that maximum power is extracted. The load resistance R_o may not be equal to the resistance R_T required to achieve MPPT. Further R_o may not be constant, but may be varying dynamically. In the face of a dynamically varying R_o, the DC-DC converter that is interposed between the PV source and the load resistance R_o will present a constant resistance R_T across the terminals of the PV source such that MPPT is achieved even when R_o is not equal to R_T.

The duty cycle of the DC-DC power converter is controlled in order to maintain or regulate its input resistance R_T such that MPPT condition is achieved. The terminal voltage and the terminal current of the PV source is measured and used to calculate the instantaneous power that is being drawn. This information is passed onto a MPPT control topology which will appropriately control the duty ratio of the DC-DC converter in order to regulate the PV source terminal resistance R_T.

From figure 7.1.1, one can see that PV power is being supplied to a resistive load

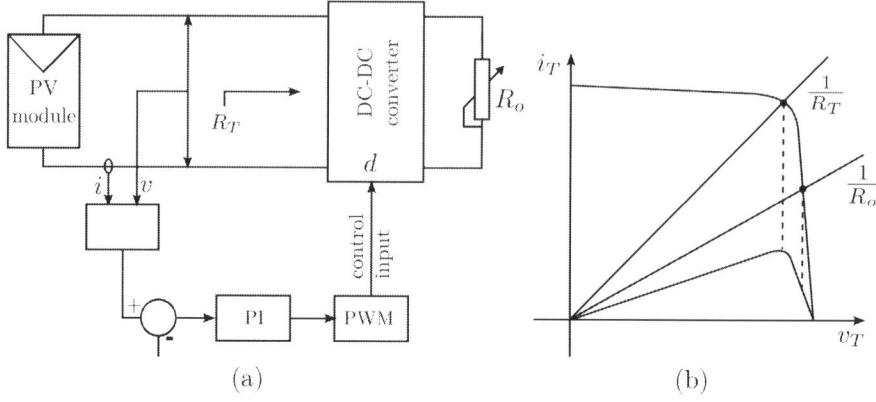

Figure 7.1.1: (a) Block diagram of MPPT system (b) i-v characteristic of the PV cell illustrating MPPT operating point

R_o. R_o is a dynamically changing quantity which can swing from 0 (short circuit) to ∞ (open circuit). Consider an arbitrary value of R_o such that the operating point is not at maximum power point. The load line is given by the $\frac{1}{R_o}$ line as shown. When $R_o = \infty$, the load line is along the v-axis and when $R_o = 0$, the load line is along the i-axis. The load line corresponding to maximum power point is when $R_o = R_T$. The main objective of maximum power point tracking algorithms is to ensure that the resistance seen across the terminals of the PV source is R_T irrespective of the value of R_o.

7.2 Input resistance of boost converter

In order to perform maximum power point tracking, the input resistance of the DC-DC converter that interfaces the PV source with the load needs to be controlled. It is meaningful only if the input resistance of the converter is a function of the duty ratio. Consider the boost converter as shown in figure 7.2.1. This is one of the primary converters.

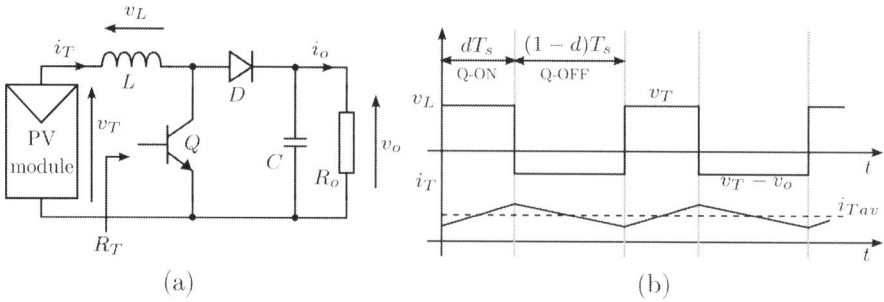

Figure 7.2.1: (a) Boost converter, (b) voltage and current waveforms of inductor

7.2. INPUT RESISTANCE OF BOOST CONVERTER

The PV source is interfaced to the load R_o through a DC-DC converter popularly known as the boost topology. It consists of an inductor on the input side, a power semiconductor switch Q, diode and output capacitor as shown in figure 7.2.1(a). The power semiconductor switch Q can be any one of the controlled switches like BJT, MoSFET, IGBT or SiC type of devices. The device Q is switched at a frequency f_s having a switching time period $T_s = \frac{1}{f_s}$. The device Q is ON for a period dT_s wherein d is the duty ratio of operation of the device Q. During the period $(1-d)T_s$ the device Q is OFF.

The waveforms of the voltage across the inductor v_L and the current through the inductor i_T are shown in 7.2.1(b). These are important waveforms that provide insight into the operation of boost circuit topology. During the period dT_s, the power switch Q is ON. The voltage across the inductor $v_L = v_T$, where v_T is the terminal voltage of the PV source. v_T is a dc quantity and can be considered to be constant for the duration of the switching period. As a consequence, the inductor current will rise linearly at the rate of $\frac{v_T}{L}$.

During the period $(1-d)T_s$, the power switch Q is OFF and the inductor current free wheels through the diode and the output capacitor-load combination. The inductor voltage during this period is $(v_T - v_o)$. The inductor current will fall with a rate of $\frac{(v_T - v_o)}{L}$. The average voltage across the inductor is always zero. Therefore, the volt-sec across the inductor in a switching period T_s should be zero in the steady state. Using the inductor volt-sec balance condition,

$$v_T dT_s + (v_T - v_o)(1-d)T_s = 0$$

$$v_o = v_T \cdot \frac{1}{(1-d)} \qquad (7.2.1)$$

Considering the current through the output capacitor, the amp-sec in a switching period should be zero in steady state. Using this amp-sec balance condition in a capacitor,

$$(-i_o)dT_s + (i_{Tav} - i_o)(1-d)T_s = 0$$

$$i_o = i_{Tav} \cdot (1-d) \qquad (7.2.2)$$

Dividing equation 7.2.1 by equation 7.2.2 one obtains

$$\frac{v_o}{i_o} = \frac{v_T}{i_{Tav}} \cdot \frac{1}{(1-d)^2}$$

$$R_o = \frac{R_T}{(1-d)^2}$$

or

$$R_T = R_o \cdot (1-d)^2 \quad (7.2.3)$$

Equation 7.2.3 gives the PV source terminal resistance as a function of load resistance R_o and duty ratio d for a boost converter interfacing PV source and load. When $d = 0$ then $R_T = R_o$. The load line has the slope of $\frac{1}{R_o}$. When $d = 1$ then $R_T = 0$. The load line has the slope of ∞ which implies that it is along the y-axis or i-axis. By controlling the value of duty ratio d between 0 and 1, the load line can be positioned any where between the extreme limits of y-axis and R_o load line.

7.3 Input resistance of buck converter

Consider now a buck converter as shown in figure 7.3.1(a). This is also one of the primary converters.

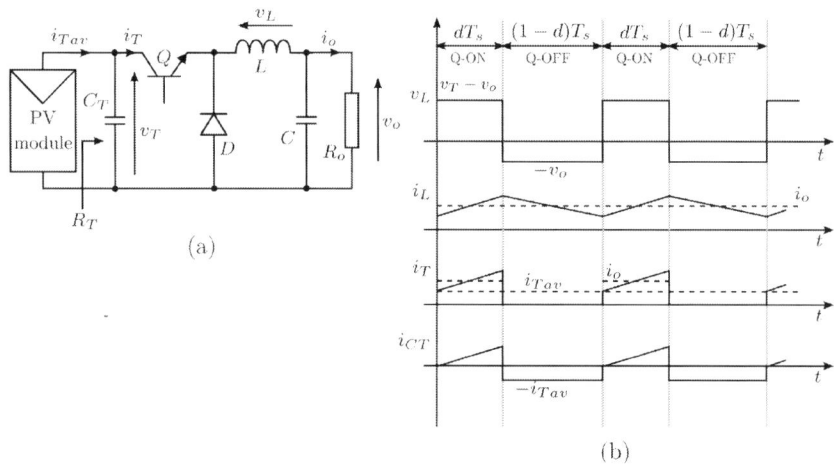

Figure 7.3.1: (a) Buck converter (b) voltage and current waveforms

The PV source is interfaced to the load R_o through the buck DC-DC converter. It consists of an inductor on the output side, a power semiconductor switch Q on the input side and diode as shown. It has a capacitor on the output side and a capacitor on the input side too as shown in figure 7.3.1(a). The power semiconductor switch Q can be any one of the controlled switches like BJT, MoSFET, IGBT or SiC type of devices. The device Q is switched at a frequency f_s having a switching time period $T_s = \frac{1}{f_s}$. The device Q is ON for a period dT_s wherein d is the duty ratio of operation of device Q. During the period $(1-d)T_s$ the device Q is OFF.

The waveforms of the voltage across the inductor v_L and the current through the inductor i_L are shown which provide the necessary information for extracting the input-output relationship. The waveforms of current through the switch Q, the current

7.3. INPUT RESISTANCE OF BUCK CONVERTER

through input capacitor C_T and the current through the PV terminals i_{Tav} are indicated in the figure 7.3.1(b).

During the period dT_s, the power switch Q is ON. The voltage across the inductor $v_L = (v_T - v_o)$, where v_T is the terminal voltage of the PV source. v_T and v_o are dc quantities and can be considered to be constant for the duration of the switching period. As a consequence, the inductor current will rise linearly at the rate of $\frac{(v_T - v_o)}{L}$. During this period, the diode D is OFF and the current i_T through the switch Q will be same as that through the inductor.

During the period $(1-d)T_s$, the power switch Q is OFF and the inductor current freewheels through the output capacitor-load combination and diode. The inductor voltage during this period is $(-v_o)$. The inductor current will fall with a rate of $\frac{(-v_o)}{L}$. The current through the switch Q will be zero and the diode current will be same as the inductor current.

The input capacitor C_T has an important role. If it were not present, then the terminal current of the PV source would have been a switched current like the waveform of i_T. During period $(1-d)T_s$, as the PV source current is zero, there can be no power drawn from the PV source. During period dT_s the current amplitude value is dependent on only the load. Therefore, MPPT cannot be performed. By placing the input capacitor C_T, the PV source terminal current is continuous and equal to the average of the current through switch Q. This current is a function of the duty ratio, thereby making MPPT operation possible.

In the steady state the average inductor current will be i_o. The average voltage across the inductor is always zero under steady state conditions. Therefore, the volt-sec across the inductor in a switching period T_s should be zero. Using the inductor volt-sec balance condition,

$$(v_T - v_o)dT_s + (-v_o)(1-d)T_s = 0$$

$$v_o = v_T \cdot d \tag{7.3.1}$$

Considering the current through the output capacitor, the amp-sec in a switching period should be zero. One can consider the amp-sec balance condition in a capacitor or one could consider that in an ideal situation the input and output powers are conserved. Therefore

$$i_{Tav} \cdot v_T = i_o \cdot v_o$$

$$i_{Tav} = i_o \cdot \frac{v_o}{v_T} = i_o \cdot d \tag{7.3.2}$$

Using equations 7.3.1 and 7.3.2 one obtains

$$\frac{v_o}{i_o} = \frac{v_T}{i_{Tav}} \cdot d^2$$

$$R_o = R_T \cdot d^2$$

or

$$R_T = \frac{R_o}{d^2} \quad (7.3.3)$$

Equation 7.3.3 gives the PV source terminal resistance as a function of load resistance R_o and duty ratio d for buck converter interfacing PV source and load. When $d = 0$ then $R_T = \infty$. The load line has the slope of 0. It is along the x-axis or the v-axis. When $d = 1$ then $R_T = R_o$. The load line has the slope of $\frac{1}{R_o}$. By controlling the value of duty ratio d between 0 and 1, the load line can be positioned any where between the extreme limits of x-axis and R_o load line.

7.4 Input resistance of buck-boost converter

Consider the buck-boost converter as shown in figure 7.4.1(a). This is the third of the three primary converters.

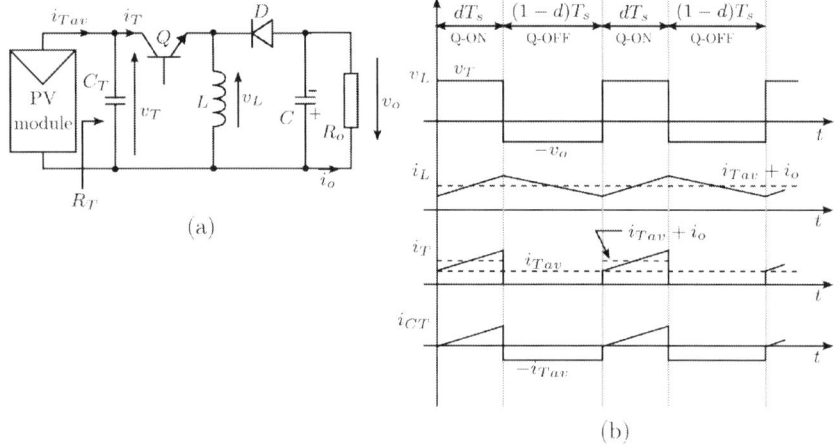

Figure 7.4.1: (a) Buck-Boost converter (b) voltage and current waveforms

The PV source is interfaced to the load R_o through the buck-boost DC-DC converter. It consists of an inductor, a power semiconductor switch Q on the input side and diode as shown. It has a capacitor on the output side and a capacitor on the input side as well as shown in figure 7.4.1(a). The power semiconductor switch Q can be any one of the controlled switches like BJT, MoSFET, IGBT or SiC type of devices. The device Q is switched at a frequency f_s having a switching time period $T_s = \frac{1}{f_s}$. The device Q is ON for a period dT_s wherein d is the duty ratio of operation of device Q. During the period $(1-d)T_s$ the device Q is OFF.

7.4. INPUT RESISTANCE OF BUCK-BOOST CONVERTER

The waveforms of the voltage across the inductor v_L and the current through the inductor i_L are shown which provide the necessary information for extracting the input-output relationship. The waveforms of current through the switch Q, the current through input capacitor C_T and the average current through the PV terminals i_{Tav} are indicated in the figure 7.4.1(b).

During the period dT_s, the power switch Q is ON. The voltage across the inductor $v_L = v_T$, where v_T is the terminal voltage of the PV source. v_T and v_o are dc quantities and can be considered to be constant for the duration of the switching period. The inductor current will rise linearly at the rate of $\frac{v_T}{L}$. During this period, the diode D is OFF and the current i_T through the switch Q will be same as that through the inductor.

During the period $(1-d)T_s$, the power switch Q is OFF and the inductor current freewheels through the output capacitor-load combination and diode. The inductor voltage during this period is $(-v_o)$. The inductor current will fall with a rate of $\frac{(-v_o)}{L}$. The current through the switch Q will be zero and the diode current will be same as the inductor current.

The input capacitor C_T has an important role like in the case of buck converter. If it were not present, then the terminal current of the PV source would have been a switched current like the waveform of i_T. During period $(1-d)T_s$, as the PV source current is zero, there can be no power drawn from the PV source. During period dT_s the current amplitude value is dependent on only the load. Therefore, MPPT cannot be performed. By placing the input capacitor C_T, the PV source terminal current is continuous and equal to the average of the current through switch Q. This current is a function of the duty ratio and thereby making MPPT operation possible.

In the steady state the average inductor current will be $(i_o + i_{Tav})$. The average voltage across the inductor is always zero under steady state conditions. Therefore, the volt-sec across the inductor in a switching period T_s should be zero. Using the inductor volt-sec balance condition,

$$(v_T)dT_s + (-v_o)(1-d)T_s = 0$$

$$v_o = v_T \cdot \frac{d}{(1-d)} \qquad (7.4.1)$$

Considering the current through the output capacitor, the amp-sec in a switching period should be zero. Using this amp-sec balance condition in a capacitor or input and output powers balance, one obtains

$$i_{Tav} \cdot v_T = i_o \cdot v_o$$

$$i_o = i_{Tav} \cdot \frac{(1-d)}{d} \qquad (7.4.2)$$

Using equations 7.4.1 and 7.4.2 one obtains

$$\frac{v_o}{i_o} = \frac{v_T}{i_{Tav}} \cdot \frac{d^2}{(1-d)^2}$$

$$R_o = R_T \cdot \frac{d^2}{(1-d)^2}$$

or

$$R_T = R_o \cdot \left(\frac{1-d}{d}\right)^2 \tag{7.4.3}$$

Equation 7.4.3 gives the PV source terminal resistance as a function of load resistance R_o and duty ratio d for buck-boost converter interfacing PV source and load. When $d = 0$ then $R_T = \infty$. The load line has the slope of 0. It is along the x-axis or the v-axis. When $d = 1$ then $R_T = 0$. The load line has the slope of ∞. It is along the y-axis or the i-axis. By controlling the value of duty ratio d between 0 and 1, the load line can be positioned any where between the extreme limits of x-axis and y-axis.

7.5 Input impedance of isolated converters

This section discusses few popular isolated converters from the perspective of input impedance. These converters are derived from one or other of the primary converters.

7.5.1 Forward converter

The circuit of the forward converter is shown in figure 7.5.1. It consists of a high frequency transformer wherein the primary winding is connected in series with the PV source and a power semiconductor switch. It is assumed that a capacitor is connected across the PV source for reasons similar to that explained in the primary buck and buck-boost converters. The dot polarities of the transformer is such that when the switch Q is ON, the dot end of the transformer primary is positive. The power is transferred to the secondary circuit. The secondary side circuit is similar to a buck converter in operation. During the time when switch Q is OFF, the core of the transformer needs to demagnetise. For this a diode-resistor series circuit is placed in parallel across the primary winding. This will help in providing a freewheeling path for the magnetising current.

The input-output voltage relationship for this converter is given as

$$v_o = n \cdot v_T \cdot d \tag{7.5.1}$$

where n is the secondary to primary turns ratio of the forward converter transformer. Assuming a lossless converter, the output power P_o will be equal to the input power P_{in}. Therefore,

7.5. INPUT IMPEDANCE OF ISOLATED CONVERTERS

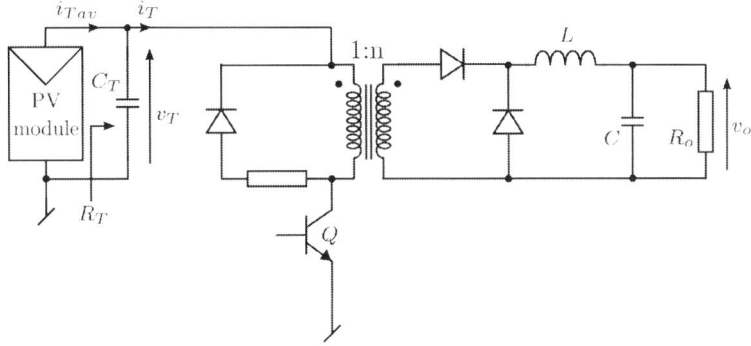

Figure 7.5.1: Forward converter circuit

$$v_T \cdot i_T = v_o \cdot i_o$$

$$i_T = \left(\frac{v_o}{v_T}\right) \cdot i_o = n \cdot d \cdot i_o \tag{7.5.2}$$

From equations 7.5.1 and 7.5.2, the terminal resistance across the PV source on the input side can be deduced as

$$R_T = \frac{v_T}{i_T} = \frac{R_o}{(n \cdot d)^2} \tag{7.5.3}$$

From equations 7.3.3 and 7.5.3, one can see that the input resistance for the forward converter has an additional degree of freedom with respect to buck converter in terms of the turns ratio. When turns ratio is unity, both the input resistances are same. When $d = 0$, the input resistance $R_T = \infty$ and the load line is aligned along the voltage axis of the i-v characteristics of PV source. When $d = 1$, the input resistance is $R_T = \frac{R_o}{n^2}$. By controlling the duty ratio, the loadline can be positioned any where between the voltage axis and $\frac{n^2}{R_o}$ line. The turns ratio n can be used as a design parameter to extend or reduce the span of the MPPT operation.

7.5.2 Flyback converter

Flyback converter is probably one of the most popular isolated converter for low power applications due to its low component count. Figure 7.5.2 gives the circuit schematic of this converter. The circuit is directly derived from the primary buck-boost topology.

In the buck-boost topology one simple modification is made. The inductor is converted into the device that provides galvanic isolation. The inductor comprises of a primary part and a secondary part. When the switch Q is ON, energy is stored within the permeance of the inductor. During this time no current flows in the secondary part

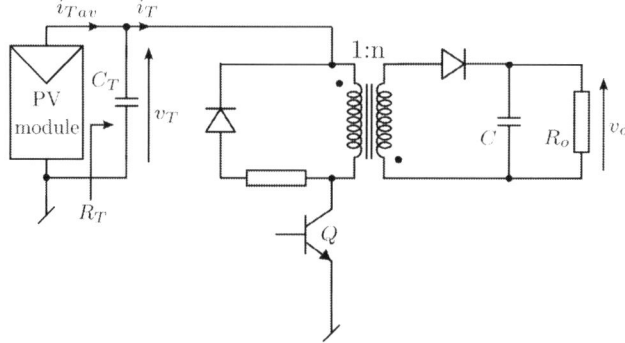

Figure 7.5.2: Flyback converter circuit

of the inductor. When the switch Q goes OFF, the energy stored in the permeance of the inductor is released in the secondary part through the diode, capacitor and load.

The input-output voltage relationship for this converter is given as

$$v_o = n \cdot \left(\frac{d}{1-d}\right) \cdot v_T \tag{7.5.4}$$

where n is the secondary number of turns to the primary number of turns of the inductor. Assuming a lossless converter, the output power P_o will be equal to the input power P_{in}. Therefore,

$$v_T \cdot i_T = v_o \cdot i_o$$
$$i_T = \left(\frac{v_o}{v_T}\right) \cdot i_o = n \cdot \left(\frac{d}{1-d}\right) \cdot i_o \tag{7.5.5}$$

From equations 7.5.4 and 7.5.5, the terminal resistance across the PV source on the input side can be deduced as

$$R_T = \frac{v_T}{i_T} = \frac{R_o \cdot (1-d)^2}{(n \cdot d)^2} \tag{7.5.6}$$

From equation 7.5.6, the terminal resistance across the PV source is given as a function of duty ratio and turns ratio. When $d = 0$ then $R_T = \infty$. The load line has the slope of 0. It is along the x-axis or the v-axis. When $d = 1$ then $R_T = 0$. The load line has the slope of ∞. It is along the y-axis or the i-axis. By controlling the value of duty ratio d between 0 and 1, the load line can be positioned any where between the extreme limits of x-axis and y-axis. The turns ratio will not change the span of MPPT operation. However, turns ratio n is an additional variable that provides design freedom in deciding load voltage and current levels.

7.5.3 Pushpull converter

This converter topology is derived from the buck converter. The operation is similar to that of two back to back forward converters. Figure 7.5.3 gives the circuit schematic of the converter. Q1, D1 and one half of the pushpull transformer will form one of the back to back forward converter part and Q2, D2 and the other half of the pushpull transformer forms the other back to back forward converter.

When Q1 is ON and Q2 is OFF, the dot ends of the windings of the pushpull transformer are positive. D1 will conduct. D2 is reverse biased and therefore OFF. D1, L, C and the load operate like an averaging circuit. When Q2 is ON and Q1 is OFF, the non-dot ends of the windings of the pushpull transformer are positive. D2 will conduct. D1 is reverse biased and therefore OFF. D2, L, C and the load act as filter. There will be a duration of time within a switching period when neither Q1 nor Q2 is ON. During this time, the inductor current will freewheel through both D1 and D2. Due to this, the freewheeling inductor current will split and flow through the dot end winding and non-dot end winding of the pushpull secondary. This leads to cancellation of mmf and no voltage is induced across the winding a that time.

The voltage across the inductor will be an ORing of the pulses due to Q1 ON and Q2 ON. This will be double the frequency of the individual switching frequency of the switches Q1 and Q2. The duty ratio for output LC averaging circuit is double that of the duty ratio of the primary side switching operation.

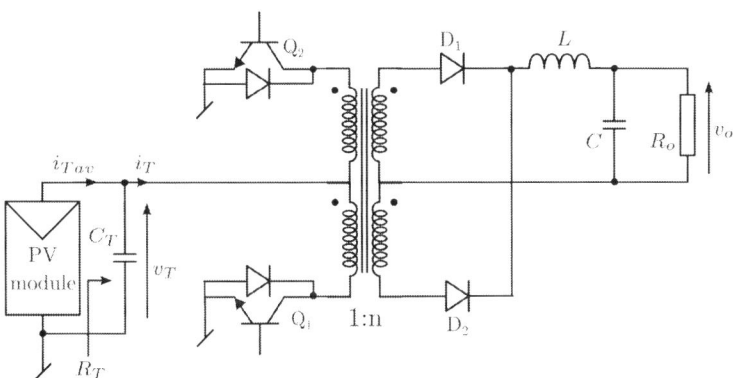

Figure 7.5.3: Pushpull converter circuit

The pushpull transformer has a better utilisation factor compared to the forward converter transformer. In the case of the forward converter transformer, the flux swing is only in the positive side of the B-H characteristics. However, in the case of pushpull transformer, the flux swings on both positive and negative sides of the B-H characteristics, thereby resulting in better utilisation. The off state voltage stress across the primary side switches of the pushpull converter is at least twice the applied terminal voltage v_T. The input-output voltage relationship for this converter is given as

$$v_o = 2 \cdot n \cdot v_T \cdot d \qquad (7.5.7)$$

where n is the secondary to primary turns ratio of the converter transformer and d is the duty ratio of the primary switches. Assuming a lossless converter, the output power P_o will be equal to the input power P_{in}. Therefore,

$$v_T \cdot i_T = v_o \cdot i_o$$

$$i_T = \left(\frac{v_o}{v_T}\right) \cdot i_o = 2 \cdot n \cdot d \cdot i_o \tag{7.5.8}$$

From equations 7.5.7 and 7.5.8, the terminal resistance across the PV source on the input side can be deduced as

$$R_T = \frac{v_T}{i_T} = \frac{R_o}{(4 \cdot n \cdot d)^2} \tag{7.5.9}$$

From equations 7.3.3 and 7.5.9, one can see that the input resistance for this converter has an additional degree of freedom in terms of the turns ratio. When $d = 0$, the input resistance $R_T = \infty$ and the load line is aligned along the voltage axis of the i-v characteristics of PV source. When $d = 1$, the input resistance is $R_T = \frac{R_o}{4 \cdot n^2}$. By controlling the duty ratio, the loadline can be positioned any where between the voltage axis and $\frac{4 \cdot n^2}{R_o}$ line. The turns ratio n can be used as a design parameter to extend or reduce the span of the MPPT operation.

7.5.4 Half bridge converter

This converter is also another derivative of the primary buck converter. The schematic of the half bridge topology is shown in figure 7.5.4. The part of the circuit on the secondary side of the high frequency transformer is exactly similar to the push pull topology. The primary side bridge has two capacitor-resistor divider (R_1C_1 and R_2C_2) that divides the dc link voltage. The voltage across each capacitor is $\frac{V_T}{2}$. When the primary side switch Q1 is ON, Q2 should be OFF and vice-versa. When Q1 is ON, the voltage across the primary is $\frac{V_T}{2}$ and when Q2 is ON, it is $-\frac{V_T}{2}$. The primary winding of the transformer should necessarily have zero average voltage across it in order to prevent flux build up or flux walking. A capacitor C_{fw} is used as shown to prevent flux walking and saturation of the transformer core. The operation of the secondary side is exactly same as that described for the push pull circuit. An advantage here over the push pull circuit is in the voltage rating of the primary side switches. The voltage rating of the primary side switches should be at least V_T, as against $2V_T$ for the push pull topology.

The input-output voltage relationship for this converter is given as

$$v_o = 2 \cdot n \cdot \frac{v_T}{2} \cdot d \tag{7.5.10}$$

where n is the secondary (centre tap) to primary turns ratio of the converter transformer and d is the duty ratio of the primary switches. Assuming a lossless converter, the output power P_o will be equal to the input power P_{in}. Therefore,

7.5. INPUT IMPEDANCE OF ISOLATED CONVERTERS

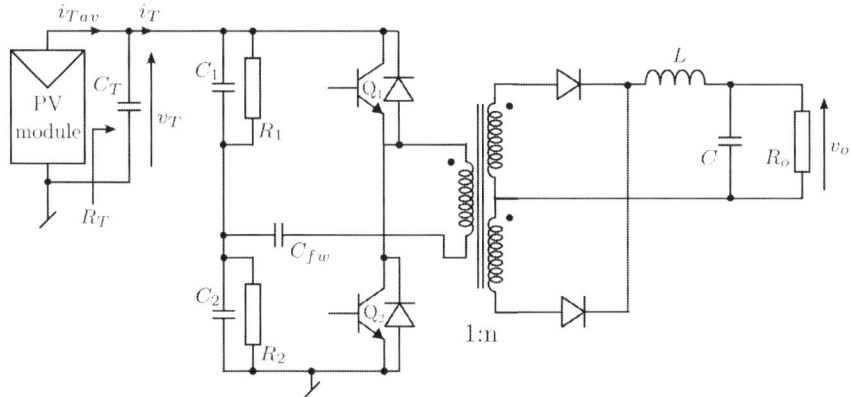

Figure 7.5.4: Half bridge converter circuit

$$v_T \cdot i_T = v_o \cdot i_o$$

$$i_T = \left(\frac{v_o}{v_T}\right) \cdot i_o = n \cdot d \cdot i_o \qquad (7.5.11)$$

From equations 7.5.10 and 7.5.11, the terminal resistance across the PV source on the input side can be deduced as

$$R_T = \frac{v_T}{i_T} = \frac{R_o}{(n \cdot d)^2} \qquad (7.5.12)$$

From equations 7.3.3 and 7.5.12, one can see that the input resistance for this converter is similar to that of the buck converter with an additional degree of freedom in terms of the turns ratio selection. When $d = 0$, the input resistance $R_T = \infty$ and the load line is aligned along the voltage axis of the i-v characteristics of PV source. When $d = 1$, the input resistance is $R_T = \frac{R_o}{n^2}$. By controlling the duty ratio, the loadline can be positioned any where between the voltage axis and $\frac{n^2}{R_o}$ line. The turns ratio n can be used as a design parameter to extend or reduce the span of the MPPT operation.

7.5.5 Full bridge converter

This converter is yet another derivative of the primary buck converter. The schematic of the full bridge topology is shown in figure 7.5.5. The part of the circuit on the secondary side of the high frequency transformer is exactly similar to the half bridge topology. The primary side has four switches as shown. The diagonally opposite switch pairs are turned ON or OFF simultaneously. When Q1 and Q4 are ON, the voltage across the primary is V_T and when Q2 and Q3 are ON, it is $-V_T$. Here too, capacitor C_{fw} is used to prevent flux walking in the transformer core. The operation of the secondary side is exactly same as that described for the push pull circuit.

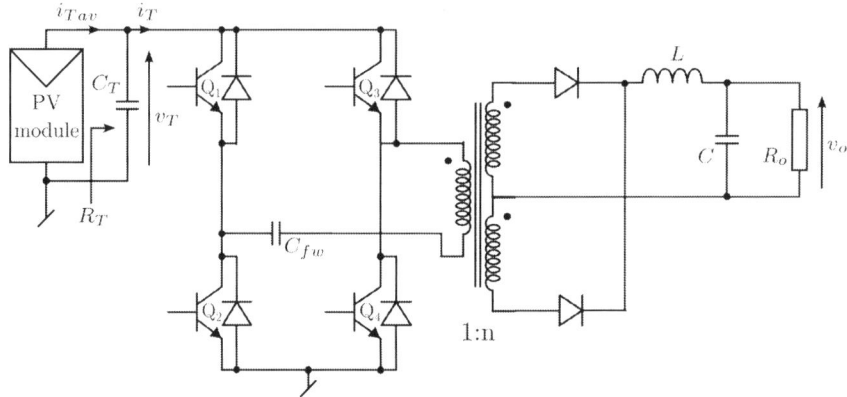

Figure 7.5.5: Circuit of full bridge converter

The input-output voltage relationship for this converter is given as

$$v_o = 2 \cdot n \cdot v_T \cdot d \qquad (7.5.13)$$

where n is the secondary (centre tap) to primary turns ratio of the converter transformer and d is the duty ratio of the primary switches. Assuming a lossless converter, the output power P_o will be equal to the input power P_{in}. Therefore,

$$v_T \cdot i_T = v_o \cdot i_o$$
$$i_T = \left(\frac{v_o}{v_T}\right) \cdot i_o = 2 \cdot n \cdot d \cdot i_o \qquad (7.5.14)$$

From equations 7.5.13 and 7.5.14, the terminal resistance across the PV source on the input side can be deduced as

$$R_T = \frac{v_T}{i_T} = \frac{R_o}{(4 \cdot n \cdot d)^2} \qquad (7.5.15)$$

From equations 7.3.3 and 7.5.15, one can see that the input resistance for this converter is similar to the push pull converter relation. When $d = 0$, the input resistance $R_T = \infty$ and the load line is aligned along the voltage axis of the i-v characteristics of PV source. When $d = 1$, the input resistance is $R_T = \frac{R_o}{4 \cdot n^2}$. By controlling the duty ratio, the loadline can be positioned any where between the voltage axis and $\frac{4 \cdot n^2}{R_o}$ line.

7.6 MPPT through input resistance control

All maximum power point tracker algorithms will essentially try to regulate the input resistance of the converters that are interfaced to the PV source. Typical block

7.6. MPPT THROUGH INPUT RESISTANCE CONTROL

schematic is as shown in figure 7.1.1. The input resistance of the converter is controlled by using the duty ratio as the control input. The terminal voltage and the current of the PV source are measured and used as feedback signals for MPPT control. Most algorithms differ in the manner by which the reference for MPPT is generated. There are several methods that are in use. They are

Using reference cell for obtaining the reference voltage

Using reference cell for obtaining the reference current

Sampling voltage or current

Methods for power slope identification

Hill climbing method

Most MPPT algorithms are based on one of the above methods or variants of them. The following subsections will discuss the concepts related to these methods.

7.6.1 Reference cell method - voltage scaling

The concept of this method is illustrated in figure 7.6.1. This is one of the simplest and reliable methods. The power versus voltage curve shows that the peak power point occurs at some voltage, say V_m. As the insolation varies over the day, the value of the peak power P_m will vary, but the terminal voltage at which it occurs will be in the close neighbourhood of V_m.

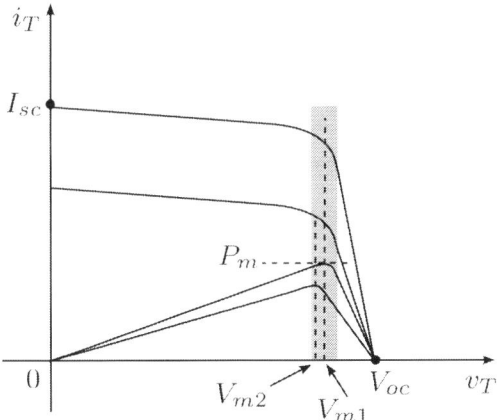

Figure 7.6.1: Voltage scaling method

This is indicated as a tolerance band in the figure 7.6.1. The ratio of the voltage at peak power V_m to the open circuit voltage V_{oc} of the PV source is assumed to be a constant. The value of this ratio is approximately around 0.7 for most types of PV sources. Thus,

$$\frac{V_m}{V_{oc}} = K = 0.7 \tag{7.6.1}$$

The challenge is to estimate the value of V_{oc} for the PV source. In this method, a reference cell of the same make is mounted in the field along with the PV panels. The reference cell is open circuited and its terminal voltage is measured. This is scaled by gain K as per equation 7.6.1 to estimate the reference or command value of voltage at peak power point V_m^*. Figure 7.6.2 illustrates the control block schematic for this MPPT method. The terminal voltage v_T of the PV source is measured and used as the feedback signal. The error from the comparison of V_m^* and v_T operates the proportional-integral (PI) controller. The output of the PI controller modulates a high frequency triangular carrier to produce a PWM signal. The PWM signal regulates the input resistance of the DC-DC converter thereby controlling the terminal resistance seen by the PV source. In the steady state, the error input to the PI controller is zero and therefore, the terminal voltage v_T of the PV source is at the desired operating point that extracts the maximum power from the PV source.

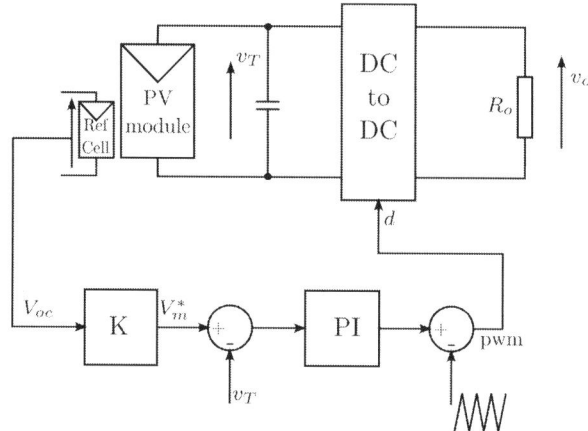

Figure 7.6.2: Control block diagram of voltage scaling method

This method ensures that the terminal voltage is always in the neighbourhood of V_m by controlling the terminal voltage to be within the tolerance band. This way, the power drawn from the PV source will be in the neighbourhood of the peak power capability.

There is some uncertainty in the ratio $\frac{V_m}{V_{oc}}$. The value may not be precisely 0.7 for MPPT. The determination of open circuit voltage V_{oc} is dependent on the reference cell only and not on the actual set of PV panels that act as the source. Determining V_{oc} of the actual PV source is a challenge. There may be partial shading on the PV source panels but not on the reference cell and vice-versa. This also will cause the operating point to be at a point that is away from the maximum power point operation. Notwithstanding these drawbacks, this is a simple and reliable method of MPPT.

7.6.2 Reference cell method - current scaling

A similar concept like that discussed in the reference cell-voltage scaling method is used with current scaling too. Figure 7.6.3 illustrates this concept. The ratio of the current at peak power I_m to the short circuit current value I_{sc} at a given insolation is considered to be a constant. For most PV cells, this value is in the neighbourhood of 0.9. Therefore the constitutive relation in this method is,

$$\frac{I_m}{I_{sc}} = K = 0.9 \qquad (7.6.2)$$

As insolation varies, the short circuit current value will change. The peak power also changes and as a result the current at peak power will change. However, the ratio of the currents will be as per equation 7.6.2.

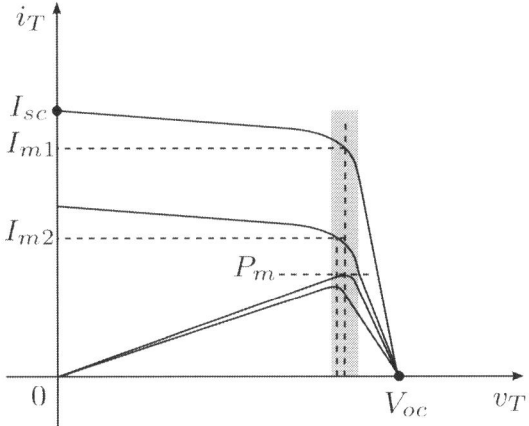

Figure 7.6.3: Current scaling method

Figure 7.6.4 gives the control block diagram for this method. The block schematic is similar to that of the voltage scaling method. Instead of obtaining the V_{oc} information from an open circuited reference cell, the I_{sc} information is extracted from a shorted circuited reference cell. The extracted I_{sc} is scaled as per equation 7.6.2 to obtain the command value or the reference value for the current at peak power I_m^*. The actual terminal current of the PV source is sensed and used as the feedback variable. This is compared with I_m^* and the error is used to drive a PI controller as shown. The output of the PI controller modulates a high frequency triangular carrier to generate the duty ratio control for the converter which in turn regulates its input resistance. The duty ratio and thereby the input resistance of the converter will vary and stabilise at an operating point where the actual terminal current $i_T = I_m^*$. This would imply MPPT operation.

In this method too, the main drawback is the uncertainty in the determination of I_m^*. The I_m^* that is determined is that for a reference cell and not of the active PV source. This may be at variance with respect to the short circuit current of the actual

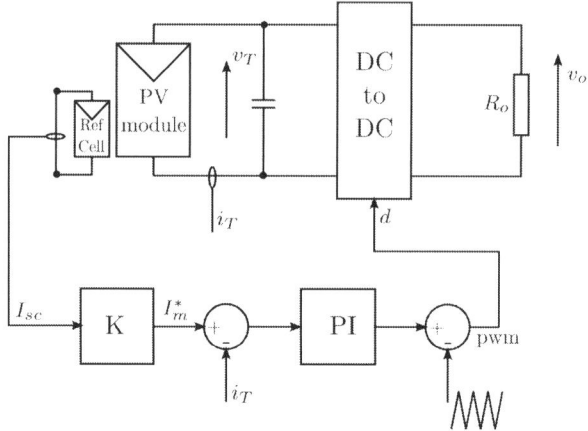

Figure 7.6.4: Control block diagram of current scaling method

PV source. The other aspect of uncertainty that get introduced into this method is the assumed value of K as per equation 7.6.2.

7.6.3 Sampling method

The sampling method addresses one of the main drawbacks of reference cell method. In the reference cell approach, the voltage at peak power point V_m or the current at peak power point I_m are derived from a reference cell and not from the actual PV source that delivers the power to the load. This would mean that the command values of the terminal voltage or current may not be correct and the actual terminal voltage or current of the PV source that are controlled, may not actually be at the peak power operating point.

In this method, reference cells are not used. The open circuit voltage or the short circuit current of the actual PV source that is delivering power to the load, are measured by sampling. These measured values are used to estimate the command values. In figure 7.6.5(a), a single pole double throw switch S is shown that is used to sample the open circuit voltage. The throw 'b' is connected to a capacitor which in turn is connected to a DC-DC converter and load. The throw 'a' is connected to a very high shunt resistance R_{sh}. The switch S is operated with a very small duty ratio like that shown in figure 7.6.5(c). In a period of 1s, the pole of the switch S is connected to throw 'a' for a duration of only $1\mu s$. The sample and hold circuit will activate and hold the value of V_{oc} of the entire PV source. As the duty ratio of this sampling $d_s = 10^{-6}$, the capacitor at the input of the DC-DC converter will buffer the main power circuit during this sampling time. Once the open circuit voltage is obtained, the command value of the voltage at peak power is estimated using equation 7.6.1 and the rest of the control process is same as in the reference cell-voltage scaling method.

In a similar manner, it can be noted from figure 7.6.5(b) that by shorting throw 'a' to the circuit ground and sampling the current in the throw, one can sample and hold

7.6. MPPT THROUGH INPUT RESISTANCE CONTROL

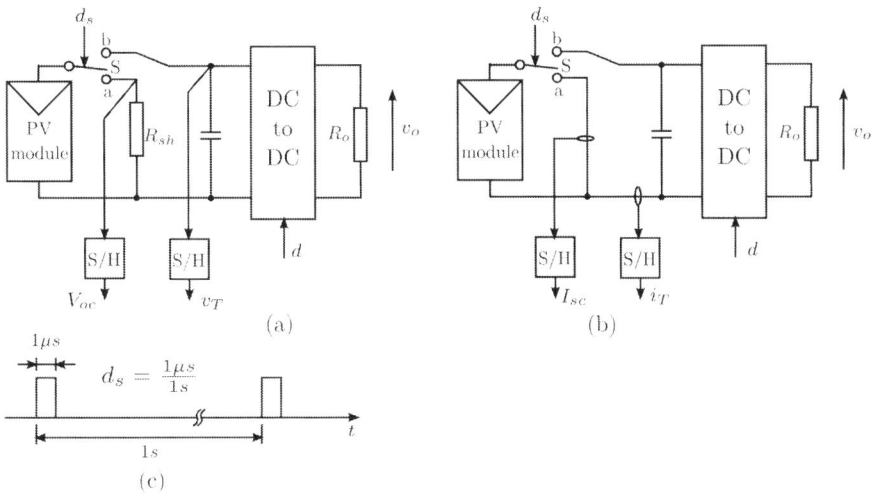

Figure 7.6.5: Sampling method for MPPT (a) Open circuit voltage sampling, (b) Short circuit current sampling, (c) Sampling duty ratio

the short circuit current value of the PV source. Similar control strategy as employed by equation 7.6.2 will follow. In this manner, the sampling method removes one of the main uncertainties introduced by the reference cell approach and will prove to be more accurate in controlling the terminal resistance of the PV source.

7.6.4 Power slope - incremental resistance method

Another method for achieving maximum power point tracking without the drawbacks of reference cell methods is the power slope method. Figure 7.6.6(a) shows the i-v characteristics of the PV source. It also shows the power versus voltage curve. The power slope method works by estimating the sign of slope of the power curve for a given operating point. If the operating point is on the low voltage side of the curve, then the power slope is positive. If the operating point is on the higher voltage part of the curve, then power slope is negative. This slope sign is an indicator that signals the direction of change for the duty ratio of the converter which in turn will change the input resistance or the terminal resistance of the PV source.

With reference to the characteristics shown in figure 7.6.6(a), the instantaneous power is given as

$$p = v_T \cdot i_T$$
$$\frac{dp}{dv_T} = i_T + v_T \cdot \frac{di_T}{dv_T}$$

At maximum power operating point, $\frac{dp}{dv_T} = 0$, and therefore

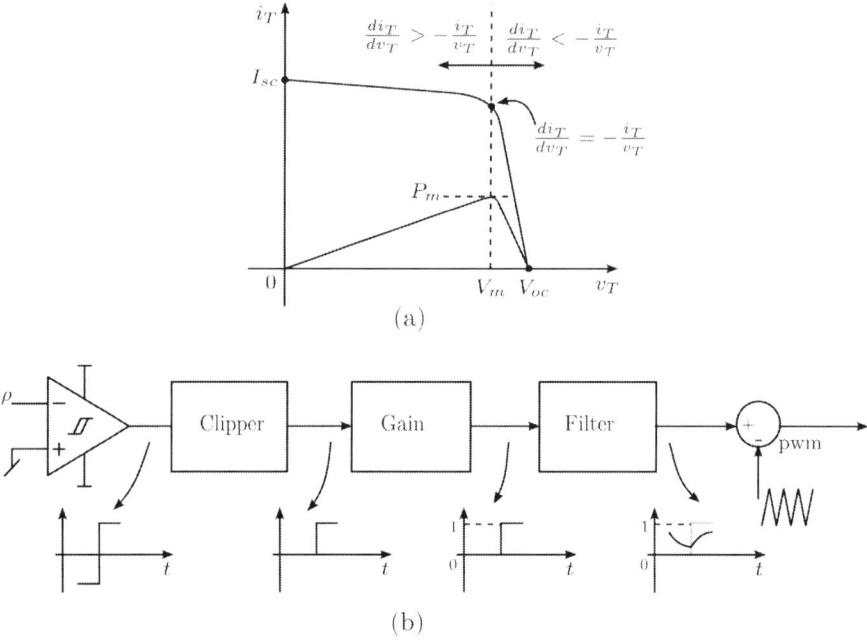

Figure 7.6.6: Using the slope of power curve for MPPT (a) concept through i-v characteristic (b) block diagram

$$\left[\frac{di_T}{dv_T}\right]_{@P_m} = -\frac{i_T}{v_T} \qquad (7.6.3)$$

Equation 7.6.3 provides the maximum power point tracking principle. $\frac{di_T}{dv_T}$ gives the inverse incremental resistance change across the terminals of the PV source. Three cases arise as

$\frac{di_T}{dv_T} + \frac{i_T}{v_T} > 0$: This implies that the slope of the power curve is positive and the operating point is to the left of the operating point corresponding to MPP. The duty ratio of the converter should be changed in such a direction as to increase the terminal resistance.

$\frac{di_T}{dv_T} + \frac{i_T}{v_T} < 0$: This implies that the slope of the power curve is negative and the operating point is to the right of the operating point corresponding to MPP. The duty ratio of the converter should be changed in such a direction as to decrease the terminal resistance.

$\frac{di_T}{dv_T} + \frac{i_T}{v_T} = 0$: This implies that the slope of the power curve is zero and the operating point coincides with that corresponding to MPP.

Define $\frac{di_T}{dv_T} + \frac{i_T}{v_T} \equiv \rho$. Then from figure 7.6.6(b) ρ is a threshold function. The block schematic for generating the modulating signal for changing the duty ratio is shown in

7.6. MPPT THROUGH INPUT RESISTANCE CONTROL

figure 7.6.6(b). ρ is passed through a schmitt comparator to invert the signal. This is clipped by using a clipper circuit. An amplifier can set an appropriate gain as shown. The output of the gain block is filtered and compared with a high frequency triangle to generate the PWM signal for the DC-DC converter. When the function $\rho > 0$, then the power slope is positive. The blocks process the signal leading to decrease in duty ratio. When $\rho < 0$, the power slope is negative and the duty ratio increases to appropriately modulate the terminal resistance of the PV source to achieve maximum power point tracking.

7.6.5 Power slope - pulse phase method

Pulse phase method is another power slope detection algorithm that can be effectively used for MPPT. Figure 7.6.7 gives the principle of operation. A small amplitude pulse is superimposed on the terminal voltage of the PV source. The x-axis or the voltage axis will have pulsation as shown. This pulsation is projected onto the i-v curve and then onto the y-axis or the current axis. It can be seen that there will be pulsations in the terminal current too. The product of the terminal current and terminal voltage will give the instantaneous power that is being extracted from the PV source. The power extracted from the PV source will also have pulsations. If the operating point lies to the left of the peak power operating point, then the power slope is positive and the power pulsations will be in phase with the pulsations of the terminal voltage. If the operating point lies to the right of the peak power operating point, then the power slope is negative and the power pulsations will be out of phase with the pulsation of the terminal voltage. Detecting the phase of the power pulses with respect to the voltage pulses will provide the relative indication on the position of the peak power operating point with respect to the current operating point. This will enable one to change the duty ratio in the proper direction.

Figure 7.6.8 gives the block schematic of the implementation of this algorithm. The power and voltage signals are passed through dc-blocking circuit to remove any dc bias levels. They are passed through zero crossing detectors and added. If the operating point lies to the left of the peak power operating point, then on adding the pulse amplitudes will increase. If the operating point lies to the right of the peak power operating point, the power and voltage pulses will be out of phase and on adding, the result will be zero. The output of the adder is passed through a rectifier or absolute value circuit. This will give a high value if operating point is on the positive part of power slope and zero value if the operating point is on the negative part of the power slope. This signal is passed through a schmitt comparator circuit having a low threshold comparison to obtain an output which gives high or low. This is followed by a filter and the output of the filter is compared with a high frequency triangular carrier to provide the PWM waveform which will drive the DC-DC converter. When filter output is decreasing the duty ratio will decrease and make the operating point move to towards open circuit voltage point and when filter output increases the duty cycle will increase and make the operating point to shift towards the short circuit point. This way the terminal resistance is regulated to achieve an operating point that delivers peak power.

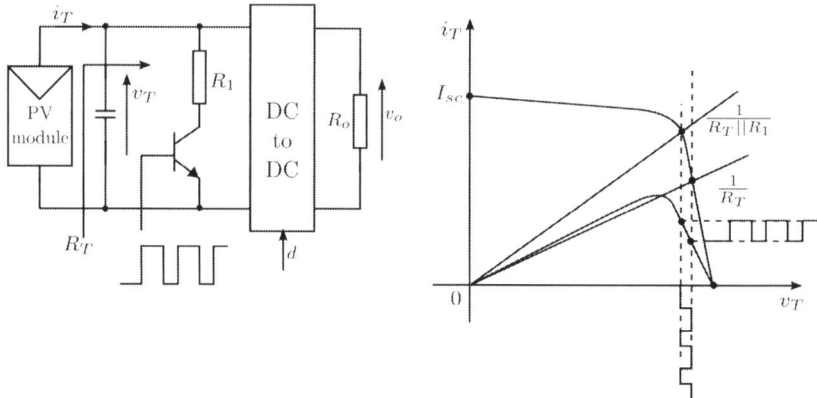

Figure 7.6.7: Alternate method of using the slope of power curve for MPPT

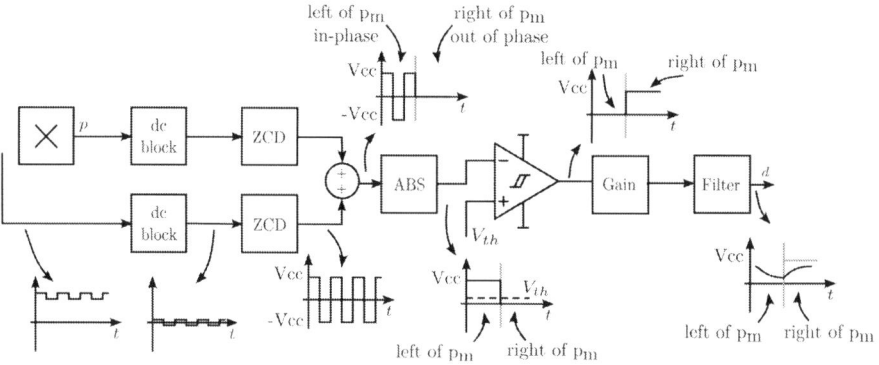

Figure 7.6.8: Implementation block diagram

7.6.6 Hill climbing method

Another very popular and robust tracking method is by the use of the hill climbing or gradient method. It works on the principle that the power versus voltage curve of the PV source is a hill and the goal is for the current operating point to climb up the slope of the hill to achieve peak power operation. Figure 7.6.9 illustrates this concept.

The terminal voltage and terminal current are sensed and their product provides the power signal. The power signal is passed through two filters, one is a fast filter and the other is a slow filter. The output of the slow filter P_{base} is visualised to be a plane which slowly moves up or down the power curve hill. The output of the fast filter $P_{current}$ is visualised to be a ball that represents the current operating point power. The ball continuously moves in one direction say from left to right climbing the slope reaching the peak and then moving down the slope on the other side till it hits the plane P_{base}. On hitting the P_{base} plane, the ball bounces back and starts moving in the reverse direction moving right to left climbing the slope, reaching the peak and then moving down the slope on the other side till it hits the plane once again. On hitting the plane, the ball once again bounces and reverses the direction of motion. This keeps happening continuously.

The base plane P_{base} is slowly rising and restricts the motion of the ball $P_{current}$ more and more till the operating point ball is more or less in the neighbourhood of the peak power point.

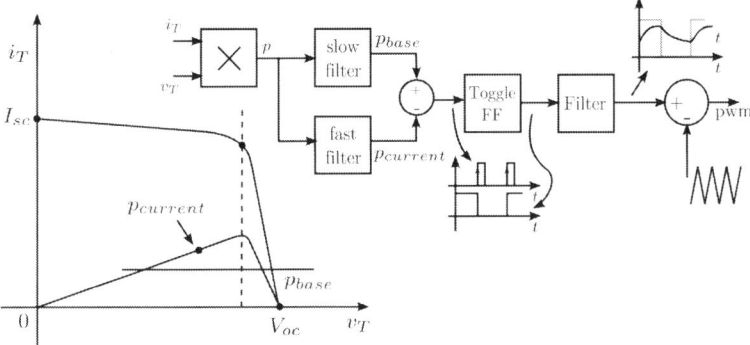

Figure 7.6.9: Hill climbing algorithm for MPPT

The block schematic is also given in the figure 7.6.9. The output of the fast and slow filters are compared using a comparator. P_{base} will act as the rising reference and is given to the positive terminal of the comparator. $P_{current}$ is given to the negative terminal of the comparator. The output of the comparator gives the visualised bounce back signal when the ball bounces on the plane. The direction reversal for the motion of the ball is done by using a toggle flip flop. Every time a bounce signal arrives, the flip flop is toggled indicating direction reversal. The toggle flip flop output is filtered and compared with a high frequency triangular carrier to produce the PWM waveform which controls the DC-DC converter that in turn regulates the terminal resistance across the PV source.

7.6.7 MPPT with shading

In a real installation, the PV source may be placed on roof tops or fields. There can be shadows of the nearby buildings and/or trees that can fall on the PV panels. This will result in partially shaded panels and their i-v characteristics will be like that discussed in chapter 2 for non-identical cells in series and parallel. Consequently, the power versus voltage curve may have more than one power peak. This would mean that the MPPT should track the global power peak for partially shaded panels. This is a challenge.

Insolation changes due to shading is a large time scale event. It occurs in the time scales of minutes. The duty cycle of the converters are at the time scale of microseconds. Once every few minutes, a duty cycle sweep from 0 to 1 is made. One complete duty cycle sweep would take at most a millisecond. During the duty cycle sweep, terminal voltage and current are measured and the power at the corresponding duty cycle operating points are calculated. This would provide the profile of the power curve versus voltage. From this profile, the global maximum power point can be extracted giving the region or zone of global maximum power point. From here on, one of the MPPT algorithms discussed earlier can be employed to track the peak power point in the immediate neighbourhood. In this manner maximum power point tracking can be enabled even with partial shading of the PV panels.

7.7 Questions

1. Which one of the following is the most appropriate DC-DC converter for maximum power point tracking of a PV panel?
   a) Boost converter
   b) Buck converter
   c) Buck-boost converter
   d) Cascaded buck converter

2. Why input capacitor is mandatory at the PV panel output for maximum power point tracking with the buck converter?
   a) The input voltage of the buck converter is discontinuous
   b) The input current of the buck converter is discontinuous
   c) The inductor current of the buck converter is discontinuous
   d) The diode current of the buck converter is discontinuous

3. What is the input impedance seen by an ideal boost converter where R_o is the output load and d is the duty ratio of the converter switch?
   a) $R_o \cdot (1-d)^2$
   b) $\frac{R_o}{d^2}$
   c) $R_o \cdot \frac{(1-d)^2}{d^2}$
   d) $R_o \cdot \frac{d^2}{(1-d)^2}$

4. For a given PV panel, at the peak power operating point,
   a) change in rate of change in current is equal to to the ratio of the operating point current to operating point voltage
   b) change in current with respect to voltage is same as the ratio of the operating point current to operating point voltage
   c) change in power with respect current is zero
   d) change in power with respect to voltage is equal to the ratio of the operating point current to operating point voltage

5. At maximum power point, dI/dV is equal to
   a) $-I/V$
   b) I/V
   c) 1
   d) 0

6. Reference Cell method based on current scaling is more accurate than the reference Cell method based on voltage scaling because

a) Sensed current is directly proportional to insolation levels
b) Sensed current is not affected due to the changes in insolation levels
c) Sensed voltage is directly proportional to insolation levels
d) Sensed voltage is not affected due to the changes in insolation levels

7. Consider a PV system wherein the PV is interfaced to a load resistance Ro by means of a dc-dc converter. If the ratio of the terminal voltage to the terminal current of the PV panel is greater than load resistance, then in order to achieve maximum power point tracking, the interfacing dc-dc converter is a
a) Buck or Boost
b) Cuk converter
c) Buck-Boost or Buck
d) Buck-Boost or Boost

8. From the i-v characteristics of a PV cell, which of the following statements is true for the operating points lying to the right of maximum power point? (P represents power)
a) ripple signals of P and V are in phase ; P is directly proportional to V
b) ripple signals P and V are in phase ; P is inversely proportional to V
c) ripple signals P and V are out of phase ; P is directly proportional to V
d) ripple signals P and V are out of phase ; P is inversely proportional to V

9. Consider a PV system wherein the PV is interfaced to a load resistance by means of a dc-dc converter. If the ratio of the terminal voltage to the terminal current of the PV panel is lesser than load resistance, then in order to achieve maximum power point tracking, the interfacing dc-dc converter is a
a) Buck or Boost
b) Cuk converter
c) Buck-Boost or Buck
d) Buck-Boost or Boost

10. A reference cell voltage scaling method is used to track maximum power point for a solar panel. The open circuit voltage is 43.6 V then the voltage at maximum power point is approximately equal to
a) 43.6 V
b) 21.8 V
c) 30.5
d) 13.6 V

11. Suppose that a boost converter is connected as an interface between PV panels and a DC bus for MPPT and the operating point is at peak power point. If the insolation now reduces, then
a) Duty ratio of converter is to be decreased
b) Duty ratio of converter is to be increased
c) Load on the converter is to be decreased
d) Load on the converter is to be increased

12. A solar photovoltaic panel has an open circuit voltage of 43.6V and short circuit current of 5.45A. The maximum power that can be drawn from the panel is 175W. The maximum power occurs at a voltage of 81% of open circuit voltage. What is the load resistance that needs to be connected to the terminals of the PV panel to extract maximum power?

13. A solar photovoltaic panel has an open circuit voltage of 43.6V and short circuit current of 5.45A. The maximum power that can be drawn from the panel is 175W. The maximum power occurs at a voltage of 81% of open circuit voltage. A boost converter is connected between the PV panel and a 10 ohm load resistance. Calculate the duty ratio at which the boost converter should operate in order to extract maximum power from the PV panel.

14. A buck converter is operated with the following specifications: Input voltage = 36V and output voltage = 9V. If the average value of current through the diode is 0.375A, then the input resistance of the converter is ⋯.

15. For most PV panels, the ratio of the voltage at peak power operating point to the open circuit voltage is around ⋯⋯.

16. For most PV panels, the ratio of the current at peak power operating point to the short circuit current is around ⋯⋯.

17. Consider a buck converter that is interfacing a PV panel to a 48V battery load. A 10A current flows through the inductor to charge the battery. The buck converter switch is switching at 20kHz with a duty ratio of 0.5. The inductor current has very low switching frequency ripple. A capacitor C_T is placed across the terminals of the PV panel to buffer the switched input current of the buck converter. What is the rms current flowing through the capacitor C_T?

18. A PV array is rated as 600 Wp. The short circuit current is 26A and the open circuit voltage is 30V. If Vmp = 24V, find the optimum load resistance that can be matched to the array.

19. A PV array is rated as 600 Wp. The short circuit current is 26A and the open circuit voltage is 30V. The voltage at peak power is 24V. A 2 ohms load needs to be supplied by the PV source. If a boost converter is used to interface the PV array and load, find the duty cycle in order to draw maximum power from the PV array.

20. A PV array is rated as 600 Wp. The short circuit current is 26A and the open circuit voltage is 30V. The voltage at peak power is 24V. A 2 ohms load is connected to the array directly, and if a buck-boost converter is used to interface the PV array and load, find the duty cycle in order to draw maximum power from the PV array.

Chapter 8

Battery Charging

8.1 Introduction

One of the most important and essential applications of PV systems is the PV-battery interface. Batteries are integral part of most photovoltaic driven applications. One of the commonly encountered PV-battery interface is the charge controller, where the charge that is put into the battery or removed from the battery is enabled or disabled by electronic switching mechanisms. Another important PV-battery interface application is the battery charger, where charging the battery is done from PV or PV driven DC-DC converters.

Yet another important challenge that needs addressing is the interconnections of several batteries. Batteries may be connected in series or parallel. When batteries are connected in series, there is a problem with charge sharing. All batteries in series do not share charge equally while being charged and discharged. Therefore charge equalization needs to be consciously built into the charging system for batteries connected in series. Likewise, batteries in parallel have the problem of current sharing and circulating currents. The circulating currents cause unwanted losses and more importantly reduce the capacity of the set of batteries in parallel. Active paralleling of batteries need to be done in order to remedy these problems. The sections to follow will discuss and provide typical solutions with regard to PV and battery interfaces.

8.2 Direct connection

Consider that a PV source is connected with a battery load. The battery is not an ideal source. It has series internal resistance R_B that affects the load line. R_B is a dynamic resistance, however it is considered as constant in the simplified model for battery. Figure 8.2.1(a) shows the loadline when the battery is charging. The x-axis is the terminal voltage across the battery and the y-axis is the terminal current. The terminal current i_T is the same as i_B in this case with the battery acting as a sink where in the battery current is flowing into the positive terminal of the battery.

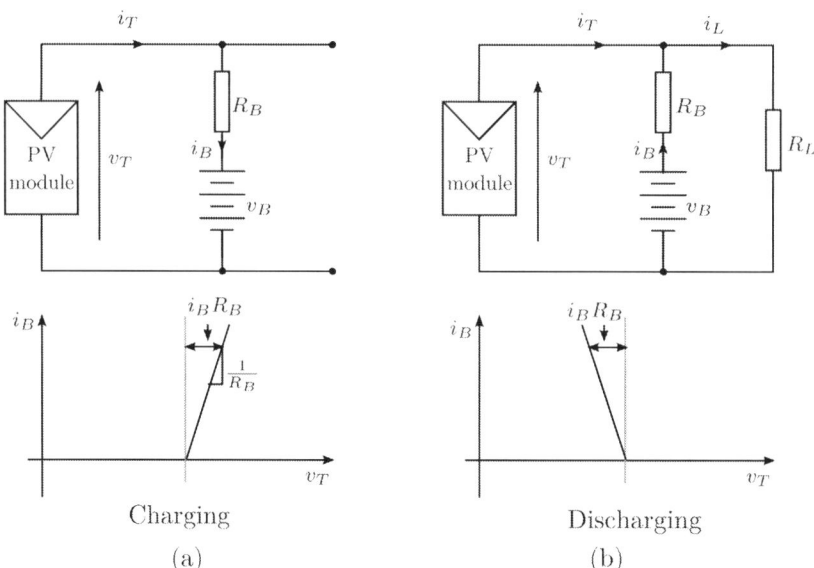

Figure 8.2.1: Battery load (a) during charge (b) during discharge

During charging conditions, the drop across R_B is such that it adds to the battery voltage to give a terminal voltage that is higher than v_B. The figure 8.2.1(a) shows the load line of the battery under charging conditions. The slope of the load line is slightly positive due to the internal series resistance R_B. The terminal voltage is given as

$$v_T = v_B + i_B R_B \qquad (8.2.1)$$

Figure 8.2.1(b) shows the load line when the battery is discharging to an external load determined by resistor R_L. The load resistor is supplied by both the battery current i_B and the PV panel current i_T. The battery current i_B flows out of the positive terminal of the battery and supplies the external load. The drop across R_B is in a direction wherein the terminal voltage is less than the battery voltage by the voltage drop across R_B. The load line is plotted on the i-v coordinate system. Here the y-axis is the terminal current which supplies to the combination of the battery and the external load. In this case, the load line has a negative slope as can be seen in figure 8.2.1(b). The terminal voltage is given as

$$v_T = v_B - i_B R_B \qquad (8.2.2)$$

The i-v characteristics of the PV panel and that of the battery load line can be superimposed as shown in figure 8.2.2. Consider the battery load line wherein the battery is in discharged condition with a voltage of v_{B1}. This load line intersects the PV source i-v characteristics as shown wherein a charging current of I_{chg} can flow into

8.2. DIRECT CONNECTION

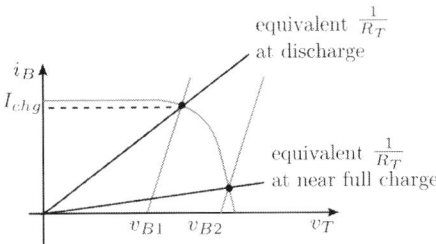

Figure 8.2.2: i-v characteristic and loadline

the battery to charge it. The load is decoupled from the battery which allows the battery to charge up. As the battery gets charged, the battery load line starts shifting towards the open circuit voltage. The battery voltage increases and the charging current will decrease as dictated by the PV i-v characteristics. At the operating point shown where in the battery voltage is v_{B2}, the charging current is at trickle charging condition. When the battery voltage reaches open circuit voltage value of PV source, then no charging current flows and battery is in float charging condition.

It can be seen from the figure 8.2.2 that the points of intersection of the battery load line with the PV i-v characteristics are operating points of the PV system. These operating points have equivalent load lines having slopes of $\frac{1}{R_T}$, where R_T is the terminal resistance at the PV source. If a converter which can handle the indicated range of resistance change is interposed between the PV source and the battery-load combination, then the power converter can be operated with MPPT operation for battery charging.

It should be noted that when the battery needs to be charged, the load should be decoupled. In this condition, the battery will not drain out through the external load while it is being charged. When the insolation is low or absent, the load is coupled to the battery and power will be supplied from the battery. This is a simple direct connection approach with minimum number of components. Applications like serial lights powered by solar cells use the direct connection method for interfacing the PV, battery and the LED serial lights. The coupling and decoupling of the load is done through a manual switch. Further this direct connection method does not permit MPPT operation.

8.2.1 Example

Consider an example, where a battery is directly connected to a PV panel. The battery is a lead acid battery type of 12 volt nominal voltage, 20 Ah capacity and with a C-rate of C10. Figure 8.2.3 shows the i-v characteristics of the PV panel superimposed with the battery load line.

Let the battery be in a discharged condition. For a lead acid battery, the voltage at discharged condition will be around 10.8 volts. The load line at battery voltage of 10.8V is shown in figure 8.2.3. When the battery is fully charged, the voltage across the battery terminals will be 14.2V. The open circuit voltage of the PV source should

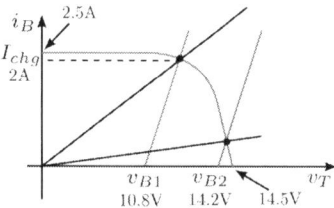

Figure 8.2.3: Direct connection example

be matched to 14.2V. It can be chosen slightly higher than 14.2V. The operating point will swing between these two extreme points from discharged to charged condition. For a C10 20Ah battery, a charging current of 2A is needed. The short circuit current of the PV source should be set at slightly above 2A as shown. As the battery charges, the voltage increases and the charging current automatically decreases as per the i-v characteristics. When the battery is fully charged, it will be in either trickle or float charge operation as the charging current will be very low or even zero.

8.3 Charge controller

Charge controller is a circuit that provides an automated PV source, battery and load interface. Consider the i-v characteristic as shown in figure 8.3.1(a). The battery voltage will be at V_{min} under discharged condition and at V_{max} under full charged condition. The battery load line will swing in between these two extreme voltage limits. In order to maintain the operating point such that the voltage is within the V_{min} and V_{max} window, two switches S_1 and S_2 are used to interface the PV source, battery and load as shown in figure 8.3.1(b). Switch S_1 is a unidirectional current flow switch which will allow current to flow only in the direction of PV source to battery-load combination. On the other hand, switch S_2 is a bidirectional switch which will allow current to flow in both directions. The battery may discharge and can also charge-up due to regenerative effects from the load.

When the battery voltage $v_B > V_{max}$, then switch S_1 is made open to prevent further charging of the battery and S_2 is closed to allow discharging of the battery to the load. This is shown in figure 8.3.1(c). When $v_B < V_{min}$, then switch S_2 is opened to disallow any further discharging of the battery through load and S_1 is closed in order to focus only on charging the battery. This mode is shown in figure 8.3.1(d). There are applications that can be driven directly by a PV source if there is sufficient insolation. Therefore, during the time when the PV source is charging up the battery, it can also deliver power to the load if the PV source is appropriately sized. A unidirectional switch S_3 may be placed as shown in figure 8.3.1(d) which will also allow a direct path for the power to flow from the PV source to the load while the battery is getting charged.

There are times when the battery voltage is in between the two extreme limits i.e. $V_{min} \leq v_B \leq V_{max}$. During this condition, both the switches S_1 and S_2 are closed. This allows a charging path from the PV source to battery and also a discharging path from

8.4. CHARGE CONTROLLER CIRCUIT

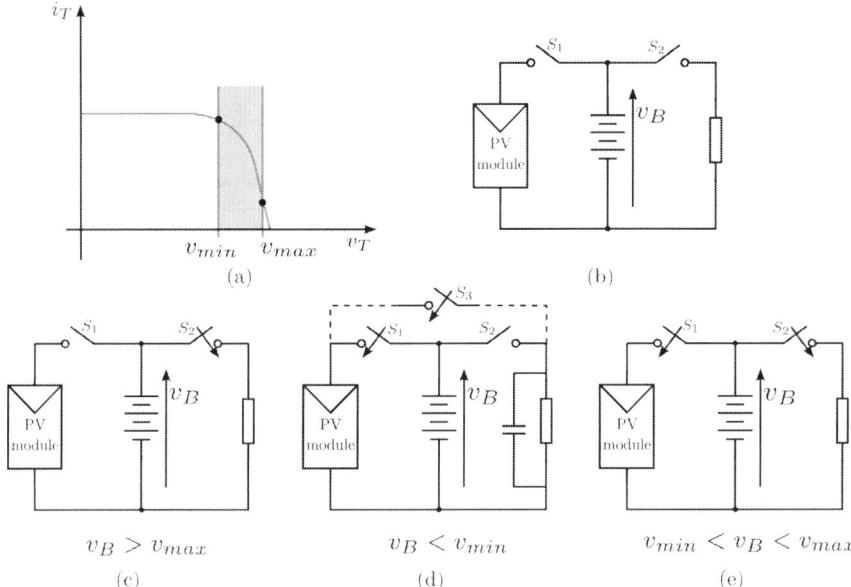

Figure 8.3.1: Charge controller modes (a) battery voltage limits, (b) 2-switch charge controller, (c) operational mode1 (d) operational mode 2 and 3-switch controller, (e) operational mode 3

battery to load as shown in figure 8.3.1(e). Most charge controllers are designed to perform the functional logic as described.

8.4 Charge controller circuit

The functional description of the charge controller discussed in the previous section is incorporated as a circuit that is shown in figure 8.4.1. The function of the switches S_1 and S_2 are being performed by two relays. The relay contacts are driven by the current flow in the relay coils. Each relay coil is switched ON or OFF with the help of power semiconductor switches Q_1 and Q_2 which can be BJT, MoSFET or IGBT semiconductor switches. A diode is placed across the relay coil to provide a freewheeling path for the coil current when the switch is turned OFF.

The battery voltage is sensed by using a simple resistive divider attenuator. A zener regulator is used as a reference to compare with the battery voltage. A hysteresis comparator compares these battery voltage with the reference to determine the output state. When $v_B < V_{max}$, then the comparator output goes high and turns on the switch Q_1. This will in-turn cause the relay coil current to flow and turn ON the contacts of switch S_1. A similar circuit is also used for driving the relay S_2. Here S_2 is switched ON when $v_B > V_{min}$. Both V_{min} and V_{max} are derived from zener regulators.

The function of the relay switches S_1 and S_2 can also be performed by semiconductor devices. In figure 8.4.2, the switch S_1 is replaced by a PNP BJT switch. The

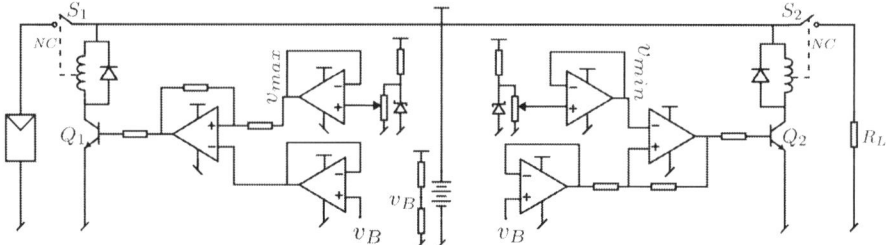

Figure 8.4.1: Charge controller circuit using relays

BJT S_1 switch is turned ON and OFF by the driver switch Q_1.

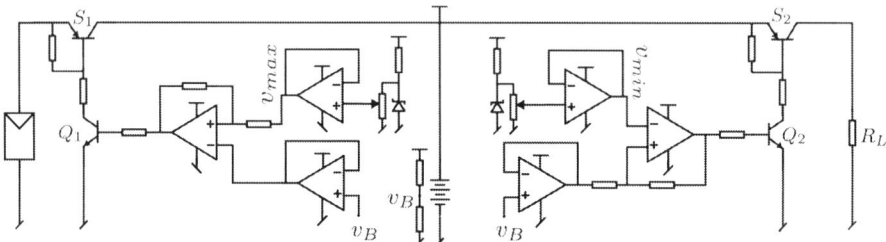

Figure 8.4.2: Charge controller circuit using semiconductor switches

8.5 Switched mode battery charger with current control

Charge controllers route the flow of power from the PV source to battery and load. However, it does not control the charging profiles. The batteries in most cases will not be directly charged from the PV source. They will be interfaced through a DC-DC converter in order to maintain the operating point in the neighbourhood of the maximum power point. Figure 8.5.1 shows the block schematic of a typical charger wherein a DC-DC converter is used to interface the PV source with the battery-load combination. The DC-DC converter is controlled in a manner to provide both MPPT and charge profiling.

Any DC-DC converter can in principle be used for charging the battery. The choice of the converter is made based on the impedance matching for MPPT and requirements like galvanic isolation. Current control of the charger is desirable for any type of DC-DC converter. The schematic of the buck DC-DC converter with current control is shown in figure 8.5.2(a). The battery is charged through a buck converter. The power semiconductor switch Q is switched ON and OFF based on the value of the current in the inductor L.

The control circuit consists of a clock generator which sets the output of a S-R latch. The clock determines the switching period of the converter. The clock generator

8.5. SWITCHED MODE BATTERY CHARGER WITH CURRENT CONTROL

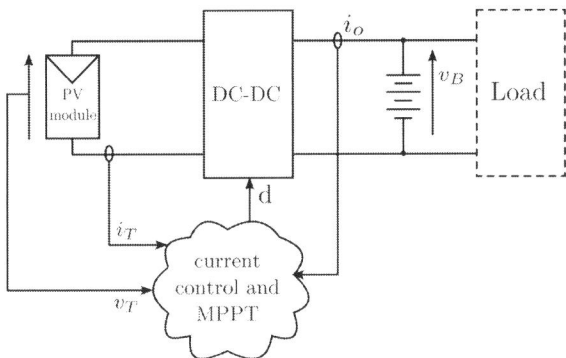

Figure 8.5.1: Block schematic of the battery charger

provides small width pulses every clock period as shown in the figure 8.5.2(b). At the beginning of every clock period, the clock pulse will set the S-R latch and turns ON the switch of the converter. The PV source current magnetically charges up the inductor energy. The voltage across the inductor is $(v_T - v_B)$. The voltage across the inductor when switch is ON is constant and the inductor current rises linearly at the rate $\left(\frac{v_T - v_B}{L}\right)$ as shown.

The inductor current i_L is compared with the desired or reference value of the inductor current i_{Lref} and the output of it is passed through a zero crossing comparator. When the error $(i_{Lref} - i_L)$ goes negative, then the zero crossing comparator output also goes high. The comparator output will reset the S-R latch and the S-R latch output will go low thereby turning the switch OFF. The inductor current will freewheel through the battery and the diode of the buck converter and start to decrease linearly at the rate $\left(-\frac{v_B}{L}\right)$. The switch will remain in OFF condition till the end of clock period. At the start of the next clock period, the clock pulse will again set the S-R latch and turn ON the switch. The cycle then repeats.

8.5.1 Slope Compensation

A practical current controlled converter has a stability issue that needs to be addressed for it to be an useful charger. Consider the current through the inductor i_L. During the ON period (dT_s) of the switch, the inductor current rises with a slope of $m_1 = \left(\frac{v_T - v_B}{L}\right)$. And during the OFF period of the switch, the inductor current decreases linearly with a slope of $m_2 = \left(-\frac{v_B}{L}\right)$. If Δi_L is the change in the inductor current during either ON time dT_s or OFF time $(1-d)T_s$, then

$$m_1 \cdot dT_s = m_2 \cdot (1-d)T_s$$

If δ_1 is a perturbation at the start of the clock period, the perturbed inductor current will increase linearly in a path parallel to the non-perturbed current having a slope m_1 until it reaches the reference current limit level as decided by i_{Lref} as shown in figure

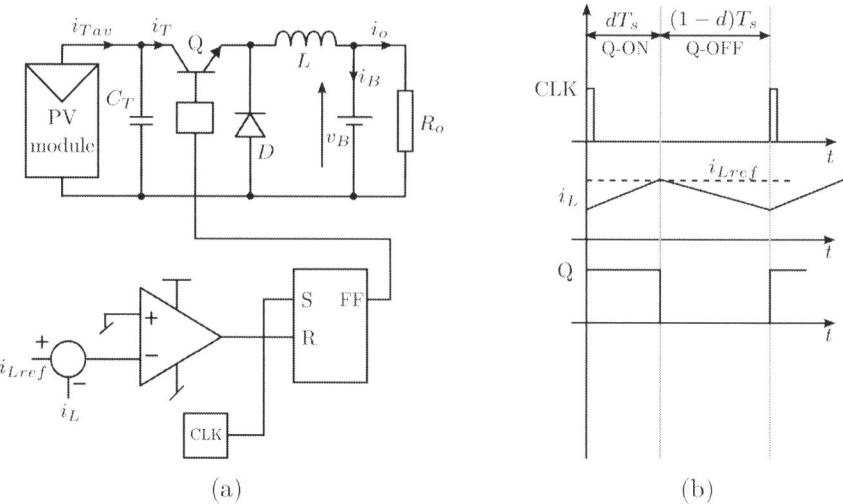

Figure 8.5.2: Current controlled charger (a) Buck type charger circuit (b) operational waveforms

8.5.3(a). During the OFF time of the switch, the perturbed current path will decrease linearly and take a path parallel to the non-perturbed current having slope m_2 until the end of clock period. Consider δ_2 to be the resulting perturbation at the end of the clock period. Then by simple trigonometry, it can be seen that

$$\frac{\delta_1}{m_1} = \frac{\delta_2}{m_2}$$

and

$$\delta_2 = \left(\frac{m_2}{m_1}\right) \cdot \delta_1 \qquad (8.5.1)$$

It can be seen from equation 8.5.1 that the resulting perturbation δ_2 at the end of each clock cycle would converge only if $m_2 < m_1$. For this buck converter, this implies that any perturbation would converge towards zero if the duty ratio is less than 0.5. The system is considered to be stable under such conditions. However, if the duty ratio is more than 0.5, then $m_2 > m_1$ and perturbations will diverge at the end of the clock cycle leading to unstable operation like pulse skipping as shown in the waveforms of figure 8.5.3(b).

In figure 8.5.3(c), the reference current i_{Lref} is not a constant but has a sloped sawtooth type of waveshape. If the slope of i_{Lref} is made same as m_2, then it can be seen that whatever be the perturbation in the inductor current and whatever be the duty ratio, the inductor current perturbation will converge to zero within one clock cycle. Therefore, i_{Lref} is always given a compensating slope to handle perturbations in current for any operative duty ratio. This is called slope compensation.

8.5. SWITCHED MODE BATTERY CHARGER WITH CURRENT CONTROL

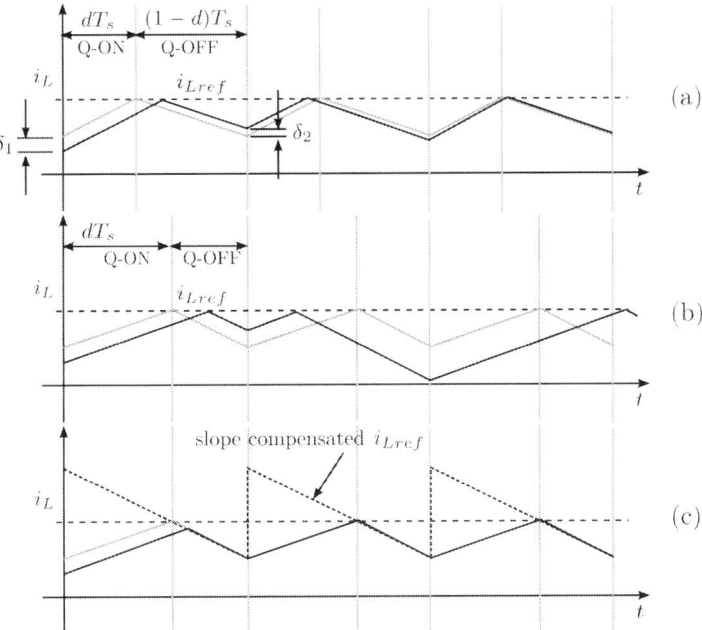

Figure 8.5.3: Waveforms of current control (a) error propagation for perturbations with d¡0.5 (b) error propagation for perturbations with d¿0.5 (c) error propagation with slope compensation

Figure 8.5.4(a) shows the waveforms related to generation of a slope compensated reference. The uncompensated reference i_{Lref} is a constant value as shown. A current controlled signal i_{slope} is generated as a sawtooth waveform as indicated in figure 8.5.4(a). The slope of the signal i_{slope} is set as a positive slope with the magnitude of the slope equal to the magnitude of m_2. For the case of the buck converter example, $m_2 = \left(-\frac{v_B}{L}\right)$ and therefore, the slope of the sawtooth signal is $\left(\frac{v_B}{L}\right)$. The signal i_{slope} is generated by the structure as shown in figure 8.5.4(b). This can be implemented either in analog domain using op-amps and switches or in the digital domain as a program.

The voltage across the inductor is measured during the period when it is discharging. In the buck converter examples, the voltage across the inductor during the period $(1-d)T_s$ is the battery voltage v_B itself. This value is scaled using the value of the inductor and C_s value. An equivalent value of current $\left(\frac{v_B C_s}{L}\right)$ is pumped into the capacitor C_s. The voltage across the capacitor C_s will be a ramp with a slope of $|m_2|$ which is given by

$$v_{Cs} = \frac{1}{C_s}\int I \cdot dt = \frac{v_B}{L}\int dt$$

The capacitor is reset using a semiconductor switch on every rising edge of the clock signal. This will provide the required i_{slope} signal. The i_{slope} signal is subtracted

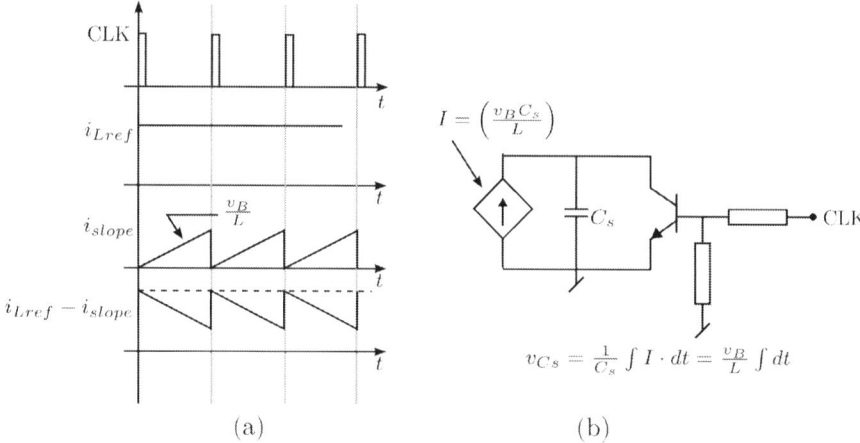

Figure 8.5.4: Slope compensation (a) slope compensated reference generation waveforms, (b) implementation schematic

from the uncompensated reference i_{Lref} to obtain the slope compensated reference signal that is given to the current controller.

8.6 Charging with MPPT

The control scheme shown in figure 8.5.2 does not include MPPT. However it is easy to include MPPT also into the current control scheme. Any one of the MPPT algorithms can be incorporated. The block schematic shown in figure 8.6.1 includes a reference current MPPT algorithm. The MPPT controller output is connected to the uncompensated inductor reference point i_{Lref} in the current controller. The output voltage of the charger is the battery voltage v_B. The voltage is wholly determined by the battery. Therefore, the power that is being pumped into the battery is decided by the amount of current that is fed into it. The MPPT algorithm tracks the operating point to the maximum power point of the PV source. The output of the MPPT algorithm is a signal that correlates directly with the maximum power point of the PV source. As the output voltage is decided by the battery, the MPPT output can decide the current reference that needs to be injected into the battery to achieve maximum power point operation. The MPPT output sets the current reference i_{Lref} and the current controller does the job of injecting this value of current into the battery. Thus the battery is charged with PV source operating at maximum power point.

8.7 Batteries in Series

A single cell has a voltage ranging from 2V to 4V and the current capability of few amperes. However, most applications for renewable energy systems and electric drives need much larger battery voltages and much larger current capabilities. Therefore, it

8.7. BATTERIES IN SERIES

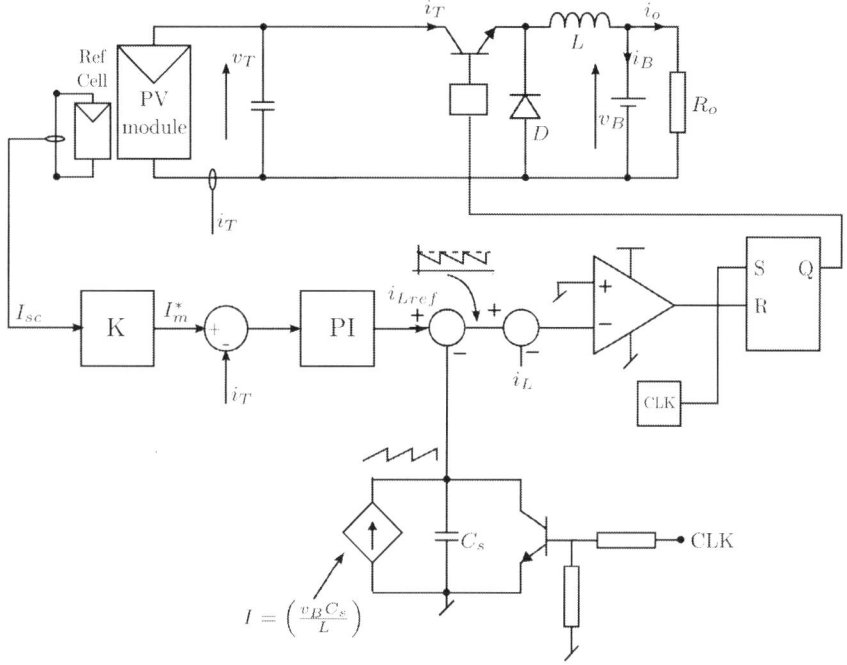

Figure 8.6.1: Current control along with MPPT

is necessary to have battery cells in series and parallel combinations to achieve the desired voltage and current capabilities.

The voltage capability can be increased by placing battery cells in series. For best utilization of the battery, all the cells in series should have equal charge sharing during discharging to load and while charging up. However, the situation is not ideal in practice and the cells connected in series can have differing levels of charges. In order to obtain the best performance from the battery pack wherein the cells are connected in series, charge equalisation must be performed continuously and dynamically. Therefore, cells in series must be accompanied by charge equalisation circuits as well.

8.7.1 Resistive equaliser

Figure 8.7.1 shows the circuit where a resistor is connected across each cell. The resistors act like attenuator circuit with resistors in series. If all the resistors are equal, then the voltage across the resistors are equal. The charges in the cell will eventually settle in such a manner that the voltages across the cells are equal. The equalisation currents will flow through the resistors and therefore this method is lossy. Empirically, around 2% of the load current is allowed to flow through the resistors for charge equalisation. If there are n cells in series which is supposed to cater to a load having maximum discharge current as $I_{discharge}$, then value of R is given as

$$R = \frac{V_{dclink}}{n \cdot \left(I_{discharge} \cdot \frac{2}{100}\right)}$$

where V_{dclink} is the nominal voltage of the series connected battery pack. The maximum power dissipation P_R in the equalisation resistors is given as

$$P_R = \frac{V_{dclink}^2}{n^2 \cdot R}$$

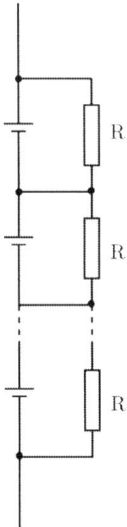

Figure 8.7.1: Charge equalisation with resistive equalisers

The resistors that are connected across the cells can be active devices that are operated in the linear region. BJTs and MoSFETs can be used for the purpose with appropriate biasing.

8.7.2 Charge pump based active equaliser

The resistive equaliser is very simple and effective method for achieving charge equalisation of the battery cells. However, this method is lossy. In order to address this, switched mode converters are used to perform charge equalisation across the cells.

Figure 8.7.2 show n cells connected in series. The cell voltages are v_{B1}, v_{B2} and v_{Bn} as shown. Across each cell, a buck-boost converter is connected. It is desired that the series connected cells be balanced and have equal charge distribution. The voltages across each cell are supposed to be equal and $\frac{1}{n}$ the terminal voltage v_T under charge equalised condition.

8.7. BATTERIES IN SERIES

Let say $n = 2$, then the voltage across each cell is half the terminal voltage. If $v_{B1} > \frac{v_T}{2}$, then as a consequence $v_{B2} < \frac{v_T}{2}$. This will make the comparator to enable the clock pulses to the gate drive of the power switch present in the buck boost converter across battery B1. During this gated period, high frequency fixed duty cycle pulses are applied to the gate of the power switch. Energy is drawn from battery B1 and stored in the inductor. When the power switch is off, the energy in the inductor freewheels through the other cells downstream and charges up the reservoir capacitor C_R. In this manner, the buck boost converter across every cell will operate and place the extra charge removed into the reservoir capacitor C_R. A boost converter is used to pump the energy in the reservoir capacitor to the top of the battery string whenever the reservoir reaches a threshold energy level. In the steady state, all the charges in the cells will be equalised and $v_{B1} = v_{B2} = \frac{v_T}{2}$. The same concept is extended to n cells in series where in at steady state $v_{B1} = v_{B2} = \cdots v_{Bn} = \frac{v_T}{n}$.

The voltage that Q1 should withstand is $(v_T + v_{CR} - v_{B1})$. The voltage rating of Q2 should be at least $(v_T + v_{CR})$. For n cells in series, it can be seen that the voltage rating of the power switches progressively increases from the ground-side end to the rail-side end. However, the current ratings of all the devices will be same and it is dependent on the value of the inductors.

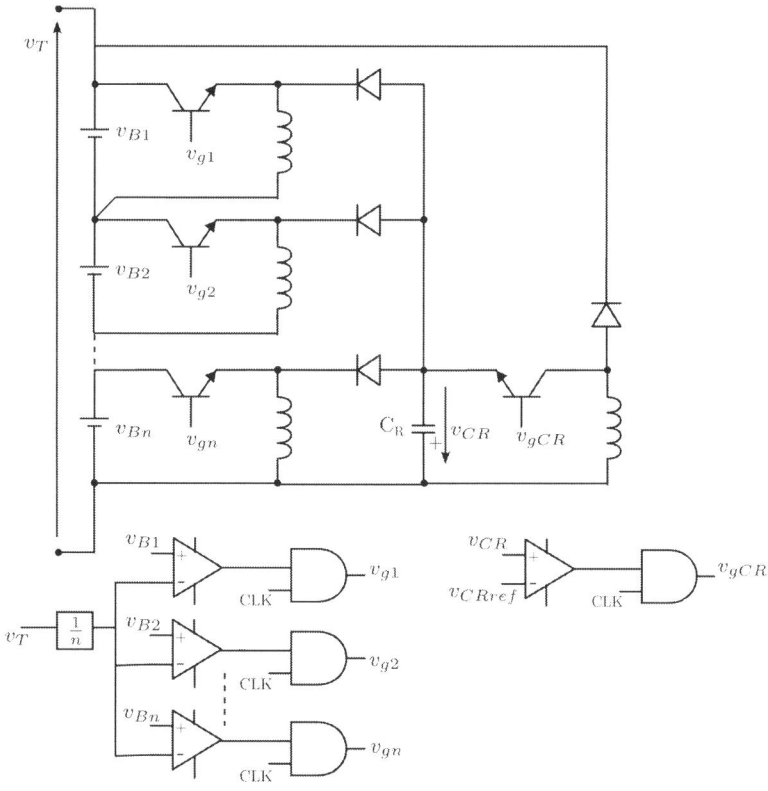

Figure 8.7.2: Charge equalisation with buck-boost converter

The active charge equalisation system basically consists of a switched mode converter that draws charge from the battery cell when the cell has excess charge and dumps this excess charge into a reservoir or buffer like the capacitor C_R that has been used in figure 8.7.2. The switched mode converter can also use topologies that includes high frequency galvanic isolation to achieve a measure of modularity in design and component selection. Figure 8.7.3 uses a flyback converter to achieve galvanic isolation. In the flyback configuration, the voltage rating of the power switches are all the same. Therefore, one can use the same inventory power switches for all the cells. The turns ratio also affects the off-state voltage of the switches. This provides another degree of freedom in design and selection of the power switches. The power switch voltage withstanding capability is given as

$$V_{CEO-Q_i} = v_{Bi} + \frac{v_{CR}}{n} \tag{8.7.1}$$

where

V_{CEO-Q_i} is the voltage rating of the ith power switch

v_{Bi} is the cell voltage of the ith cell of the series string

n is the turns ratio of the high frequency transformer which is to ratio of secondary turns to primary turns

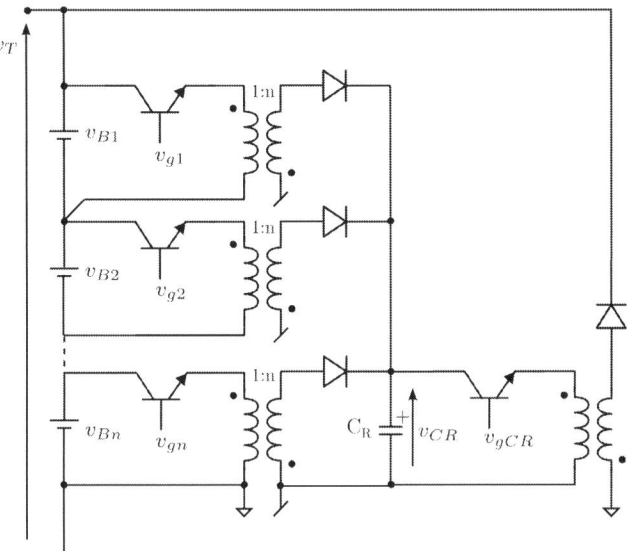

Figure 8.7.3: Charge equalisation with isolation

With use of high frequency transformers, one can further simplify the circuit by removing the reservoir portion of the equalisation system. Figure 8.7.4 gives such a circuit topology. The device ratings here too are same for all devices and is given as

8.8. BATTERIES IN PARALLEL

$$V_{CEO-Q_i} = v_{Bi} + \frac{v_T}{n}$$

Each charge pump module is rated for $\left(\frac{v_T}{n} \cdot I_Q = \frac{P_{eq}}{n}\right)$, where P_{eq} is the power for charge equalisation and I_Q is the average current of the flyback switch. It is also the current drawn for equalisation of each battery cell. The charge equalisation power P_{eq} is normally taken as less than 10% of the maximum load power delivered from the series string.

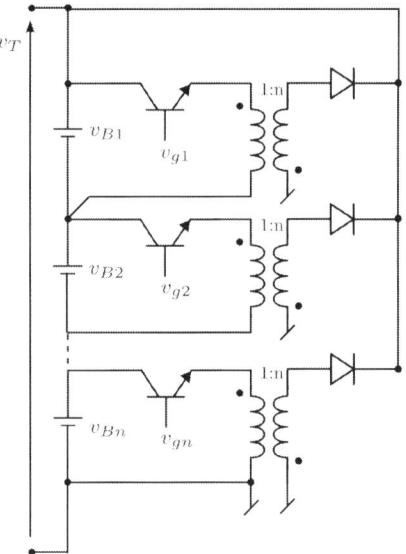

Figure 8.7.4: Charge equalisation with isolation and no reservoir

8.8 Batteries in Parallel

Cells and batteries are required to be connected in parallel for enhancing the current capabilities. However, connecting batteries in parallel is more challenging than connecting them in series. Batteries and cells are potential based sources and therefore their series impedance is negligibly small. If two batteries are connected in parallel, it is like connecting two voltage sources in parallel. As their series impedances are negligible, even small voltage differences can cause very large circulating current between them resulting in catastrophic failures. This can lead to thermal runaway eventually resulting in battery fires. Therefore, batteries should be connected in parallel with the aid of active circuits that will ensure that circulating currents do not flow through them. Active paralleling circuits for interconnecting batteries and battery strings in parallel to eliminate circulating currents are essential to any battery package system.

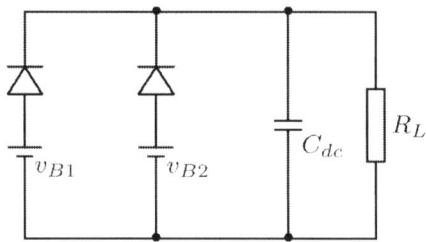

Figure 8.8.1: Paralleling through diode ORing

When batteries are connected in parallel, circulating current results if even one of the batteries in the pack has deteriorated. This will make the deteriorated battery a sink and the other batteries need to supply this sink current apart from the regular load current. As a result the batteries will be supplying current that is more than the load requirement especially as the deteriorated battery is also acting as an extra load. One of the main objectives of active circuits when paralleling is to isolate the deteriorated battery from the circuit. Figure 8.8.1 has two batteries v_{B1} and v_{B2} that are supplying a dc-link supported by the capacitor C_{dc}. A load R_L is connected across the capacitor which loads the two batteries. The batteries are connected in parallel through diode ORing as shown in figure 8.8.1. The diodes ensure that there can be no circulating current between the two batteries. If one of the batteries has deteriorated, then its voltage will be lower than the good battery and will automatically get isolated by the diode. However, one drawback of this method of paralleling is that the batteries cannot sink current. This can be a serious disadvantage if the load is of regenerative type which transfers energy back to the batteries.

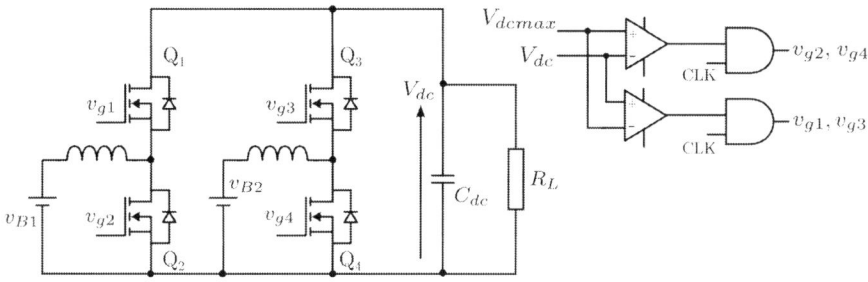

Figure 8.8.2: Bidirectional boost active parallel circuit

Another drawback of using only diode ORing is that the dc-link voltage and the battery voltage are strongly coupled. The dc-link voltage will be a diode drop less than the highest battery voltage of the pack at a given point in time. This means that the battery voltages should be matched to the dc-link voltage for a given application. A measure of decoupling of voltages between batteries and dc-link can be achieved by the topology shown in figure 8.8.2. Each battery is interfaced to the dc-link V_{dc} through a bidirectional boost converter. The inductor along with a half bridge arm

8.8. BATTERIES IN PARALLEL

forms a bidirectional boost converter. In the figure 8.8.2, Q_1 and Q_2 are switches of a half-bridge arm which combined with the inductor forms the boost converter for battery v_{B1}. Likewise, Q_3 and Q_4 along with the inductor forms the boost converter for battery v_{B2}.

In the forward power flow operation, power flows from the batteries to the dc-link. As long as the dc-link voltage is below a maximum threshold voltage V_{dcmax}, the bottom switches of the half bridge arms Q_2 and Q_4 are enabled for pulse width modulated switching. This means that the gates of Q_2 and Q_4 will receive pulse width modulated gate pulses that will drive each half bridge boost converter section to draw power from the batteries at lower voltage and to deliver power to the dc-link capacitor C_{dc} at higher voltage. When Q_2 and Q_4 are ON, the inductors charge up magnetically and when Q_2 and Q_4 turn OFF, the inductor discharges through the body diodes of Q_1 and Q_3 into the dc-link capacitor C_{dc}.

When V_{dc} is greater than the maximum threshold voltage V_{dcmax}, the dc-link capacitor should be discharged so that the surplus energy can charge up the batteries. During this time, the power flows from the dc-link to the batteries. The batteries now act as sinks. During this mode, the switches Q_1 and Q_3 are enabled for pulse width modulation and the switches Q_2 and Q_4 are disabled. The body diodes of the bottom switches are used for inductor current free-wheeling action. The topology in the reverse power flow mode operates as a buck converter.

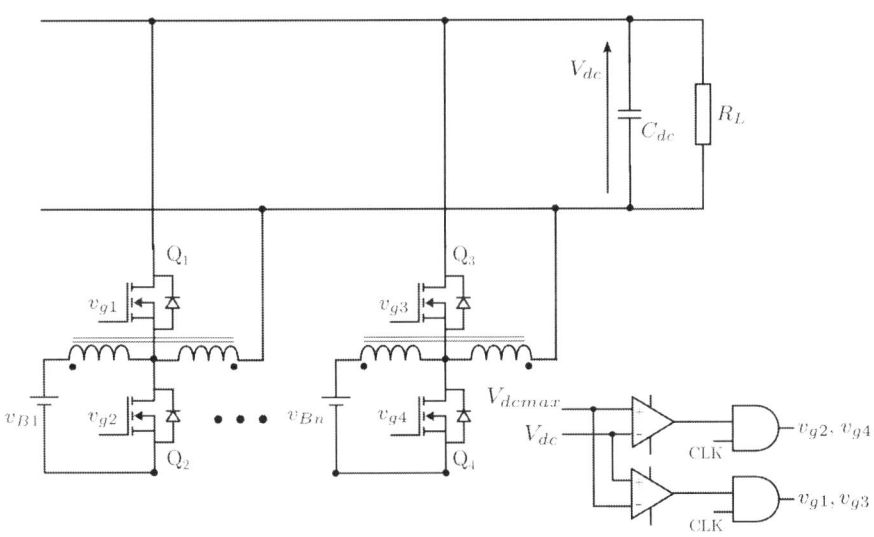

Figure 8.8.3: Bidirectional flyback active parallel circuit

The battery voltages v_{B1} and v_{B2} can be quite different. The duty cycle of the boost converter is controlled such that each battery voltage is made compatible with the dc-link. This implies that one may even use an eutectic combination of batteries that may even have different chemistries. However, the boost converter has a practical maximum boost ratio of about 3 to 4 and the range of voltage differences among the batteries should be selected such that they are addressable within the practical boost ratio range. If the battery voltages are widely differing in value, then the lowest battery voltage may not be boosted up to the dc-link voltage value by the boost converter. This will make the low voltage batteries redundant.

The circuit in figure 8.8.3 uses a flyback topology to address the limited boost ratio of the boost topology. Each battery is interfaced to the dc-link through a bidirectional flyback converter. The converter is innovatively conceptualised such that a regular half bridge topology can be used. This is a significant advantage as the half bridge modules are commercially available more easily. In this circuit too, during forward power flow operation, power flows from the batteries to the dc-link. As long as the dc-link voltage is below a maximum threshold voltage V_{dcmax}, the bottom switches of the half bridge arms are enabled. With reference to figure 8.8.3, Q_2 and Q_4 are enabled. The gates of Q_2 and Q_4 will receive pulse width modulated gate pulses. When Q_2 and Q_4 are ON, the flyback primary inductors charge up magnetically and when Q_2 and Q_4 turn OFF, the secondary coupled inductor discharges through the body diodes of Q_1 and Q_3 into the dc-link capacitor C_{dc}.

When V_{dc} is greater than the maximum threshold voltage V_{dcmax}, the dc-link capacitor should be discharged to allow the surplus energy to charge up the batteries. During this time, the power flows from the dc-link to the batteries. During this mode, the switches Q_1 and Q_3 are enabled for pulse width modulation and the switches Q_2 and Q_4 are disabled. The power flow in reverse direction also happens through flyback action.

The circuit of figure 8.8.3 can handle any degree of voltage differences between batteries and dc-link voltage. This is due to another degree of design freedom given by the turns ratio of the flyback coupled inductors. This implies that one can use vastly different batteries of differing voltages and chemistries to interface to the dc-link. The maximum expected load current are equally divided among the battery-flyback interfaces. If there are n batteries in parallel, then each flyback converter will handle only $\frac{1}{n}$ of the load power and current.

8.9 Questions

1. A photovoltaic module has a open circuit voltage of 16V and short circuit current of 3A. This PV module is directly connected to a battery which has a no-load voltage of 12V. The battery is getting charged from the PV module. The most appropriate operating point is
   a) (13.5V, 2.7A)
   b) (14.5V, 1A)
   c) (9.5V, 2.9A)
   d) (12V, 2.8A)

2. A photovoltaic module has a open circuit voltage of 16V and short circuit current of 3A. This PV module is directly connected to a battery which has a no-load voltage of 12V. Both the PV module and the battery are supplying power to a load resistor of 1.64 ohms connected across the terminals of the PV module.

8.9. QUESTIONS

The most appropriate operating point is
a) (13.5V, 2.7A)
b) (14.5V, 1A)
c) (9.5V, 2.9A)
d) (12V, 2.8A)

3. A photovoltaic module has a open circuit voltage of 16V and short circuit current of 3A. This PV module is directly connected to a battery which has a no-load voltage of 12V. The battery is getting charged from the PV module. The internal resistance of the battery is ....
   a) 0.55 ohms
   b) 2.5 ohms
   c) 0.86 ohms
   d) 0 ohms

4. A photovoltaic module has a open circuit voltage of 16V and short circuit current of 3A. This PV module is directly connected to a battery which has a no-load voltage of 12V. Both the PV module and the battery are supplying power to a load resistor of 1.64 ohms connected across the terminals of the PV module. The most appropriate operating point is
   a) 0.55 ohms
   b) 2.5 ohms
   c) 0.86 ohms
   d) 0 ohms

5. Five batteries each of 12V nominal are connected in series. A charge pump-based equalization circuit is employed. The reservoir capacitor voltage can be
   a) 84V
   b) 72V
   c) 60V
   d) 48V

6. A 20Ah battery with a nominal voltage of 12V is connected to the terminal of a PV module for charging. The PV modules has a short circuit current of 2.5A and open circuit voltage of 14.5V. The battery has a c-rate of C/10. If the internal resistance of the battery is 0.5 ohm, then the input resistance seen by the PV module across the terminals is .....

7. A buck converter is used to charge a battery in a current controlled manner. The buck converter is being supplied by a PV source. The battery has a nominal voltage of 12V. The buck converter uses a MOSFET power semiconductor switch which is switched at 50kHz. The buck converter uses a 4mH inductor with a maximum current of 3A. The slope in the current reference should be .....

8. A buck converter is supplied by a 20V PV source. The semiconductor switch in the buck converter is being switched at 50kHz with a duty ratio of 0.4. The buck converter is operating such that there is no discontinuity in the inductor current.

The buck converter is connected to a load comprising 5V battery having an equivalent series resistance of 3 ohms. Calculate the battery charging current?

9. A boost converter is used to charge a battery in a current controlled manner. The boost converter is being supplied by a PV source at a PV panel terminal voltage of 36V. The battery has a nominal voltage of 60V. The boost converter uses a MOSFET power semiconductor switch which is switched at 50kHz. The boost converter uses a 4mH input inductor with a maximum current of 3A. The slope in the current reference should be ⋯⋯.

10. Consider an application where 4 batteries of 12V each are to be connected in series to achieve a 48V dc bus. If a resistive charge equalisation circuit is employed, then what is the loss in the charge equalisation circuit if 1k ohm is placed across each battery?

Chapter 9

Peltier Cooling with Photovoltaic

9.1 Introduction

Peltier effect was discovered by a watch maker physicist Jean Charles Peltier. He found that by passing a current through a junction or dissimilar metals there results a temperature difference between the junctions. The peltier elements are made using bi-metallic components or also by semi-conductor material junctions.

The semiconductor Peltier element is the more popular type that is available. Interfacing the Peltier junction and the photovoltaic cell will lead to an interesting product called the heat pump. A heat pump transfers heat energy from a cold body to a hot body or in other words from a body at lower temperature to a body at higher temperature.

This mechanism of heat pump by use of electrical energy at its terminals has several important applications. It is an important part of portable refrigeration equipment that are mainly used to carry and store medicines in remote locations. The peltier elements are also used as thermal management components and are used to pump out heat in collaboration with heatsinks or forced coolers.

9.2 Peltier device

Consider two dissimilar metals as shown in figure 9.2.1. One of the metals is copper and the other is nickel. Iron-constantan pair or copper-alumel pair can also be used with similar effect. There are two connections between the metals as shown. The connections are done by spot welding without adding any other material like solder. The two copper ends are connected to a voltage source through a resistor R in series.

On energising the circuit, a current will flow from the positive terminal of the voltage source, through the resistor R, through the bi-metallic junction marked T_1, through the nickel wire, through the second bi-metallic junction marked T_2, and back

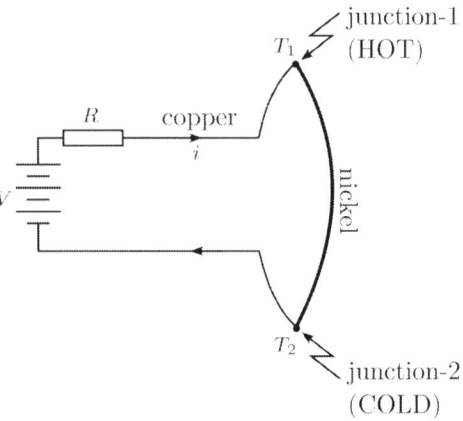

Figure 9.2.1: Peltier principle

to the source. Due to peltier effect, there will be a temperature difference between the two bi-metallic junctions. The junction T_1 will be hot and is called the *hot junction*. The direction of current flow is important. Here the current flows from copper to nickel.

At junction T_2, the current flows from nickel to copper. This results in T_2 being colder than T_1. This is called the *cold junction*. Thus, a current flow through bi-metallic junctions will lead to temperature gradient which leads to the concept of the heat pump.

Figure 9.2.2(a) shows a visualisation of a semiconductor based peltier junction. The actual device is shown in figure 9.2.2(b). It consists of alternate P and N blocks with copper interconnects as shown. The copper contacts based P and N blocks are mounted on ceramic substrate. Wires are brought out from the end copper contacts. A voltage source with a resistor in series is connected. The P and N blocks connected to copper make up the thermal junctions. Current flowing in the direction of copper to N block will be a hot junction and current flowing from N block to copper will be a cold junction. Current flowing from P block to copper will be a hot junction and current flowing from copper to P block will be a cold junction. The N and P blocks are placed alternately and copper acts as the interconnecting element between the N and P blocks. The copper interconnections are made in such a way that all hot junctions will be on one surface and all cold junctions will be on the opposite surface as illustrated in figure 9.2.2. Depending on the direction of current flow, hot and cold junctions get formed resulting in one surface being hot and the other being cold. In a given peltier element, there will be multiple copper clad interconnections for the cold junction and so also for the hot junction.

The heat absorbed or generated is proportional to the magnitude of the current flowing through the peltier junction. The constitutive heat equation for the peltier device is given as

9.3. HEAT PUMP

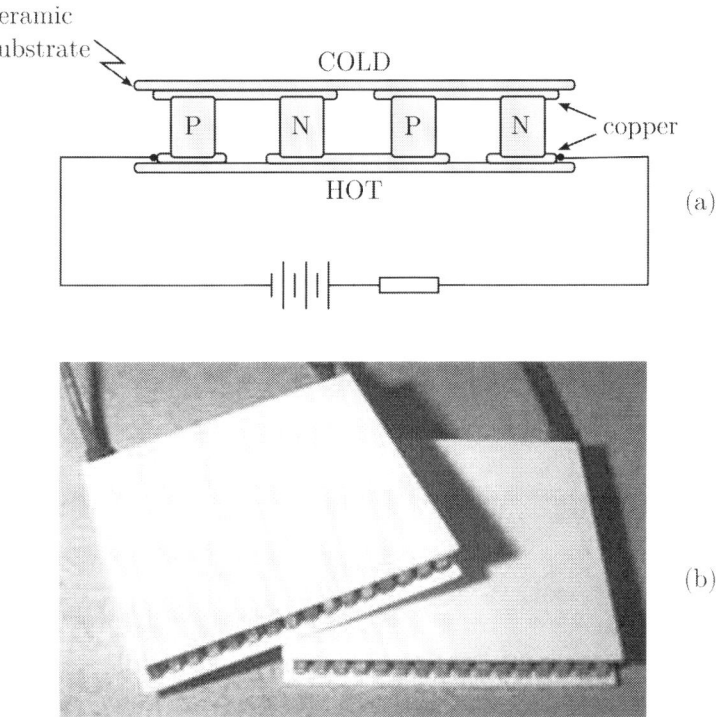

Figure 9.2.2: (a) Semiconductor peltier section (b) peltier device

$$Q = \rho \cdot i \qquad (9.2.1)$$

wherein Q is the amount of heat that flows out of the cold junction or the amount of heat that flows into the hot junction when a current i flows through the peltier junctions. In equation 9.2.1, the proportionality factor ρ is called the peltier coefficient.

9.3 Heat pump

The heat pump equivalent model is shown in figure 9.3.1. There is a body that is at temperature T_1 and a body which is at temperature T_2. Let $T_2 > T_1$. Then the body at temperature T_1 is called the cold body and that at temperature T_2 is called the hot body. For the heat pump to operate, it must pump heat energy from the cold body to the hot body. This can happen only by supplying some external energy to perform the pumping action. Referring to figure 9.3.1, an amount of heat energy Q_C is drawn from the cold body. An amount of energy E is supplied to perform the pumping action. Q_H is the amount of energy needed to transfer Q_C amount of energy from the cold body to hot body.

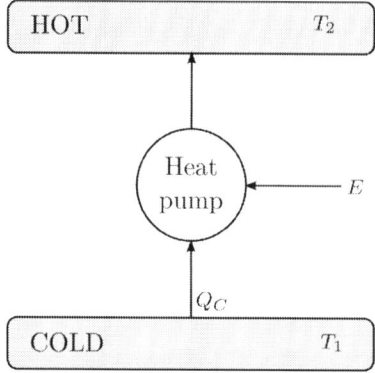

Figure 9.3.1: Heat pump equivalent model

$$Q_H = Q_C + E$$

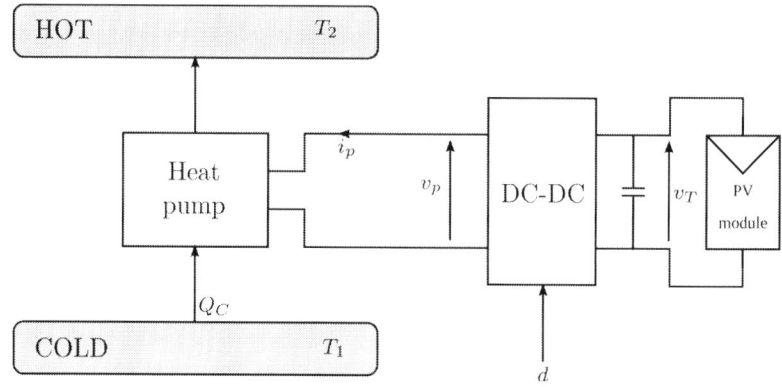

Figure 9.3.2: Peltier cooling system

The heat pump needs an external energy input of E, in order to pump Q_C amount of heat energy from the cold body to the hot body. Figure 9.3.2 shows the cooling system wherein the external energy is derived from an electrical source. A PV source is interfaced through a DC-DC converter to supply the external energy needed to pump up the heat energy. In figure 9.3.2, a peltier device can be used as a heat pump. The external energy is supplied from a PV source through a DC-DC converter. The DC-DC converter is connected to the terminals of the peltier device in order to supply the external electrical energy E.

One should note that external energy is required to pump up heat from the cold body to hot body. However, there is also a natural heat flow from the hot body to the cold body that does not require any external energy. This is also occurs at the same time. It is analogous to water pumping wherein the water is pumped from a low level to a higher level and if there is a large leak in the over head tank there will be significant drop in the level of the water in the overhead tank. It is important that the thermal resistance of the heat flow path from the hot body to cold body be significantly increased by design in order to minimize the loss of heat through this natural heat flow. This is done be improving the thermal insulation around the cold and hot bodies.

Coefficient of performance (CoP) is a parameter which can be considered as a figure of merit in order to compare various types of heat pumps. CoP is the ratio of the amount of heat that is removed from the cold body in contact with the cold junction of the peltier to the amount of external energy supplied for the purpose of heat pumping. In the case of peltier heat pumps, the external energy supplied is electrical in nature. Therefore, CoP is given as

$$CoP = \frac{Q_C}{E} \qquad (9.3.1)$$

where Q_C is the amount of heat in joules that is removed from the cold junction and E is the amount of external electrical energy in joules that is supplied to the peltier junction to perform the heat pump action.

In equation 9.3.1, if no external energy is required for pumping the heat, then the external energy $E = 0$. This means that the coefficient of performance CoP is infinite. Higher the CoP, better is the heat pump.

In equation 9.3.1, the rate of energy flow would give the average power that is pumped from the cold junction to the hot junction. Therefore, one may also estimate the CoP using the average power flows. in such a case, Q_C and E should both be expressed in watts instead of joules.

9.4 Datasheet

There are several parameters listed in the datasheet of the peltier element. One of the important parameters is Q_{max} which is the maximum amount of heat energy flow rate in watts that one can pass through the peltier device at maximum external power supplied and with zero temperature difference across the two surfaces. The wattage of the peltier device is basically the Q_{max} value. For example, a 25W peltier device implies that the Q_{max} value is 25W. This means energy can be removed from the cold junction side and transferred to the hot junction side at the energy flow rate of 25W at maximum current injected into the device and also having zero temperature difference between the surfaces. Heat pumping at zero temperature difference condition occurs predominantly near phase change conditions. For example, during distillation of water when the latent heat in steam needs to be transferred to boiling water, this heat transfer occurs at zero temperature difference condition.

Another absolute maximum value parameter usually given is the $\triangle T_{max}$. This is the maximum temperature difference allowed between the hot and cold junctions and is expressed in ^oC. Most peltier devices have $\triangle T_{max}$ less than 100^oC. This is specified at zero heat flow rate condition and for a given hot side temperature as can be seen in the nomograph of figure 9.4.1.

The maximum current that can flow through the peltier element is another absolute max parameter and is denoted as I_{max}. This gives a measure of the maximum amount of heat power that can be pumped at a specified temperature difference across the two surfaces. Likewise the maximum voltage across the peltier is given by the V_{max} parameter. The peltier element has an equivalent series impedance and is given in the datasheet as module resistance in ohms.

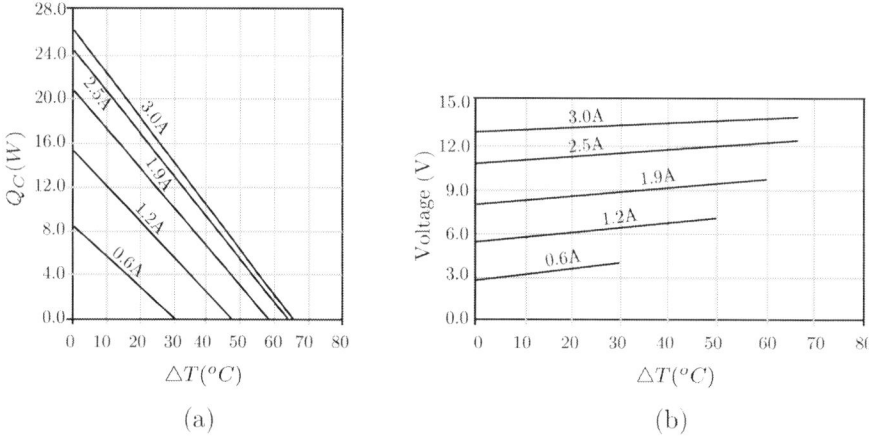

Figure 9.4.1: Performance curves at hot side temperature of 25^oC (a) Heat flow versus temperature difference, (b) Peltier voltage versus temperature difference

Most datasheets of peltier devices give two important nomographs that are useful for design of peltier heat pumps. Figure 9.4.1 show the typical nomographs of $\triangle T$ vs Q_C and $\triangle T$ vs V. Both nomographs are a family of curves with the peltier current as the parameter. In figure 9.4.1(a), it can be seen than when $\triangle T = 0$ maximum rate of energy flow can be achieved at a given peltier current. The voltage across the peltier device being constant, the amount of peltier current gives an indication of the amount of external power drawn for the pumping action. This implies that more the peltier current, higher will be the rate of energy removal from the cold junction. This is evident from the figure 9.4.1(a). The nomograph of figure 9.4.1(b) gives the voltage that needs to be applied across the peltier device in order for the specific amount of current to flow through the peltier such that a specific temperature difference is maintained across the cold and hot junctions.

Consider an example wherein it is required to pump 20W of power from the cold junction to the hot junction. From the sample nomograph of figure 9.4.1(a), it can be seen that the 20W horizontal intercept cuts the three curves related to peltier currents 1.9A, 2.5A and 3.0A. Any one of these three curves can be used for design. Let the curve corresponding to 2.5A peltier current be selected. From the figure 9.4.1(b) consider the voltage curve corresponding to the 2.5A peltier current. It can be seen that for a 20W Q_C a temperature difference of around 12^oC results. For the same $\triangle T$, one obtains that about 10V needs to be applied across the peltier device to achieve this heat pump action.

9.5 Peltier cooling

Referring to figure 9.5.1(a) consider the application wherein a component is generating heat by virtue of the electrical current through it and voltage across it. This power loss needs to removed or pumped out of the component so that the temperature of the components core is within rated limits. Let the component be mounted on a printed circuit board as shown. The inner core or junction of the component J is the hottest part where the heat is generated. The heat will naturally flow out to the ambient, external to the component. Let the temperature of the component core or junction be T_j and that of the external ambient be T_a. The power loss within the component is the rate at which the heat needs to be removed in order to maintain the core temperature of the component within rated value in equilibrium conditions. The power loss can in fact be denoted as Q_C in watts.

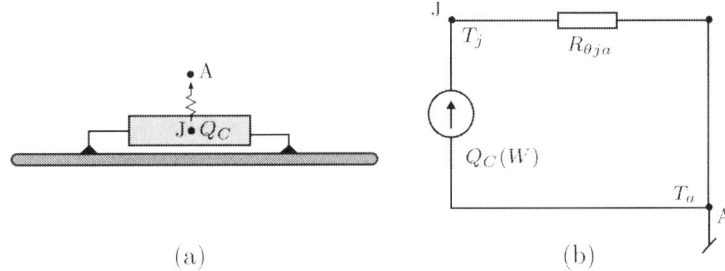

Figure 9.5.1: (a) heat flow from electronic component (b) thermal equivalent circuit

The figure 9.5.1(b) gives an electrical equivalent of the thermal domain. It is easier for electrical engineers to visualize this equivalent circuit. Heat flow or the rate of heat energy flow is considered as equivalent to current flow in electrical circuits. Therefore, the power that is dissipated is the rate at which the heat needs to be removed. Q_C which is the rate of heat flow is the equivalent current source in the electrical equivalent circuit. The heat energy flows from the core or junction which is at temperature T_j to the external ambient which is at temperature T_a. The temperature is considered equivalent to voltage or potential in the electrical equivalent. The rate of heat flow is offered a resistance to its flow and is called thermal resistance. The thermal resistance from the junction of the device to the ambient is called $R_{\theta ja}$. Thus the ohms law for thermodynamics can be expressed as

$$R_{\theta ja} = \frac{(T_j - T_a)}{Q_C} \qquad (9.5.1)$$

The thermal resistance is expressed in the units of $^oC/W$ or K/W.

As an example consider that an amount of 20W of power is dissipated by the electronic component. If the thermal resistance from the junction to ambient is $10^{\circ}C/W$ and the ambient is at a temperature of $40^{\circ}C$, then what is the core or junction temperature?

From equation 9.5.1, one obtains

$$T_j = T_a + R_{\theta ja} \cdot Q_C$$
$$= 40^{\circ}C + 10^{\circ}C/W \cdot 20W$$
$$= 240^{\circ}C$$

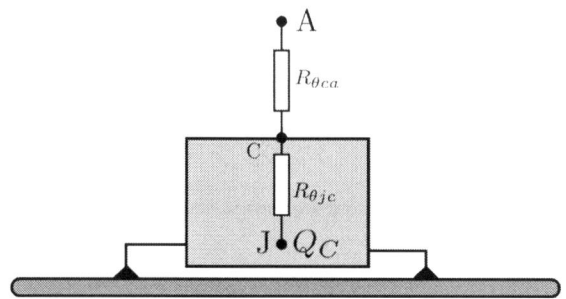

Figure 9.5.2: Component thermal resistance

Consider figure 9.5.2 wherein the dissipating component is zoomed-in from the point of view of thermal resistances. Along the path of the heat flow, one can identify a thermal resistance from the core or junction to the case of the component and another thermal resistance from the case of the component to the ambient. Both these thermal resistances are in series. The junction to case thermal resistance is indicated as $R_{\theta jc}$ and the case to ambient thermal resistance is indicated as $R_{\theta ca}$. The power loss generated at the core of the component Q_C will need to overcome both these resistances in order to flow into the ambient which is considered equivalent to the circuit ground for the thermal domain. In order to overcome these resistances, analogous to voltage drops, there will be temperature drops across these thermal resistances. They are $\triangle T_{jc}$ across the thermal resistance $R_{\theta jc}$ and $\triangle T_{ca}$ across the thermal resistance $R_{\theta ca}$. The governing equation is given as

$$(T_j - T_a) = (T_j - T_c) + (T_c - T_a)$$
$$= (Q_C \cdot R_{\theta jc}) + (Q_C \cdot R_{\theta ca}) \quad (9.5.2)$$
$$= (R_{\theta jc} + R_{\theta ca}) \cdot Q_C$$

Equation 9.5.2 gives the temperature drop across the two thermal resistances in series for a power loss of Q_C in the dissipative component.

9.5.1 Heatsink

In equation 9.5.2, $R_{\theta jc}$ is the thermal resistance of the component material from the core or junction up to the surface or case of the component. The value of this thermal resistance is dependent on the material properties of the component. Likewise, the thermal resistance $R_{\theta ca}$ is dependent on the convection properties of air in the immediate neighbourhood of the component surface. The designer of the peltier cooling system will not have any flexibility in deciding the value of the thermal resistance of either $R_{\theta jc}$ or $R_{\theta ca}$. Use of heatsink gives design flexibility for improving the thermal resistances especially from the case to the ambient.

A heatsink is a good conductor of heat, typically a metal like aluminum which has low thermal resistance. The heatsink acts as an interface between the component surface and the ambient as shown in figure 9.5.3(a). It conducts the heat more easily with low thermal resistance deeper into the ambient zone. The heatsink also is designed such that it has a much larger surface area of contact with the ambient to facilitate the heat removal process. That is the reason the heatsinks are designed with several different fin arrangement in order to increase the surface contact area with the ambient thereby significantly reducing the thermal resistance.

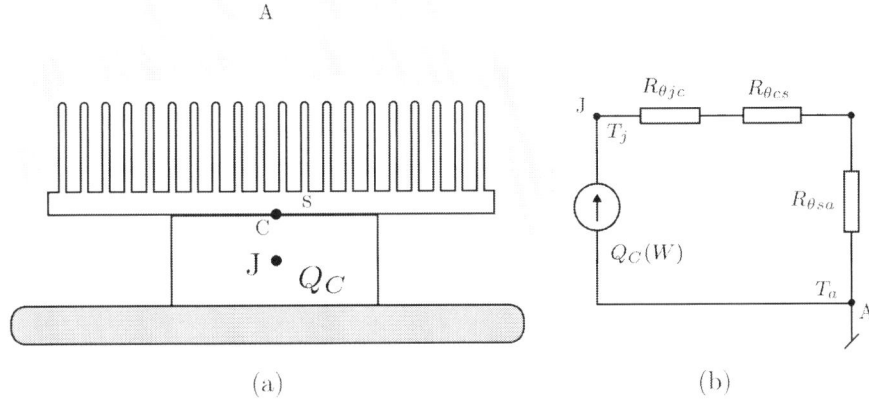

Figure 9.5.3: (a) Component cooling with heatsink (b) thermal equivalent circuit

The electrical equivalent circuit of the thermal system is shown in figure 9.5.3(b). As before, Q_C is the power loss which results in heat flow from the component core or junction to the ambient. The component junction is at temperature T_j and the ambient is at temperature T_a. The case or the surface of the component is at temperature T_c and the heatsink is at temperature T_s. From the equivalent circuit, one obtains

$$\begin{aligned}(T_j - T_a) &= (T_j - T_c) + (T_c - T_s) + (T_s - T_a) \\ &= R_{\theta jc} \cdot Q_C + R_{\theta cs} \cdot Q_C + R_{\theta sa} \cdot Q_C \\ &= (R_{\theta jc} + R_{\theta cs} + R_{\theta sa}) \cdot Q_C \end{aligned} \qquad (9.5.3)$$

In equation 9.5.3 the thermal resistance of junction to case $R_{\theta jc}$ is in the neighbourhood of $0.5^\circ C/W$, the thermal resistance of case to sink $R_{\theta cs}$ is around $0.1^\circ C/W$. Consider the example discussed in the previous sub-section. If the thermal resistance of the heat sink is $1^\circ C/W$ then the junction temperature can be calculated as

$$T_j = T_a + R_{\theta jc} \cdot Q_C + R_{\theta cs} \cdot Q_C + R_{\theta sa} \cdot Q_C$$
$$= 40^\circ C + (0.5 + 0.1 + 1) \cdot 20W$$
$$= 72^\circ C$$

Alternately, the problem can be rephrased as a design problem. What should be the thermal resistance of the heatsink such that the junction temperature does not exceed $100^\circ C$ given that all other parameters are as before. From equation 9.5.3, one obtains

$$R_{\theta sa} = \frac{(T_j - T_a)}{Q_C} - R_{\theta jc} - R_{\theta cs}$$
$$= \left(\frac{100^\circ - 40^\circ}{20W}\right) - 0.5 - 0.1$$
$$= 1.5^\circ C/W$$

Thus the use of heatsinks can provide significant design flexibility in thermal design of the system.

9.5.2 Heatsink with peltier

Further improvement to the thermal design can be conceived with the integration of peltier element or elements. Figure 9.5.4 shows a single peltier element that is placed in between the hot component surface and the heatsink surface. Here an active heat pump is being used to improve the heat removal process.

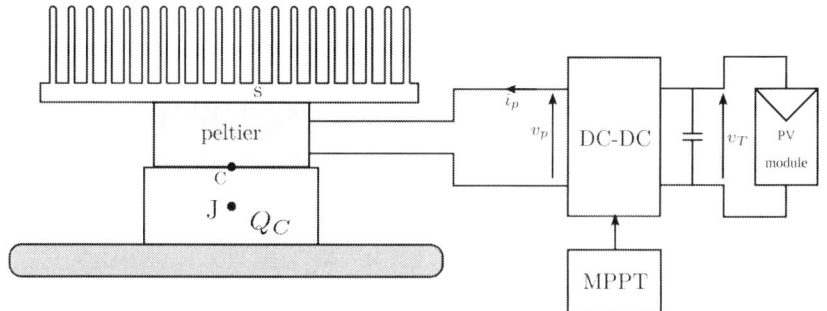

Figure 9.5.4: Peltier and heatsink cooling system

Figure 9.5.4 shows the method of interfacing the peltier to the component surface and the heat sink surface. The cold junction of the peltier is in contact with the component surface and the hot junction of the peltier is in contact with the heatsink. For

9.5. PELTIER COOLING

the peltier element to act as a heat pump, external electric power is being supplied by a PV source operated through a DC-DC converter as shown. The DC-DC converter is operated to extract the maximum power from the PV source.

If the coefficient of performance of the peltier element is CoP, then the electric power that the peltier needs to draw for it to pump Q_C amount of power from the component surface to the heatsink is given as

$$E = \frac{Q_C}{CoP}$$

The heatsink temperature and the component case temperature can be decided based on the heat flow and the thermal resistances. The case to sink thermal resistance $R_{\theta cs}$ is the one that will be affected by the introduction of the peltier element. As the peltier element is an active heat pump, the $R_{\theta cs}$ will become negative and improve the heat removal process. The $\triangle T$ for the peltier junction is $(T_c - T_s)$. This is a design decision and can be set at a prescribed value. From the datasheet, the thermal nomographs can be used to obtain the current that needs to be allowed into the peltier terminal in order to transport Q_C amount of power at the prescribed $\triangle T = (T_c - T_s)$.

9.5.3 Forced cooling

The peltier device that is interposed between the component surface and the heatsink is a active heatpump and will pump out the heat energy from the component at the Q_C rate of energy. The heatsink should also allow the flow of heat energy from the heatsink to the ambient at the same rate. If the rate of heat flow from the heatsink to the ambient is slower, then the heatsink temperature will rise and diminish the effect of the peltier pump. Forced cooling of heatsinks is a popular solution wherein a fan is used to blow the air around the heatsink fins to make the heat removal process more effective.

A combination of several effects are seen during the heat removal process from a component. Conduction, convection, radiation and mass transport processes are all at play while removing heat from a hot source. The heat that is generated due to power loss in the component at its core, flows by conduction up to heatsink and its immediate ambient. The air surrounding the fins of the heatsink remove the heat through the process of convection. The air molecules with some velocity are able to remove the heat much better and faster than static air molecules. It is popular and useful to use a fan as shown in figure 9.5.5 to cause turbulence in the air molecules surrounding the heatsink fins so that heat removal through convection can be more efficiently achieved.

The fan together with the heatsink will provide a much lower thermal resistance from sink to ambient as compared to that without the fan. The fan is run from another output of the DC-DC converter. The peltier element, PV source, DC-DC converter with maximum power point tracking and heatsink will form the peltier cooling system.

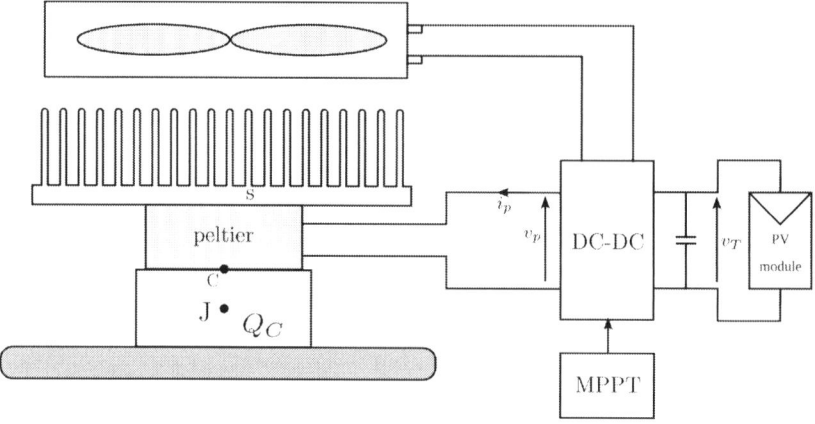

Figure 9.5.5: Peltier with forced cooling

9.6 Thermal aspects

In the peltier cooling system, the heat is pumped from the cold junction of the peltier which is in contact with the heat source, to the hot junction of the peltier. The heat from the hot junction has to be transported to the ambient. This heat removal from the hot junction to ambient needs to be performed at a rate at least equal to the rate at which the peltier element is pumping up the heat energy. Only then will the peltier cooling system be effective. This section discusses on the thermal aspects that will provide some insight into effective thermal design.

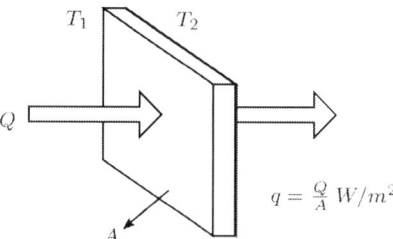

Figure 9.6.1: Specific power

Consider a solid block that can conduct heat through it. Let the cross section area of the block normal to the heat flow be A. One side of the block is at T_1 and the other side is at T_2. The heat flow Q is from the side at higher temperature T_1 to the side at lower temperature T_2. This is illustrated in figure 9.6.1. Specific power is defined as

$$q = \frac{Q}{A} \, W/m^2 \qquad (9.6.1)$$

The temperature difference between the two surfaces is $\triangle T = (T_1 - T_2)$. The ther-

9.6. THERMAL ASPECTS

mal resistance is given as

$$R_\theta = \frac{\Delta T}{Q} \; ^\circ C/W \tag{9.6.2}$$

and thermal resistivity is given as

$$r_\theta = \frac{\Delta T}{q} \; ^\circ Cm^2/W \tag{9.6.3}$$

There is another term called thermal coefficient that is also quite popular in literature. It is denoted by h. The thermal coefficient is the inverse of the thermal resistivity. Therefore,

$$q = \frac{\Delta T}{r_\theta} = h \cdot \Delta T \tag{9.6.4}$$

Thermal conductivity is yet another parameter that various literature use in the context of heat flow. Referring to figure 9.6.1 where the thickness of the solid block is Δx, the thermal conductivity k is expressed as

$$k = h \cdot \Delta x \tag{9.6.5}$$

The thermal conductivity is expressed in $W/K/m$.

9.6.1 Conduction

Conduction is a heat transfer mechanism wherein heat energy flows from the hotter part of a solid medium to the colder part of the solid medium. Consider the heat flow through a solid block as shown in figure 9.6.1. The specific power can be calculated as

$$q = k \cdot \frac{\Delta T}{\Delta x} \tag{9.6.6}$$

Equation 9.6.6 is the heat flow rate equation and is also called the Fourier law. If one multiplies the equation 9.6.6 by the cross section area A, then one obtains the heat power Q. This can be written as

$$Q = kA \cdot \frac{\Delta T}{\Delta x} \tag{9.6.7}$$

and the thermal resistance can be written in the following form

S.No.	Material	Thermal conductivity, $W/K/m$
1	Silver	410
2	Copper	385
3	Aluminum	211
4	Steel	47.6
5	Glass	1.05
6	Still air	0.026
7	Polyurethane	0.025

Table 9.1: Thermal conductivity of common materials

$$R_\theta = \frac{\triangle x}{kA} \qquad (9.6.8)$$

From equation 9.6.8 it can be seen that the thermal resistance is inversely proportional to the cross section area that is orthogonal to the flow of the heat power. Therefore, greater the cross sectional area, smaller will be the thermal resistance and better will be the heat conduction through the material. This is precisely the reason for the large number of fins on heatsinks. It increases the area of contact to the ambient that is orthogonal to the flow of heat power. As a consequence the thermal resistance is reduced.

Thermal conductivity is a material property. Table 9.1 gives a list of thermal conductivity of few common materials. Silver at $410W/K/m$ is one of the best conductors of heat. Copper and aluminum are also pretty good conductors of heat. Observe that glass has a low thermal conductivity of $1.05W/K/m$. Compare this with silver. Apparently glass is a good heat insulator. Polyurethane has a very low thermal conductivity of $0.025W/^oK/m$. It is the material which is goes by the brand name of thermocol that is used as an insulation packing material. This is a very good thermal insulator and used in containers to retain heat or cold. Likewise, still air is also a very good insulator. In fact, thermocol is a foam type material with lots of air trapped within it. It is the trapped still air that gives the good thermal insulation properties for thermocol.

9.6.2 Convection

Convection is a heat transfer mechanism wherein the heat flow occurs between solid and fluid boundaries. The heat flows from a hot solid surface to a colder fluid surrounding it. This is the typical situation with a solid heatsink and the air fluid surrounding it. The air molecules in contact with the hot solid will take in the heat energy with consequent decrease in density. The hot air molecules will rise up and the colder air molecules will take their place and will retrieve heat from the hot solid surface. This process of heat removal from the solid by the surrounding fluid is called convection.

9.6. THERMAL ASPECTS

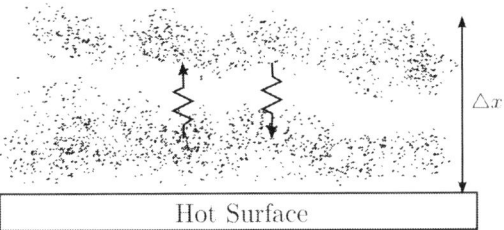

Figure 9.6.2: Heat transfer by convection

The movement of fluid in the neighbourhood of the solid hot surface can be visualised as shown in figure 9.6.2. One can improve the rate at which the fluid molecules carry away the heat by increasing the velocity of the surrounding fluid. For air, a fan can be used to force the convection to happen at a faster rate. This is called forced convection cooling.

Let the convection currents happen in the neighbourhood of the hot solid surface. As shown in figure 9.6.2, the equivalent thickness of the fluid where convection is mostly prevalent is shown as $\triangle x$. Unlike in the case of solids, the precise value of $\triangle x$ is not easy to estimate. Referring to heat rate equation 9.6.7, it can be seen that $\triangle x$ is the only quantity that is uncertain. All other parameters are known or can be estimated.

$\triangle x$ depends on the geometry and position of the solid surface. A parameter X is identified which is called the characteristic dimension of the solid body. It is associated with the surface through which heat energy transfers to the fluid medium. The characteristic dimension of the solid body is found to be proportional to $\triangle x$ or the equivalent thickness of the fluid that is in action during the convection process. The constant of proportionality is called the Nusselt's number which is a dimensionless quantity. Thus,

$$N = \frac{X}{\triangle x} \tag{9.6.9}$$

The parameter $\triangle x$ in the heat rate equation 9.6.7 can be expressed in terms of the Nusselt's number as follows.

$$Q = k \cdot A \cdot N \cdot \frac{\triangle T}{X} \tag{9.6.10}$$

and the thermal resistance for convection $R_{\theta v}$ is given as

$$R_{\theta v} = \frac{X}{k \cdot A \cdot N} \tag{9.6.11}$$

The characteristic dimension X is dependent on the geometry and position of the solid body. It could be the length of the solid, the breadth or width of the solid or the diameter of a cylindrical solid body. The characteristic dimension has been identified for several standard objects through experimentation and is given in table 9.3.

There are two variants in convection. One is called the free convection wherein the heat flow happens through natural flow of the fluid molecules or air molecules in the case of heatsink to ambient air. This is based on the natural circulation of fluid molecules that is attained by the movement of fluid molecules due to density gradients. The other variant is called forced convection wherein a fan or blower is used to increase the velocity of fluid or air molecules that are in contact with the hot solid surface. The Nusselt's number is different for free and forced convection. The Nusselt's number for various standard geometries and positions are listed in table 9.3.

9.6.2.1 Nusselt's number for Free convection

In the case of free convection, the Nusselt's number is determined as a function of the Rayleigh number. The Rayleigh number gives a measure of the type of flow. Below a specific threshold value of the Rayleigh number, the flow is said to be laminar in nature. Above the threshold, the flow is said to be turbulent. The Rayleigh number R_a for free convection is given as

$$R_a = \frac{g \cdot \beta \cdot X^3 \cdot \triangle T}{\delta \cdot \upsilon} \quad (9.6.12)$$

where
R_a = Rayleigh number
$g = 9.81 \, m/s^2$
β = co-efficient of thermal expansion of fluid
   = $\frac{1}{330}$ for air
X = characteristic dimension
δ = thermal diffusivity of fluid
   = $2.6 \times 10^{-5} \, m^2/s$ for air at $50^{\circ}C$
υ = kinematic viscosity
   = $1.8 \times 10^{-5} \, m^2/s$ for air at $50^{\circ}C$

The parameters in equation 9.6.12 like gravitational acceleration, coefficient of thermal expansion, thermal diffusivity and kinematic viscosity are properties of the fluid and are known for a given fluid. Based on the solid surface from which heat flows, the characteristic dimension is ascertained from table 9.3 and the Rayleigh number can be calculated. Using the Rayleigh number, one can determine whether the fluid flow is laminar or turbulent and appropriately calculate the Nusselt's number as given in table 9.3. Once the Nusselt's number is estimated, the thermal resistance for free convection can be easily obtained. In the table, the flow type is indicated using 'L' for laminar flow and 'T' for turbulent flow. Note that in table 9.3, for the case of parallel plates, only turbulent flow is considered. However, if the Rayleigh number is below 10^5, the flow is laminar and the plates can be considered as independent horizontal or vertical plates and the Nusselt's number arrived at appropriately.

Consider a horizontal flat plate. Several of the heatsinks are flat plates either mounted horizontally or vertically. The flat plate can have rectangular or circular

9.6. THERMAL ASPECTS

S.No.	Solid body	Flow	Range of R_a	Nusselt's number
1	$X = \frac{a+b}{2}$	L	$10^2 < R_a < 10^5$	$0.54 \cdot R_a^{0.25}$
		T	$R_a > 10^5$	$0.14 \cdot R_a^{0.33}$
2	$X = d$	L	$10^2 < R_a < 10^5$	$0.54 \cdot R_a^{0.25}$
		T	$R_a > 10^5$	$0.14 \cdot R_a^{0.33}$
3	$X = b$	L	$10^4 < R_a < 10^9$	$0.56 \cdot R_a^{0.25}$
		T	$10^9 < R_a < 10^{12}$	$0.2 \cdot R_a^{0.4}$
4	$X = d$	L	$10^4 < R_a < 10^9$	$0.56 \cdot R_a^{0.25}$
		T	$10^9 < R_a < 10^{12}$	$0.2 \cdot R_a^{0.4}$
5	$X = d$	L	$10^4 < R_a < 10^9$	$0.47 \cdot R_a^{0.25}$
		T	$R_a > 10^9$	$0.1 \cdot R_a^{0.33}$
6	$X = l$	L	$10^4 < R_a < 10^9$	$0.56 \cdot R_a^{0.25}$
		T	$10^9 < R_a < 10^{12}$	$0.2 \cdot R_a^{0.4}$
7	$X = d$	T	$R_a > 10^5$	$0.062 \cdot R_a^{0.33}$
8	$X = d$ for $\theta < 50^o$	T	$R_a > 10^5$	$0.062 \cdot R_a^{0.33}$

Table 9.3: Nusselt's number for free convection

S.No.	Solid body	Flow	Range of R_e	Nusselt's number
1	$X = a$	L	$R_e < 5 \times 10^5$	$0.664 \cdot R_e^{0.5} \cdot \left(\frac{v}{\delta}\right)^{0.33}$
		T	$R_e > 5 \times 10^5$	$0.37 \cdot R_e^{0.8} \cdot \left(\frac{v}{\delta}\right)^{0.33}$
2	$X = d$	L	$0.1 < R_e < 1000$	$\left(0.35 + 0.57 \cdot R_e^{0.52}\right)\left(\frac{v}{\delta}\right)^{0.3}$
		T	$1000 < R_e < 5 \times 10^5$	$0.26 \cdot R_e^{0.6} \cdot \left(\frac{v}{\delta}\right)^{0.3}$

Table 9.4: Nusselt's number for forced convection

geometry. The characteristic dimension of rectangular and circular flat plates placed horizontally is given in the table 9.3. Calculate the Rayleigh number and check if the flow is laminar or turbulent from the thresholds given in the table 9.3. After determining the type of flow, the Nusselt's number can be calculated from the empirical relations given in the table. The Nusselt's number can be calculated in a similar manner for vertical plates, vertical and horizontal cylinders and parallel plates too. If the solid is of non-standard geometry, then the Nusselt's number is determined by performing convection heat flow experiment on the solid of non-standard geometry.

9.6.2.2 Nusselt's number for forced convection

In the case of forced convection, another number called the Reynolds number is used to determine the Nusselt's number. The Reynolds number is written as

$$R_e = \frac{u \cdot X}{v} \qquad (9.6.13)$$

where
R_e = Reynolds number
u = mean velocity of forced fluid flow m/s
X = characteristic dimension
v = kinematic viscosity
   = $1.8 \times 10^{-5}\ m^2/s$ for air at 50^oC

The Reynolds number for the flat plate and cylindrical solid are given in table 9.4.

The table 9.4 gives the case for two popular geometries for forced convection. One of the commonly used heatsinks is the flat plate with a fan or blower mounted as shown in the figure given in the table. The other is a cylindrical geometry with the fan placed as shown. The laminar and turbulent flows are determined by estimating the Reynolds number. The Nusselt's number is again based on empirical relationship comprised of the Reynolds number, the kinematic viscosity and the thermal diffusivity of the fluid.

9.7 Peltier refrigeration

Consider an example of refrigeration within a closed box container. Let the closed box container be a cylindrical box placed vertically. The peltier cooled cylindrical box refrigerator has several application as a portable box cooler. This peltier cooling example can be considered with either free convection or forced convection mechanism of heat removal. Both these methods will be discussed and compared.

9.7.1 Free convection method

The refrigeration container is as shown in figure 9.7.1. A cross section of the cylindrical box is shown. The container is enclosed on all sides by thermal insulator like polyurethane material such that the thermal resistance of the walls of the container is high. This will minimise the heat flow from the higher temperature external ambient to flow into the lower temperature volume within the container. Heat exchange between the container volume within and the external ambient can happen only through the top circular plate of the cylindrical box. Peltier element or elements are fitted onto the top surface with the cold junction in contact with the container and the hot junction in contact with a solid cylindrical aluminum block which will act as the heatsink. The Peltier element is powered by a PV source to obtain the external electrical energy for pumping the heat from within the chamber to the aluminum block heatsink. The heat flows out to the external ambient from the aluminum heatsink by the free convection mechanism.

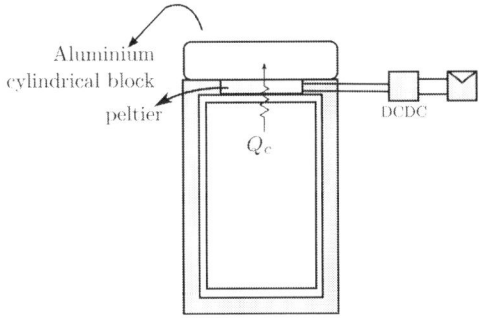

Figure 9.7.1: Peltier refrigeration using free convection

Q_C amount of heat is being pumped out of the container into the external ambient. Let the aluminum block or heatsink have a diameter of 22cm and a height of 11cm. The external ambient temperature surrounding the container T_a is at 30^oC. It is required to maintain the inside of the container at 10^oC. If one considers a 40^oC temperature difference across the peltier junctions, then it implies that the aluminum block which is in contact with the hot junction of the peltier will be at a temperature not exceeding 50^oC. Let us assume for this example that a peltier device is chosen which requires 10W of electric power in order to pump heat through a 40^oC temperature potential.

The aluminum block which acts as the heatsink is the part from which heat flows into the external ambient. Heat energy flows out to the ambient from the top surface of the cylindrical aluminum block and also from the sides. One can apply the free convection model to the aluminum block. Consider first the top surface of the aluminum block as a flat plate. Calculate the Rayleigh number using the following parameters,

$$R_a = \frac{g \cdot \beta \cdot X^3 \cdot \triangle T}{\delta \cdot \upsilon}$$

and the values of the various parameters are
$g = 9.81 \ m/s^2$
$\beta = \frac{1}{330}$ for air
$X = 0.22m$ diameter of flat surface
$\delta = 2.6 \times 10^{-5} \ m^2/s$ for air at $50°C$
$\upsilon = 1.8 \times 10^{-5} \ m^2/s$ for air at $50°C$
$\triangle T = 50° - 10° = 40°K$

Using these parameter values, the Rayleigh number is calculated and $R_a = 2.7054 \times 10^7$. Referring to table 9.3 s.no.2, $R_a > 10^5$ and therefore this implies that the flow is turbulent and one uses the appropriate empirical relation to obtain the Nusselt's number. Therefore

$$N = 0.14 \cdot R_a^{0.33} = 39.7$$

Using the heat flow rate equation, one obtains the amount of heat power that is removed from the aluminum block top surface. Denoting this as Q_{top}, one has

$$\begin{aligned} Q_{top} &= k \cdot A \cdot N \cdot \frac{\triangle T}{X} \\ &= \frac{(0.026 \ W/°K/m) \cdot \left(\frac{\pi}{4} \times 0.22^2 \ m^2\right) \cdot 39.7 \cdot 40°K}{0.22 \ m} \\ &= 7.13 \ W \end{aligned}$$

Heat flows to the ambient from the sides of the aluminum block. Let this be denoted as Q_{side}. Considering the heat flow from the sides of the aluminum block as equivalent to the free convection for vertically placed cylinder, one can compute the Rayleigh number as before. Only change will be in the characteristic dimension which is $X = 0.11m$. Thus,

$$R_a = 3.38 \times 10^6$$

Referring to table 9.3 s.no.6, the condition $10^4 < R_a < 10^9$ is met and the flow type is laminar. Therefore, the Nusselt's number is given as

$$N = 0.56 \cdot R_a^{0.25} = 24.015$$

9.7. PELTIER REFRIGERATION

Using the heat flow rate equation, one obtains the amount of heat power that is removed from the aluminum block side surface. This is

$$Q_{side} = k \cdot A \cdot N \cdot \frac{\Delta T}{X}$$

$$= \frac{(0.026\,W/^oK/m) \cdot (\pi \times 0.22 \times 0.11\,m^2) \cdot 24.015 \cdot 40^oK}{0.11\,m}$$

$$= 17.26\,W$$

The total heat flow to the ambient from the aluminum block is $7.13 + 17.26 = 24.4W$. This includes the 10W of electrical power that is drawn by the peltier to perform the heat pump action. This amount also is delivered to the ambient. Therefore the amount that is drawn from the refrigeration container is $24.4W - 10W = 14.4W$. Thus $Q_C = 14.4W$ and the coefficient of performance for the peltier unit is computed as

$$CoP = \frac{Q_C}{E} = \frac{14.4}{10} = 1.44$$

9.7.2 Forced convection method

The same example of the previous sub-section will be modified for a forced convection heat removal mechanism. Figure 9.7.2 shows the refrigeration system wherein a fan has been included in such a manner that there is air flow on the top surface and sides of the cylindrical aluminum block which acts as a heatsink. Rest of the system is identical to that described in the previous sub-section.

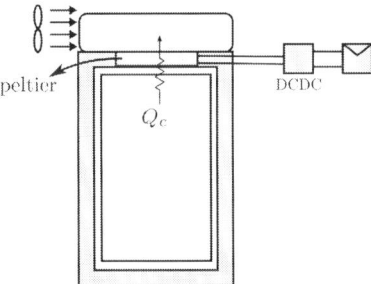

Figure 9.7.2: Peltier refrigeration using forced convection

Consider first the top surface of the aluminum block as a flat plate. Calculate the Reynolds number using the following parameters,

$$R_e = \frac{u \cdot X}{\upsilon}$$

and the values of the parameters are
$u = 3\ m/s$
$X = 0.22m$ diameter of flat surface
$v = 1.8 \times 10^{-5}\ m^2/s$ for air at $50^\circ C$
$\Delta T = 50^\circ - 10^\circ = 40^\circ K$

The air flow velocity is obtained from the datasheet of the fan or the blower. At a specific speed of rotation of the fan, the air velocity is provided from the nomograph in the fan datasheet. The Reynolds number is calculated to be 0.3667×10^5. Referring to table 9.4, it can be seen that $R_e < 5 \times 10^5$ and the flow is laminar. Therefore, the Nusselt's number can be computed as

$$N = 0.664 \cdot R_e^{0.5} \left(\frac{v}{\delta}\right)^{0.33}$$
$$= 112.62$$

Using the heat flow rate equation, one obtains the amount of heat power that is removed from the aluminum block top surface under forced convection. This is

$$Q_{top} = k \cdot A \cdot N \cdot \frac{\Delta T}{X}$$
$$= \frac{(0.026\ W/^\circ K/m) \cdot \left(\frac{\pi}{4} \times 0.22^2\ m^2\right) \times 112.62 \times 40^\circ K}{0.22\ m}$$
$$= 20.24\ W$$

Heat flows to the ambient from the sides of the aluminum block also. And like before, let this be denoted as Q_{side}. Considering the heat flow from the sides of the aluminum block as equivalent to the forced convection for vertically placed cylinder, one can compute the Reynolds number with the characteristic dimension $X = 0.22m$. Thus,

$$R_e = 0.3667 \times 10^5$$

Referring to table 9.4, the condition $10^3 < R_e < 5 \times 10^5$ is satisfied and the flow type is turbulent. Therefore, the Nusselt's number is given as

$$N = 0.26 \cdot R_e^{0.6} \left(\frac{v}{\delta}\right)^{0.3} = 127.54$$

Using the heat flow rate equation, one obtains the amount of heat power that is removed from the aluminum block side surface. This is

$$Q_{side} = k \cdot A \cdot N \cdot \frac{\Delta T}{X}$$
$$= \frac{(0.026\ W/^\circ K/m) \cdot (\pi \times 0.22 \times 0.11\ m^2) \times 127.54 \times 40^\circ K}{0.22\ m}$$
$$= 45.83\ W$$

The total heat flow to the ambient from the aluminum block is $20.24 + 45.83 = 66.07W$. This includes the 10W of electrical power that is drawn by the peltier to perform the heat pump action. Therefore the amount of heat power that is drawn from the refrigeration container is $66.07W - 10W = 56.07W$. Thus $Q_C = 56.07W$ and the coefficient of performance for the peltier unit is computed as

$$CoP = \frac{Q_C}{E} = \frac{56.07}{10} = 5.6$$

One can see that forced convection is significantly better wherein the CoP is 5.6 as compared to 1.44 for free convection method. Except for the fan, all other aspects being identical, one can see that 56W is being drawn out of the refrigerator container with forced cooling as against only 14.4W in the case of free convection. This would imply that for the same amount of heat power flow, the heatsink size will be significantly smaller in the case of forced convection. It should also be noted that if the power used to run the fan is also included then the CoP of the system including the fan will be slightly lower than 5.6 but still significantly better than the free convection case.

9.8 Radiation

In the case of Peltier based cooling systems, mostly the conduction and convection mechanisms of heat removal is primarily prevalent. However, there will always be some amount of radiation that will happen in any system as this mechanism will occur even in the absence of any medium. A body at temperature T_1 will radiate heat in the form of electromagnetic waves. The radiation mechanism is visualised in figure 9.8.1.

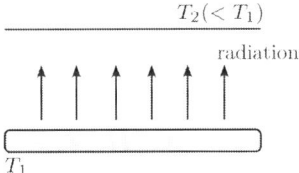

Figure 9.8.1: Radiation mechanism of heat transfer

Consider an imaginary boundary at some distance from the hot body at temperature T_1. Let the temperature at this imaginary boundary be $T_2 < T_1$. The heat energy received at this imaginary boundary of cross section area A normal to the incident radiation is given as

$$Q = h_r \cdot A \cdot \triangle T \qquad (9.8.1)$$

In equation 9.8.1, h_r is the radiation thermal coefficient and $\triangle T = (T_1 - T_2)$. The radiation thermal coefficient is given by the following relation

$$h_r = 4\sigma\varepsilon(1-\phi)\left(\frac{T_1+T_2}{2}\right)^3 \tag{9.8.2}$$

where $\sigma = 5.67 \times 10^{-8}$ is the Boltzmann's constant, and ϕ is called the shielding factor. It is zero for single flat plates and two parallel plates. ε is called the emittance which is a heat emitting property of specific surface textures. The emittance for some common surface textures are given in table 9.5.

S.No.	Surface texture	Emittance
1	Anodised Aluminum (black)	0.8
2	Polished Aluminum	0.095
3	Rough surface Aluminum	0.18
4	Tungsten at $1500^\circ C$	0.33
5	Polished copper	0.07
6	Wood	0.9

Table 9.5: Emittance of common surface textures

9.9 Mass transport

The fourth heat transfer mechanism is that by mass transport. Here the heat transfer occurs through the fluid mass. A mass of fluid that is flowing will receive or capture heat from a hot body and transport the heat energy along with it and release it at a different point. This is a phenomenon that can be commonly observed in vehicle and refrigerator cooling systems where a fluid is used to move heat energy from one place to another. In a refrigerator, a fluid like Freon is used for transferring heat from the chamber to an evaporator and then exchanges heat energy with the ambient through a radiator. Similarly, even in a car, the heat from the engine is removed by the flow of water and then get removed out into the external ambient through a radiator.

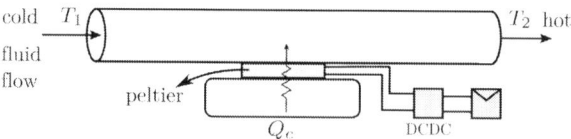

Figure 9.9.1: Mass transport mechanism of heat transfer

In the context of Peltier element being used for mass transport, consider a long pipe wherein fluid enters the pipe at temperature T_1 and leaves the pipe at temperature T_2. A Peltier element is mounted as shown in figure 9.9.1 such that its cold junction is in contact with the component surface that is dissipating heat, and the hot junction is in contact with the pipe through which the fluid flows.

A mass of fluid enters the pipe at a temperature T_1. As the fluid passes over the part where the Peltier junction is in contact with the pipe, the Peltier element which acts like a heat pump, will pump the heat power from the component that is dissipating heat and into the mass of fluid that is flowing in the pipe. The mass of fluid will continue its flow and exit from the pipe at a hotter temperature T_2. The amount of heat power that is carried by the mass of fluid is given by

$$Q = \frac{dm}{dt} \cdot c_p \cdot (T_2 - T_1) \tag{9.9.1}$$

In equation 9.9.1, $\left(\frac{dm}{dt}\right)$ is the mass flow rate and c_p is the specific heat of fluid in $J/kg/^oK$. It can also be seen from the equation 9.9.1 that the thermal resistance for mass flow $R_{\theta m}$ is given by

$$R_{\theta m} = \frac{(T_2 - T_1)}{Q} = \frac{1}{\left(\frac{dm}{dt} \cdot c_p\right)} \tag{9.9.2}$$

From equation 9.9.2, it can be seen that if the specific heat of the fluid is high, then the thermal resistance will be low. If one uses fluid with a very high specific heat capacity like Freon, the heat power transfer will be very efficient. Another important factor that is contributing to heat removal is mass flow rate itself, unlike the cases of conduction, convection and radiation which are more by thermal temperature difference. Here the thermal resistance is inversely proportional to the mass flow rate. By controlling the mass flow rate i.e. by controlling the speed of flow of the fluid, one can regulate the thermal resistance of the system. Additionally in the case of Peltier junction, the thermal resistance of the system can be further controlled by controlling the electrical power drawn by the Peltier. Thus, this thermal mechanism has the unique property of dynamically changing the fluid speed and Peltier power to achieve thermal resistance regulation.

9.10 Question

1. Heat transfer process through solids is called
    a) conduction
    b) convection
    c) radiation
    d) mass transport

2. Heat transfer process through solids by means of fluids is called
    a) conduction
    b) convection
    c) radiation
    d) mass transport

3. Heat transfer process without any medium is called
   a) conduction
   b) convection
   c) radiation
   d) mass transport

4. Which of the following has the lowest thermal conductivity?
   a) Glass
   b) Steel
   c) Copper
   d) Air

5. Which of the following has the highest thermal conductivity?
   a) Glass
   b) Steel
   c) Copper
   d) Air

6. Nusselt number is a
   a) fluid flow measure
   b) dimensionless quantity
   c) characteristic dimension
   d) thermal impedance

7. The emittance of polished aluminum is
   a) more than anodized aluminum
   b) more than glass
   c) more than wood
   d) more than polished copper

8. Which of the following statements is true?
   a) As the temperature difference between the hot and cold sides of the Peltier cooler increases, the voltage across the cooler decreases, under the condition of constant current.
   b) As the temperature difference between the hot and cold sides of the Peltier cooler increases, the removal of heat from the cold side decreases, under the condition of constant current.
   c) The coefficient of performance can never be greater than 1.
   d) The current through the Peltier cooler is allowed to exceed the rated current.

9. A peltier element is operated from a PV source. It is drawing 1A at 12V from a PV source. The peltier element is extracting heat from an object at the rate of 20 joules per second. The object from which heat is extracted is at 15 deg C and the hot surface is at 30 deg C. The co-efficient of performance of the peltier element is ....

10. Consider a peltier cooling system where in heat needs to be removed from an object at the rate of 20 joules/sec and transferred to the ambient which is at 50

9.10. QUESTION

deg C. It is required that the temperature of the hot surface of the peltier element is below 70 deg C. What should be the thermal resistance from the hot surface of the peltier element to the ambient?

11. A slab is made up of aluminum having an area of $1m^2$ and thickness of 10mm. The heat flow is normal to the $1m^2$ area of the slab. The thermal resistance is ....

12. A slab is made up of glass having an area of $1m^2$ and thickness of 5mm. The heat flow is normal to the $1m^2$ area of the slab. The thermal resistance is ....

13. In a convective peltier cooling system, an air blower is used to produce an air speed of 5 m/s over the hot surface of the peltier element which has a square section of side length 3 cm. The Reynolds number is ....

14. The specifications of a thermoelectric Peltier module from a datasheet are as given in the table below. A 20V battery is connected to this Peltier module using a series resistance R_s. Based on the data given in the table, what should be the value of R_s?

Hot side temperature	25^oC	50^oC
Q_{max} (W)	25.7	30.4
ΔT_{max} (^oC)	68	75
I_{max} (A)	3.0	3.0
V_{max} (V)	14.5	16.5
Module resistance (Ω)	0.84	0.95

15. A flat plate aluminum heat sink has a thickness of 3.5mm and $40mm \times 40mm$ square cross section to the flow of heat. One surface of the aluminum heat sink is at 50^oC and the hotter surface is at 50.1^oC. What is heat power flow through the aluminum flat plate? (The thermal conductivity of aluminum is $211W/^oK/m$)

Chapter 10

Water Pumping with Photovoltaic

10.1 Introduction

Water pumping is one of the more common of applications. In a home, it is used to pump water from an underground sump to an overhead tank to store water. This is a very familiar and common application that one encounters every day. Another application is to lift water from a well to irrigate an agricultural field. The water pump consists of two component parts, one is the prime mover which is a DC or AC motor and the other one is the pump that creates the pressure difference to lift the water. In most cases the pump-motor is powered from the main grid. In this chapter the discussion will be on using photovoltaic sources for operating the pump-motor.

10.2 Water pumping system

A typical PV based water pumping system is illustrated in figure 10.2.1. Water needs to be pumped up from a tank or reservoir at a lower height to another tank or reservoir that is placed at a higher level. In order to pump up the water, one needs a pump and a motor to drive the pump. The pump can be a dynamic pump like a centrifugal pump or a positive displacement type pump like a reciprocating pump. The shaft of the pump needs to be driven by a prime mover. Generally an induction motor is used as the prime mover. However for the PV based application, a dc motor is used to drive the shaft of the pump. The dc motor is powered from a PV source through a DC-DC converter or chopper drive.

The pump will have an inlet from which the water is drawn from the lower level tank into the pump by a pipe called the suction pipe. The water exits the pump outlet and is discharged into the higher level tank through a pipe called the delivery pipe. In most homes, the suction pipe connects the centrifugal pump to the underground sump and the delivery pipe connects the pump to the overhead tank. A pump acts as

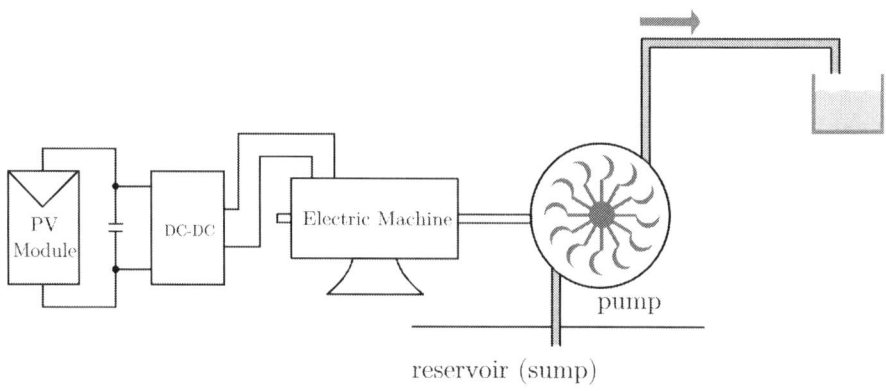

Figure 10.2.1: Water pumping system

a pressure source for the hydraulic system which lifts the water from the underground sump to the overhead tank. The pump is driven by a motor which draws power from an AC grid in most cases.

The water pumps in most homes are driven by a single phase AC induction motor and pumps used for irrigation in agricultural fields are driven by three phase AC induction motors. However, when using a photovoltaic source, the output power of the PV source should be processed through a DC-AC inverter if an AC motor drives the pump or should be processed through a DC-DC converter if a DC motor drives the pump.

10.3 Pressure heads

The pump in the hydraulic system is the pressure source that works against the opposing pressures in the system in order to lift the water from a lower level to a higher level. Mathematically, the pressure is written as

$$p = \rho g h \qquad (10.3.1)$$

where p is the pressure or force per unit area and is expressed in pascals, ρ is the density of fluid and is $1000 kg/m^3$ for water, g is the gravitational acceleration which is $9.81 m/s^2$, and h is the height of the column of fluid or water in this case that exerts force on the base and is expressed in m. As g is a constant and for a given fluid ρ is also a constant, pressure relates directly to the height of the column of fluid h. Therefore, h is called the pressure head or static pressure head.

Consider the diagram of the water pump system shown in figure 10.3.1. The length from the tip of the suction pipe to the center of the centrifugal pump is called the suction head. It is also known as static lift in some literature. This is denoted as h_s. On the delivery side, the length measured from the center of the centrifugal pump to

10.4. HYDRAULIC ENERGY

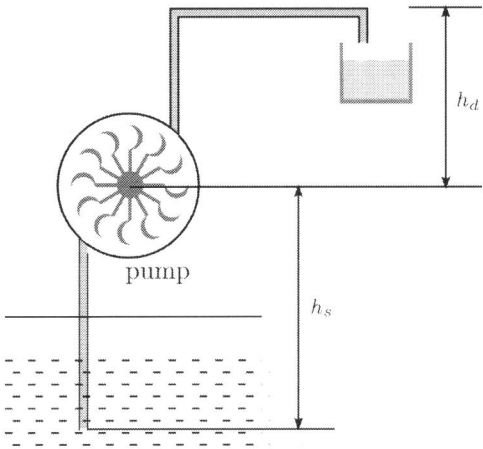

Figure 10.3.1: Dynamic heads in a water pumping system

the highest point on the delivery pipe is called the delivery head or the discharge head. It is denoted as h_d. It is also known as static height in some literature.

The pump needs to develop sufficient pressure in order to lift a mass of water against the suction head and the delivery head so that water from the lower tank can be discharged into the overhead tank. The suction head and the delivery head are both physical heads. However, there is a non-physical head too that the pump needs to overcome. Due to the flow of water in the pipes, there will be power loss due to friction by virtue of inner wall of the pipe being in contact with flowing water. The pump needs to develop sufficient pressure to overcome this friction pressure drop too. This friction pressure drop is considered as an equivalent non-physical head and is denoted as h_f the friction head. The sum of the physical and non-physical heads is called the dynamic head. Thus the total dynamic head is expressed as

$$h = h_s + h_d + h_f \qquad (10.3.2)$$

10.4 Hydraulic energy

Consider the visualisation shown in figure 10.4.1. Let a mass of water m be lifted to a height h as shown. The increase in potential energy when a mass of water is lifted through a specific height is

$$E_h = mgh \qquad (10.4.1)$$

where E_h is the hydraulic energy in joules, m is the mass of water in kg, g is the gravitational acceleration in m/s^2 and h is the height or head in m.

The mass of water is the product of density and discharge volume as given below

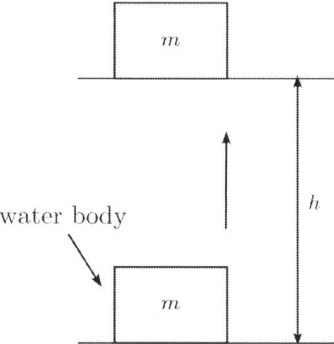

Figure 10.4.1: Visualisation for hydraulic energy calculation

$$m = \rho \cdot Q \qquad (10.4.2)$$

where ρ is the density of water and is $1000 kg/m^3$ and Q is the discharge volume in m^3. Substituting for the mass from equation 10.4.2 into equation 10.4.1, one has

$$E_h = \rho \cdot Q \cdot g \cdot h \qquad (10.4.3)$$

where E_h is the hydraulic energy in joules or watt-secs. For water pumping systems, the hydraulic energy can be written as

$$\begin{aligned} E_h &= 1000 \times 9.81 \times Q \times h \\ &= 9810 Q h \end{aligned}$$

10.5 Hydraulic power

Hydraulic power is required to rate the pump, the motor, and the power electronics downstream up to and including the PV source. The hydraulic energy is given by equation 10.4.3, and from this the hydraulic power equation can be derived. Hydraulic power is the rate of hydraulic energy and is given by

$$\begin{aligned} P_h &= \frac{dE_h}{dt} \\ &= \frac{d(\rho Q g h)}{dt} \\ &= \rho \cdot \dot{Q} \cdot g \cdot h \end{aligned} \qquad (10.5.1)$$

where P_h is the hydraulic power and is expressed in watts. $\dot{Q}$ or $\frac{dQ}{dt}$ is the discharge rate which gives the volume of water flowing in unit time. It is also called the flow rate and is expressed in m^3/s. For the case of water pumping system, the hydraulic power can be expressed as

$$P_h = 1000 \times \dot{Q} \times 9.81 \times h$$
$$= 9810 \dot{Q} h$$

wherein h is considered as the total dynamic head for the water pumping system which includes the suction head, the delivery head and the head due to friction loss.

10.6 Friction head determination

Equation 10.3.2 gives the total dynamic head which is comprised of the suction head h_s, the delivery head h_d and the non-physical head that is equivalent to the friction loss in the pipe h_f. The suction head and the delivery head are physical quantities and hence are easily measurable. However, the friction head h_f is not directly measurable. This section discusses the method to estimate the friction head for a given water pumping system.

There is an empirical relationship proposed by Weisbach. It is popularly called the Darcy-Weisbach formula, and is given as

$$h_f = f \cdot \left(\frac{L}{d}\right) \cdot \left(\frac{u^2}{2g}\right) \qquad (10.6.1)$$

where
h_f = friction head, m
f = friction factor,
L = length pipe, m
d = inner diameter of pipe, m
u = velocity of fluid, m/s
g = gravitational acceleration, $9.81 m/s^2$

L is the length of pipe which includes the suction head length, the delivery head length and also the horizontal lengths of the pipe that do not contribute to the pressure heads. Among the parameters listed above, L, d, and u are easily measurable quantities, but friction factor f is an abstract factor. It is dependent on the type of fluid flow. The Reynolds number will provide the necessary discrimination to determine whether the flow is laminar or turbulent. Therefore,

Laminar flow: if the flow is laminar, then f can be determined by the following simple empirical formula

$$f = \frac{64}{R_e} \qquad (10.6.2)$$

where R_e is the Reynolds number and is determined using equation 9.6.13. One should consider kinematic viscosity $v = 1 \times 10^{-6} m^2/s$ for water while calculating the Reynolds number. The characteristic dimension X while calculating the Reynolds number will be the inner diameter of the pipe.

Turbulent flow: if the flow is turbulent, then f is determined by another empirical relationship called the Colebrook-White formula which is given as

$$\frac{1}{\sqrt{f}} = -2log_{10}\left(\frac{\varepsilon/d}{3.7} + \frac{2.51}{R_e \sqrt{f}}\right) \quad (10.6.3)$$

where d is the inner diameter of the pipe. ε is a material property which gives an indication of the smoothness of the inner walls of the pipe. The texture of the inner wall of the pipe is determined by the minute bumps of the material. The typical values for few common materials are

Material	ε, mm
PVC	0
Asbestos cement	0.012
Steel	0.1
Smooth concrete	0.4

The ratio ε/d is called the roughness ratio as it gives a measure of the roughness of the inner walls of the pipe.

The Colebrook-White formula is a transcendental equation and therefore one must numerically solve it. This is discussed in the next sub-section. For Reynolds number $R_e < 2000$ the flow can be considered laminar and for $R_e > 4000$, the flow is considered turbulent. For values in the range ($2000 < R_e < 4000$) the flow is in the transition region between laminar and turbulent and the friction factor is not clearly defined. For the case where the Reynolds number falls in the range ($2000 < R_e < 4000$), calculate the friction factor by both equations 10.6.2 and 10.6.3 and consider the higher among the two for design purposes.

10.6.1 Moody Chart

The friction factor f can be plotted using the Colebrook-White formula. f is plotted as a function of the Reynolds number using the roughness ratio as a parameter. Such a chart which gives the nomograph of friction factor as a function of Reynolds number is called the Moody chart. This is shown in figure 10.6.1.

Referring to the figure, the x-axis gives the sweep of the Reynolds number. The R_e value is varied from 1000 to 10^7. For a given value of the roughness ratio ε/d, the friction factor is calculated as per equation 10.6.3 and plotted. A family of curves for discrete values of the roughness ratio is plotted as shown in figure 10.6.1. Each of the curve is for a different roughness index.

Observe the curve for Reynolds number less than 2000. The friction factor for this is evaluated using equation 10.6.2. For Reynolds number greater than 4000, the

10.6. FRICTION HEAD DETERMINATION

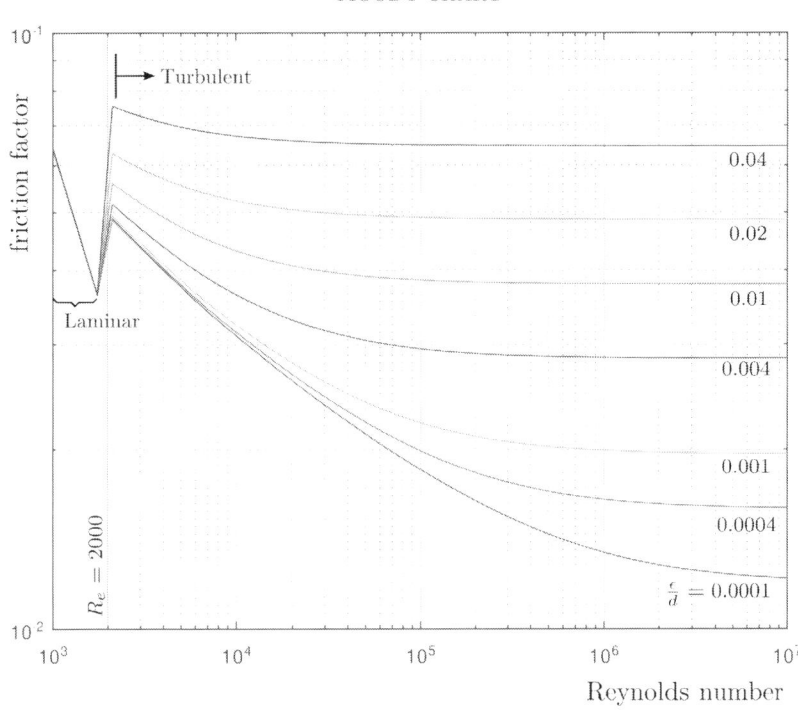

Figure 10.6.1: Moody Chart

friction factor is evaluated using equation 10.6.3. The Moody chart is used in the following manner,

Calculate the Reynolds number using equation 9.6.13.

Calculate the roughness ratio ε/d.

From the Moody chart select the curve closest to the roughness ratio calculated.

From the selected curve, for the calculated Reynolds number, read out the friction factor f.

Octave script for plotting the Moody chart

```
%----------------------------------------------------------------
% Description: This file calculates and plots the Moody Chart for
% friction factor estimation based on the Colebrook-White formula
clc
clear
```

```
%X-axis is Reynolds number
R=logspace(3,7);
%roughness factor
ed=[0.0001,0.0004,0.001,0.004,0.01,0.02,0.04];
for i=1:length(ed)
   for j=1:length(R)
f(i,j)=colebrook(R(j),ed(i));
   endfor
endfor
loglog(R,f);grid, xlabel('Reynolds number'),
ylabel('friction factor'),title('MOODY CHART')
%----------------------------------------------------------------
```

The above Octave script can be used to generate the Moody Chart. Generation of the Moody chart is based on the Colebrook-White formula. It uses a function named *colebrook()* which is discussed and listed in the next section.

10.6.2 Numerical solution to Colebrook-White formula

Another method to obtain the friction factor is to obtain a numerical solution to the Colebrook-White formula. This section will derive this solution.

Consider the following variable assignments,

$x = \frac{1}{\sqrt{f}}$, $a = \frac{\varepsilon/d}{3.7}$, and $b = \frac{2.51}{R_e}$. Applying these variable assignments to the equation 10.6.3, one has

$$x = -2\log_{10}(a+bx)$$

$$g(x) = x + 2\log_{10}(a+bx) = 0$$

It is for this equation one needs to find the root and the value of x. The Newton-Cotes numerical method can be used to obtain a solution for this equation. It is required to find the value of x such that $g(x) = 0$. Thus

$$\frac{\delta g}{\delta x} = g' = 1 + \frac{2b}{a+bx}$$

Using Newton-Cotes method, the iteration equation is

$$x_{k+1} = x_k - \frac{g(x_k)}{g'|_{x_k}}$$

10.6. FRICTION HEAD DETERMINATION

$$x_{k+1} = x_k - \left\{ \frac{x_k + 2\log_{10}(a+bx_k)}{1 + \frac{2b}{a+bx_k}} \right\}$$

Simplifying, one obtains

$$x_{k+1} = \frac{2bx_k - 2(a+bx_k) \cdot \log_{10}(a+bx_k)}{a+bx_k+2b} \qquad (10.6.4)$$

Equation 10.6.4 gives the numerical iteration equation to find the value x_k and in turn the friction factor f. In order to start the iteration one needs to have an initial starting value of x at the zeroth iteration i.e. x_0. Observe from the Moody chart that the friction factor does not go above 0.1. Assuming a high value of friction factor to start the iteration, one can set $x_0 = \frac{1}{\sqrt{0.1}}$. The iteration can be stopped by checking the value of $g(x_k)$. Compute $g(x_k)$ and exit the iteration loop if $g(x_k) < \varepsilon$ where ε is set as a very small value close to zero. After the iteration loop is completed then the friction factor is given by

$$f = \frac{1}{x_k^2}$$

Octave script for numerical solution for Colebrook-White formula

```
function [f]=colebrook(R,ed)
%-----------------------------------------------------------------
% Description: This file calculates the friction factor by using
%         the Colebrook-white equation
%
% SYNTAX: [f]=colebrook(R,ed)
% INPUTS
%      R = Reynolds number
%      ed = Roughness factor - ratio of height of surface bumps
%           to internal dia of pipe
% OUTPUT
% f = friction factor - (to be later used in Darcy-Weisbach
% formula for pipe friction loss)
%-----------------------------------------------------------------
error(nargchk(2,2,nargin));

%FLOW CHECK
if (R>0 && R<2000),
    %Flow is laminar
```

```
    f= 64/R; %friction loss for laminar flow
elseif (R>4000)
    %Flow is turbulent
a=ed/3.7;
b=2.51/R;
xk=1/sqrt(0.1); %x=1/sqrt(f) take start value of f as 0.1;
delta=xk;
while (delta>0.0001)
    xkk=(2*b*xk - (a+b*xk)*2*log10(a+b*xk))/(a +b*xk + 2*b);
delta=xkk-xk;
xk=xkk; %update xk with new value for iteration
endwhile
f=1/xk^2;

else
    %Flow is between laminar and turbulent 2000<R<4000
    %f is not defined here
    %f max of f with laminar and f with turbulent
flam= 64/R;
a=ed/3.7;
b=2.51/R;
xk=1/sqrt(0.1); %x=1/sqrt(f) take start value of f as 0.1;
delta=xk;
while (delta>0.0001)
    xkk=(2*b*xk - (a+b*xk)*2*log10(a+b*xk))/(a +b*xk + 2*b);
delta=xkk-xk;
xk=xkk; %update xk with new value for iteration
endwhile
fturb=1/xk^2;
f=max([flam,fturb]);% take the more conservative i.e.
                    % higher value
endif
endfunction
%---------------------------------------------------------------
```

10.7 Calculation steps for hydraulic power

The discussions of the previous sections are summarized here to provide a series of steps that lead to the calculation of hydraulic energy and power. This estimation of hydraulic power is essential in order to design the various components of water pumping system like the motor, DC-DC converters, and sizing of PV sources.

Step1: Set up the specifications of the water pumping system.

There are several variables and parameters that need to be defined and specified in order to calculate the hydraulic power. The variables that need to be specified are

10.7. CALCULATION STEPS FOR HYDRAULIC POWER

Q = discharge volume, m^3
$\triangle t$ = discharge time, s
d = inner diameter of pipe, m
$A = \frac{\pi}{4} d^2$, cross section area of the pipe, m^2
L = total length of the pipe, m
v = kinematic viscosity of water
ε = height of surface bumps within pipe inner wall, mm
h_s = suction head, m
h_d = discharge head, m

Step 2: Determine the discharge rate

$\dot{Q} = \frac{Q}{\triangle t}$ m^3/s

Step 3: Determine the water flow velocity in the pipes

$u = \frac{\dot{Q}}{A}$ m/s

Step 4: Determine the Reynolds number

$R_e = \frac{u \cdot d}{v}$

Step 5: Calculate the roughness ratio

roughness ratio = $\frac{\varepsilon}{d}$

Step 6: Estimate the friction factor

Moody charts - using the value of roughness ratio, select the curve and from the value of Reynolds number pick out the friction factor.

Laminar flow: for $R_e < 2000$ $f = \frac{64}{R_e}$

Turbulent flow: for $R_e < 2000$ $f = \frac{1}{x_k^2}$; where $x_{k+1} = \frac{2bx_k - 2(a+bx_k) \cdot log_{10}(a+bx_k)}{a+bx_k+2b}$, $a = \frac{\varepsilon/d}{3.7}$, and $b = \frac{2.51}{R_e}$

for $2000 < R_e < 4000$, take f as higher of that calculated for laminar flow and turbulent flow.

Step 7: Calculate the friction head

$h_f = f \cdot \left(\frac{L}{d}\right) \cdot \left(\frac{u^2}{2g}\right)$ m

Step 8: Calculate the dynamic head

$h = \left(h_s + h_d + h_f\right)$ m

Step 9: Calculate the hydraulic power

$P_h = 9810 \cdot \dot{Q} \cdot h$ $watts$

Step 10: Calculate the hydraulic energy

$E_h = 9810 \cdot Q \cdot h$ $joules$

10.8 Example 1

Consider a very common example that can be seen in most homes in the country. There will be an underground sump to store water. The water pump system will lift the water from the underground sump to the overhead tank which is placed on the roof top. Figure 10.8.1 shows the water pump system with typical dimensions. The pump is a centrifugal pump which is driven by a single phase induction motor drawing power from the main grid.

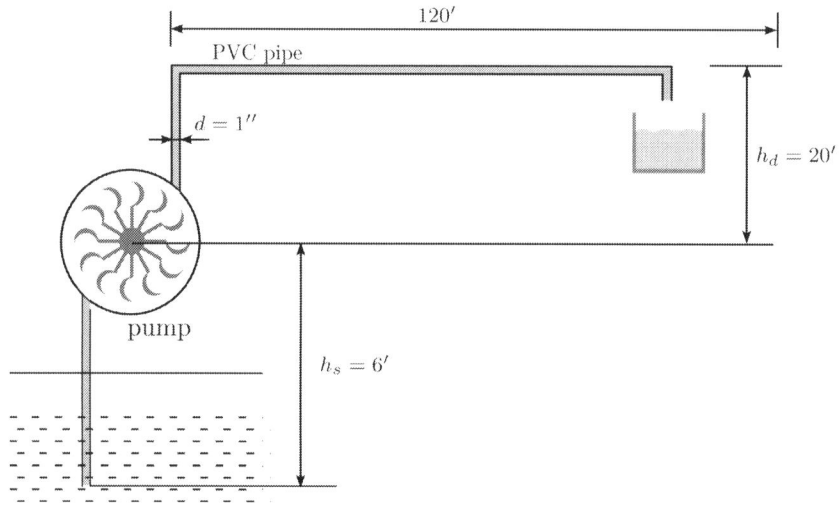

Figure 10.8.1: Example of home water pumping system

The suction head is 6ft as shown. The delivery head is 20ft. However the delivery pipe travels a length of 120ft horizontally as shown in figure. This does not contribute to the physical delivery head, but it does contribute to the friction head. The pipe is 1 inch in diameter and is made of PVC material. Let the overhead tank capacity be 1000 litres and it needs to be filled up in 10 minutes. It is required to calculate the hydraulic power so that the rating of the pump can be estimated. Following the steps outlined in the previous section, one has

Step1: Set up the specifications of the water pumping system.

There are several variables and parameters that need to be defined and specified in order to calculate the hydraulic power. The variables that need to be specified are

$Q = 1000 \text{ lt} = 1 m^3$
$\triangle t = 10*60 = 600 s$
$d = 1 \text{ inch} = 1 \times \frac{25.4}{1000} = 0.0254 m$
$A = \frac{\pi}{4} d^2 = 5.067 \times 10^{-4} m^2$
$L = \text{6ft} + \text{20ft} + \text{120ft} = 146 \times \left(\frac{12 \times 25.4}{1000}\right) = 44.5 m$
$v = 1 \times 10^{-6} m^2/s$
$\varepsilon = 0$ for PVC

$h_s = 6\ m$
$h_d = 20\ m$

Step 2: Determine the discharge rate

$\dot{Q} = \frac{Q}{\Delta t} = \frac{1}{600} = 1.67 \times 10^{-3} m^3/s$

Step 3: Determine the water flow velocity in the pipes

$u = \frac{\dot{Q}}{A} = \frac{1.67 \times 10^{-3}}{5.067 \times 10^{-4}} = 3.3 m/s$

Step 4: Determine the Reynolds number

$R_e = \frac{u \cdot d}{v} = \frac{3.3 \times 0.0254}{1 \times 10^{-6}} = 83820$

Step 5: Calculate the roughness ratio

roughness ratio $= \frac{\varepsilon}{d} = 0$

Step 6: Estimate the friction factor

Using the colebrook function in Octave, $f = 0.018683$

Step 7: Calculate the friction head

$h_f = f \cdot \left(\frac{L}{d}\right) \cdot \left(\frac{u^2}{2g}\right) = 18.05\ m$

Step 8: Calculate the dynamic head

$h = (6+20) \cdot \left(\frac{12 \times 25.4}{1000}\right) + 18.05 = 26\ m$

Step 9: Calculate the hydraulic power

$P_h = 9810 \cdot \dot{Q} \cdot h = 9810 \times 1.67 \times 10^{-3} \times 26 = 425\ watts$

The water pump should handle a load of 425W as calculated above.

Step 10: Calculate the hydraulic energy

$E_h = 9810 \cdot Q \cdot h = 9810 \times 1 \times 26 = 255060\ joules$

10.9 Example 2

In this example let the centrifugal pump be driven by a dc motor that is drawing power from a photovoltaic source. All other parameters are same as in the previous example. The system is same as shown in figure 10.8.1. A change from the previous example is the size of the overhead tank. It is now a 2000 litre capacity tank. Consider that in a day 1000 litres of water is consumed from the overhead tank by the household. This implies that 1000 litres need to be lifted up from the sump and delivered to the overhead tank in a day to make up for the 1000 litres consumption.

In this example the time for filling up the tank is linked to the insolation as the dc motor is powered from the the PV source through a DC-DC converter. The insolation at the place is plotted and shown in figure 10.9.1. The insolation in kW/m^2 is an inverted bell shaped curve as shown and it spans the daylight hours. The area under the curve is the total energy captured within the daylight hours in $kWh/m^2/day$. This is nothing but H_{at} the energy incident over the day at the place inclusive of the atmospheric effects on a tilted surface. If one considers standard insolation of $1kW/m^2$ incident at the place, then the equivalent rectangular insolation curve is also shown in the figure. The width of the rectangular insolation curve on the time scale is numerically equal to H_{at} hours.

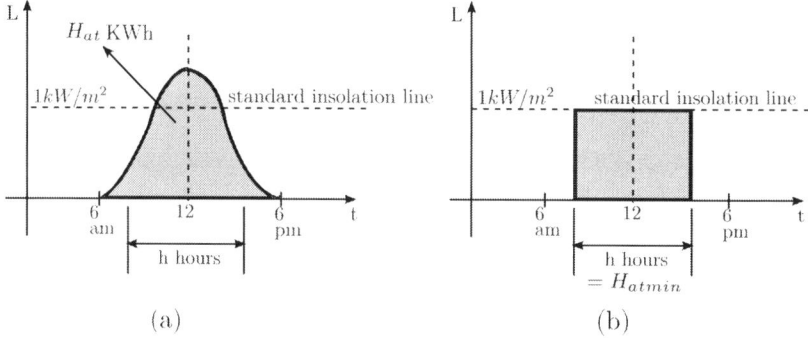

Figure 10.9.1: Example of a water pumping system

For the given place the insolation data provided gives $H_{at} = 4.58 kWh/m^2/day$. This implies that 1000 litres must be lifted up to the overhead tank in 4.58 hours. In order to be on the safe side, consider that the overhead tank should be filled up with 1000 litres in 4 hours time. Given these conditions and setup, what is the power rating of the water pump system?

Following the steps outlined in the previous section, one has

Step1: Set up the specifications of the water pumping system.

The variables that need to be specified are
$Q = 1000 \text{ lt} = 1 m^3$
$\Delta t = 4 hrs \times 3600 = 14400 \text{ s}$
$d = 1 \text{ inch} = 1 \times \frac{25.4}{1000} = 0.0254 m$
$A = \frac{\pi}{4} d^2 = 5.067 \times 10^{-4} m^2$
$L = 6ft + 20ft + 120ft = 146 \times \left(\frac{12 \times 25.4}{1000}\right) = 44.5 m$
$\upsilon = 1 \times 10^{-6} m^2/s$
$\varepsilon = 0$ for PVC
$h_s = 6 m$
$h_d = 20 m$

Step 2: Determine the discharge rate

$\dot{Q} = \frac{Q}{\Delta t} = \frac{1}{14400} = 7 \times 10^{-5} m^3/s$

Step 3: Determine the water flow velocity in the pipes

$$u = \frac{\dot{Q}}{A} = \frac{7 \times 10^{-5}}{5.067 \times 10^{-4}} = 0.1381 m/s$$

Step 4: Determine the Reynolds number

$$R_e = \frac{u \cdot d}{v} = \frac{0.1381 \times 0.0254}{1 \times 10^{-6}} = 3508.9$$

Step 5: Calculate the roughness ratio

roughness ratio $= \frac{\varepsilon}{d} = 0$

Step 6: Estimate the friction factor

Using the colebrook function in Octave, $f = 0.0415$

Step 7: Calculate the friction head

$$h_f = f \cdot \left(\frac{L}{d}\right) \cdot \left(\frac{u^2}{2g}\right) = 0.07 \ m$$

Step 8: Calculate the dynamic head

$$h = (6+20) \cdot \left(\frac{12 \times 25.4}{1000}\right) + 0.07 = 8 \ m$$

Step 9: Calculate the hydraulic power

$$P_h = 9810 \cdot \dot{Q} \cdot h = 9810 \times 7 \times 10^{-5} \times 8 = 5.45 \ watts$$

Step 10: Calculate the hydraulic energy

$$E_h = 9810 \cdot Q \cdot h = 9810 \times 1 \times 8 = 78480 \ joules$$

By spreading the discharge time period over the entire insolation time span in a day, there is significant saving in energy and wattage requirement for the pump-motor system.

¨

10.10 Example 3

Figure 10.10.1 another water pumping example. Here water is being drawn from a borewell and discharged into a reservoir at ground level. This topology is generally used for irrigation in agricultural fields. A submersible pump is used to pump water from the underground water reservoir. In this example suction head is zero as the pump is submerged in the underground water. The water pipe is a steel pipe with 4 inch inner diameter. The discharge required is 100 litres in half a minute.

Following along the same steps as before, one has

Step1: Set up the specifications of the water pumping system.

The variables that need to be specified are
$Q = 100 \ lt = 0.1 m^3$
$\triangle t = 30 \ s$
$d = 4 \ inch = 4 \times \frac{25.4}{1000} = 0.1016 m$

CHAPTER 10. WATER PUMPING WITH PHOTOVOLTAIC

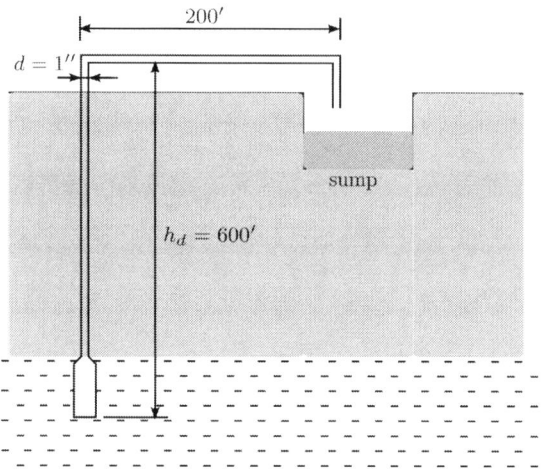

Figure 10.10.1: Example of a borewell system

$A = \frac{\pi}{4}d^2 = 8.107 \times 10^{-3} m^2$
$L = 0\text{ft} + 600\text{ft} + 200\text{ft} = 800 \times \left(\frac{12 \times 25.4}{1000}\right) = 243.84 m$
$v = 1 \times 10^{-6} m^2/s$
$\varepsilon = 0.1 mm$ for steel
$h_s = 0\ m$
$h_d = 600\ m$

Step 2: Determine the discharge rate

$\dot{Q} = \frac{Q}{\Delta t} = 3.33 \times 10^{-3} m^3/s$

Step 3: Determine the water flow velocity in the pipes

$u = \frac{\dot{Q}}{A} = 0.4112 m/s$

Step 4: Determine the Reynolds number

$R_e = \frac{u \cdot d}{v} = \frac{0.4112 \times 0.1016}{1 \times 10^{-6}} = 41773$

Step 5: Calculate the roughness ratio

roughness ratio $= \frac{\varepsilon}{d} = 9.8 \times 10^{-4}$

Step 6: Estimate the friction factor

Using the colebrook function in Octave, $f = 0.0246$

Step 7: Calculate the friction head

$h_f = f \cdot \left(\frac{L}{d}\right) \cdot \left(\frac{u^2}{2g}\right) = 0.51\ m$

Step 8: Calculate the dynamic head

10.11. PHOTOVOLTAIC SIZING

$h = (600) \cdot \left(\frac{12 \times 25.4}{1000}\right) + 0.51 = 183.4 \ m$

Step 9: Calculate the hydraulic power

$P_h = 9810 \cdot \dot{Q} \cdot h = 9810 \times 3.33 \times 10^{-3} \times 183.4 = 5997 \ watts$

10.11 Photovoltaic sizing

Consider the block schematic of a typical water pumping system as shown in figure 10.11.1. It consists of a hydraulic system composed of pipes and reservoirs that require a hydraulic power of P_h watts to lift water from a lower level to a higher level. This is the load for the pump. The pump has an efficiency of η_p. As a consequence the load requirement for the motor is $\frac{P_h}{\eta_p}$.

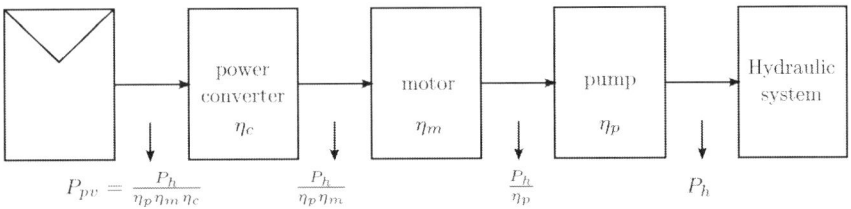

Figure 10.11.1: System blocks and efficiency

The PV or photovoltaic source powers the water pumping system. The output of the PV sources is processed through a DC-DC converter or a DC-AC inverter depending on the type of the motor being used for driving the pump. The motor and pump along with the hydraulic system will act as the load for the power converter. If η_m is the efficiency of the motor, then the load power requirement for the power converter is $\frac{P_h}{\eta_m \eta_p}$. Further, there is an efficiency associated with the power converter itself. If the efficiency of the power converter is η_c, then the load power requirement for the PV source is $P_{pv} = \frac{P_h}{\eta_c \eta_m \eta_p}$. Thus the PV source should be sized such that the losses in the power converter, motor and pump are also accounted apart from the hydraulic power requirement.

At the time of design if data is not available on the efficiencies of the power converter, motor and the pump, one can consider the pump efficiency as 70% and motor efficiency to be around 80% and the power converter efficiency to be around 90%. The the power requirement for the PV source works out to be

$$P_{pv} = \frac{P_h}{\eta_c \eta_m \eta_p} = \frac{P_h}{0.9 \times 0.8 \times 0.7} \approx 2 \cdot P_h \qquad (10.11.1)$$

Therefore from equation 10.11.1 one can conclude that after calculating the hydraulic power, the PV source should be sized for double the calculated hydraulic power so that the losses in the power converter, motor and pumps are accounted for.

10.12 Pumped hydro application

An application that extends the water pumping system is the pumped hydro. This uses the same components but the power flow is bidirectional. This is illustrated in figure 10.12.1. The system consists of the PV source followed by a DC-DC converter. This is interfaced to a DC motor which drives a centrifugal pump as shown. The centrifugal pump lifts water from a lower level reservoir to a higher level reservoir. This is the operation of a water pumping system that has been discussed in detail in this chapter. The power flow is from the PV source to the water which is pumped up. Power is drawn from the PV source.

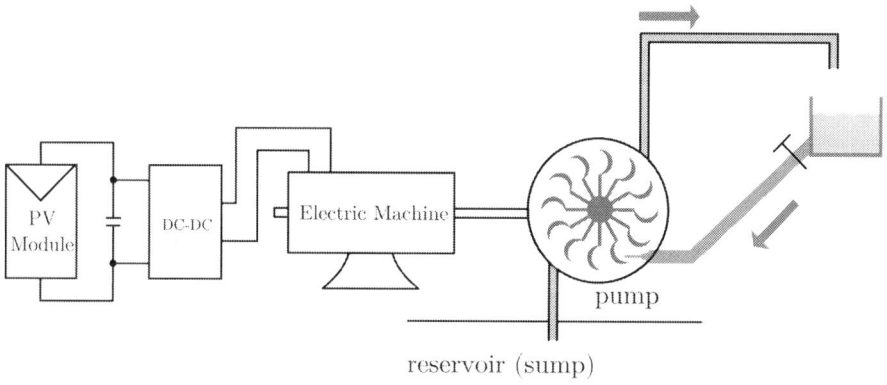

Figure 10.12.1: Single unit pumped hydro

The power can also flow in the reverse direction. In the process of pumping up the water from a lower level reservoir to a higher level reservoir, the potential energy of the mass of water lifted up is raised through the suction and delivery heads. The water that is stored in the higher level reservoir contains stored potential energy (mgh). This stored potential energy of water can be converted to kinetic energy by allowing the water to flow down a penstock and drive the centrifugal pump as a turbine. The water jet from the nozzle of the penstock will impinge on the centrifugal pump vanes and operate it as a water turbine. The water turbine is connected to the shaft of the motor and hence the motor shaft will rotate and make the motor to behave as a DC generator. If a bidirectional DC-DC converter is used, then the DC output power from the DC generator can flow back through the DC-DC converter and charge up the capacitor bank or battery that is connected at the input of the DC-DC converter. Here the pumped up water is used for generation of electricity by virtue of the stored potential energy in the pumped up water. It acts as a hydro power station when power flows in reverse. It acts as a water pumping system when power flows in the forward direction. This is the concept of the pumped hydro.

The system of figure 10.12.1 can be used for small power systems. There must be a capacitor bank or battery set connected to the output of the PV source in order to receive power from the pumped up hydro. From the battery set, power can be fed to other electrical loads. When the system is operating as a pump and is lifting water

10.12. PUMPED HYDRO APPLICATION

up to the higher level reservoir, the control valve in the penstock should be closed. During the time when the system is acting as a hydro generator, then the control valve is open. The control valve on the penstock can be regulated to control the generation power according to the load demand. During the time when the system is operating as a pumping system, it cannot operate in the hydro generation mode and vice-versa.

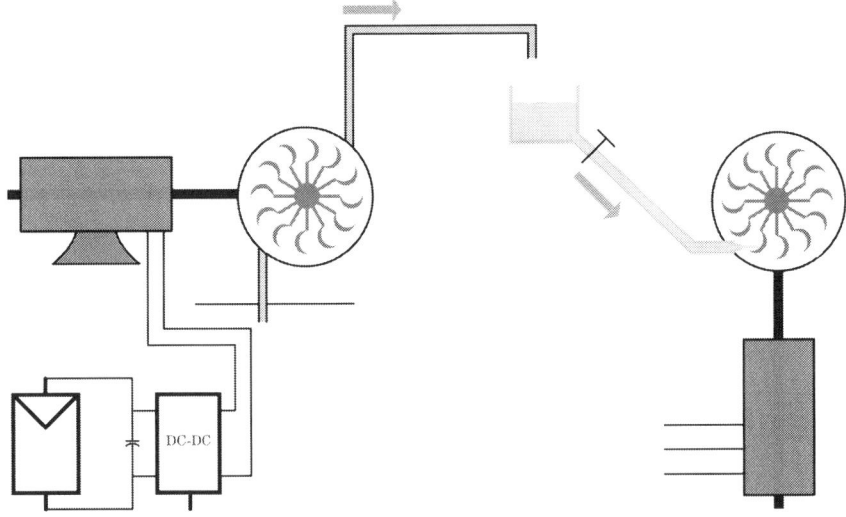

Figure 10.12.2: Dual unit pumped hydro

Consider the system shown in figure 10.12.2. This is a dual unit pumped hydro system. The water pumping portion and the hydro generation portion are decoupled. Here, the PV power is interfaced to a DC-DC converter which drives a DC motor which in turn drives a centrifugal pump in order to work as a pumping system. Water is lifted from the lower level reservoir up to to a higher level reservoir.

From the higher level reservoir, a penstock connects it to a water turbine. The water flow in the penstock is controlled by a valve as shown. The water from the jet at the end of the penstock impinges on the pelton wheels of the water turbine which drives the shaft of the motor. The motor can be a DC or AC motor depending on the application. The output of the motor drives the electrical loads. In this system, the water pumping part is decoupled from the hydro power generation part. They can act independently and simultaneously. This is more suited for higher power systems.

An advantage with the dual unit pumped hydro is that the power and rate of discharges can be differently rated for the two parts. The pumping part can be linked to the insolation. The PV power to pump up the water can be lower and the discharge can be spread over the entire daylight hours. The hydro generation part can come into action during few hours in the night time. The hydro generation part can be designed for a higher power as the night power may be needed only for a shorter duration of discharge. The hydro generation part is called a hydel system. A pico hydel system is at power levels of 5kW, a micro hydel system is at the power levels of around 50kW. For small homes and community usage, pico hydel system is an appropriate pumped hydro system to use.

10.12.1 Pico hydel system

Consider the hydro generation portion of the dual unit pumped hydro shown in figure 10.12.2. PV power is used to store the water in the higher level reservoir. The hydel part is rated for typical households and can be considered as a pico hydel system. Let the height from the higher level reservoir to the centre of the axis of the jet at the end of the penstock be H (to distinguish it from the dynamic head h). The discharge rate or the flow rate of the water at the jet end of the penstock is $\dot{Q}$. The hydraulic energy converted from potential to kinetic is $E_h = \rho \dot{Q} g H$ joules and the hydraulic power generated is $P_h = \rho \dot{Q} g H$ watts.

Consider a pelton wheel turbine as shown wherein the jet at the end of the penstock is impinging on the pelton wheel and converting the kinetic energy of water to mechanical energy. Let the velocity at the jet be u_j and the tangential turbine velocity at the circumference of the pelton wheels be u_t. The angular speed of the turbine is ω_t rad/s. Using conservation of momentum, the tangential force on the turbine is given as

$$F_t = 2\rho \dot{Q}(u_j - u_t) \tag{10.12.1}$$

The hydraulic energy E_h is converted to kinetic energy at the jet and is $\frac{1}{2}\rho Q u_j^2 = \rho Q g H$. Therefore,

$$u_j^2 = 2gH \tag{10.12.2}$$

The cross section at the jet a_j is measurable. Therefore, the jet velocity is given as

$$u_j = \frac{\dot{Q}}{a_j} \tag{10.12.3}$$

In equation 10.12.1 substituting for $\dot{Q}$ and u_j using equations 10.12.2 and 10.12.3, the tangential force can be expressed as a function of tangential velocity alone. Thus,

$$F_t = 2\sqrt{2gH}\rho a_j \left(\sqrt{2gH} - u_t\right) \tag{10.12.4}$$

The mechanical power P_m is efficiency times the jet power P_j. Thus

$$P_m = \eta P_j = \eta F_t u_t \tag{10.12.5}$$

Using equation 10.12.4 in equation 10.12.5, one can write the mechanical power in the form

$$P_m = \alpha_1 u_t - \beta_1 u_t^2 \tag{10.12.6}$$

10.12. PUMPED HYDRO APPLICATION

where α_1 and β_1 are constants.

The tangential velocity can be expressed in terms of the measurable angular velocity of the turbine shaft. If r is the radius of the turbine impeller, then $u_t = \omega_t \cdot r$. The mechanical shaft power can now be expressed as

$$P_m = \alpha \omega_t - \beta \omega_t^2 \qquad (10.12.7)$$

where

$$\alpha = 4\eta \rho a_j gHr$$
$$\beta = 2\sqrt{2gH}\eta \rho a_j r^2$$

The developed shaft torque T_d is given as

$$T_d = \frac{P_m}{\omega_t} = \alpha - \beta \omega_t \qquad (10.12.8)$$

Figure 10.12.3 shows the torque curve as a function of the shaft speed ω_t. It is a linear curve with a negative slope. The figure also shows the mechanical shaft power as a function of shaft speed. It is parabolic in shape as shown. If the generation is being used for 50Hz AC type of loads, then the peak point point should correspond to $\omega_t = \omega_s = 2\pi 50$ rad/s.

At shaft torque of zero, the angular velocity is $\frac{\alpha}{\beta}$. This should occur at the shaft torque of $\omega_t = 2\omega_s$, so that the peak power occurs at $\omega_t = \omega_s$.

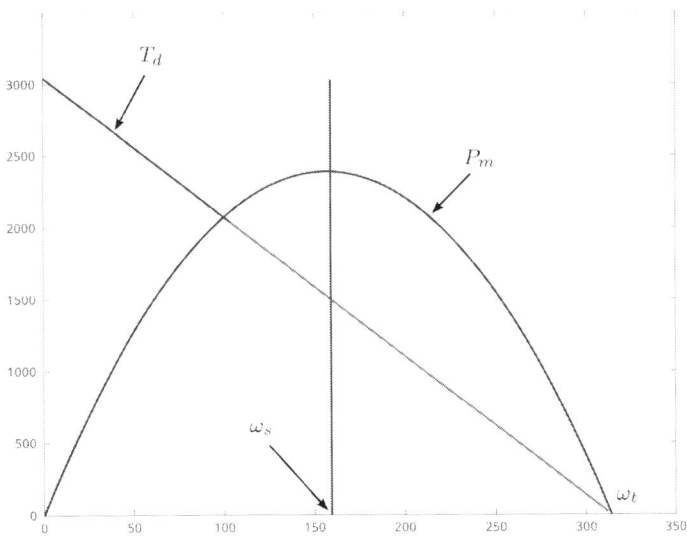

Figure 10.12.3: Pico hydel turbine torque-speed characteristic

From the peak power constraint one can note that $\omega_t = 2\omega_s$, and from the developed torque constraint, it can be noted that $\omega_t = \frac{\alpha}{\beta}$. These two may not match.

Therefore, a gearbox is need to match these two constraints. Let n_{gear} be the gear ratio of the gear box, then

$$n_{gear} \cdot \frac{\alpha}{\beta} = 2\omega_s$$

and thus,

$$n_{gear} = \frac{2\omega_s}{\frac{\alpha}{\beta}} \quad (10.12.9)$$

The torque at the generator shaft will be $\left(\frac{T_d}{n_{gear}}\right)$.

For pico hydel system, the commonly used generator for household electrical load applications is an induction generator. The power that flows into the shaft of the induction generator is electrically coupled to the AC loads using a simple open delta configuration as shown in figure 10.12.4.

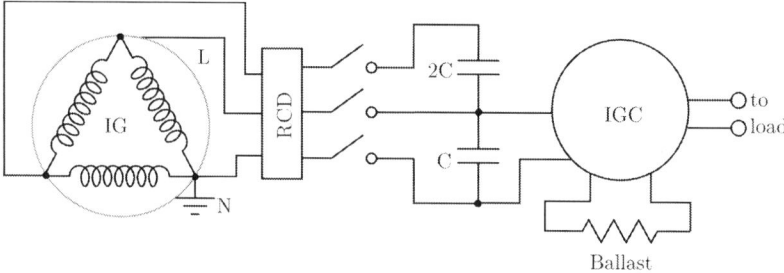

Figure 10.12.4: Induction generator interface to load

The induction generator is wound in delta as shown. The output of the induction generator are connected to a residual current detector and a multi-pole switch. Two capacitors of value C and 2C are connected in open delta configuration as shown. The output is passed through a induction generator control unit and then to the load.

The function of the induction generator controller (IGC) is to ensure that the load that is presented to the induction generator is approximately constant. A resistive ballast is usually used to perform the load balancing. When the output load is less then more of the ballast load is brought in and when the output load is high, then some part of the ballast load is cut out in order to present a constant load to the induction generator. If this IGC is not performed, then the frequency and the voltage amplitudes of the induction generator can swing over a wide range which could potentially damage the loads. Therefore, IGC is essential.

The ballast load can become hot. The ballast load is cooled by the water from the penstock that flows down into the stream or the lower level reservoir. The water is routed to flow over the ballast load to remove the heat generated by the ballast to maintain load balance.

10.13 Questions

1. A water tank (A) is on a building at 10m above the ground. Another water tank (B) is on another building at a height of 25m above ground. A motor pump set placed on the ground is supposed to pump water from tank A to tank B. The suction head for this system is
   a) 0 m
   b) 10m
   c) 15m
   d) -15m
   e) 25m

2. A water tank (A) is on a building at 10m above the ground. Another water tank (B) is on another building at a height of 25m above ground. A motor pump set placed on the ground is supposed to pump water from tank A to tank B. The total physical head for this system is
   a) 0 m
   b) 10m
   c) 15m
   d) -15m
   e) 25m

3. A ballast load is used along with stand-alone induction generator
   a) for startup
   b) for better flux control
   c) for maintaining constant frequency irrespective of load
   d) for maintaining frequency inversely proportional to load

4. Capacitor bank is connected to the terminals of a stand alone induction generator (IG)
   a) To supply magnetizing current to IG
   b) To keep IG frequency constant
   c) To improve transient stability of IG
   d) To maintain the voltage constant

5. Which of the following Reynolds number is a perfect range for turbulent liquid flow?
   a) $[0, \infty)$
   b) $[0, 2000]$
   c) $(2000, 4000)$
   d) $[4000, \infty)$

6. Which of the following represents correct analogy between electrical variables and hydraulic variables?
   a) (voltage : rate of discharge), (current: head)
   b) (voltage : discharge), (current : head)
   c) (voltage : pressure), (current : flow rate)
   d) (voltage : head), (current: flow)

7. What is the roughness factor for a steel pipe with a 0.8 inch diameter?
   a) 0
   b) 0.0049
   c) 0.0006
   d) 0.02

8. 10 litres of water has to be lifted to a height of 10m. The energy needed for the task is ······.

9. The potential energy of a body of water is increased by 10000 joules by lifting it through 100m height. The mass of the water body is ······.

10. Water has to be lifted to a height of 30m from an underground sump at the rate of 10 litre/s. Consider that the system is lossless. The power required by the prime mover is ·····.

11. In a water pumping system, the water is being pumped from a sump to an overhead tank situated 25m above ground. The sump bottom is 2m below ground. The motor-pump system is located at ground level. The water is being pumped at the rate of 24.6 litres/sec. The pipe inner diameter is 10 cm. The pipe is placed completely vertical with no horizontal part. If the friction factor is 0.037, the head equivalent of friction loss is ········.

12. In a water pumping system, the water is being pumped from a sump to an overhead tank situated 25m above ground. The sump bottom is 2m below ground. The motor-pump system is located at ground level. The water is being pumped at the rate of 24.6 litres/sec. The pipe inner diameter is 10 cm. The pipe is placed completely vertical with no horizontal part. If the friction factor is 0.037, the total dynamic head is ········.

13. In a water pumping system, the water is being pumped from a sump to an overhead tank situated 25m above ground. The sump bottom is 2m below ground. The motor-pump system is located at ground level. The water is being pumped at the rate of 24.6 litres/sec. The pipe inner diameter is 10 cm. The pipe is placed completely vertical with no horizontal part. The friction factor is 0.037. The efficiencies of the pump, motor and dc-dc converter are 70%, 80% and 90% respectively. If the system is being powered by a PV source, what is the output power requirement for the PV panels?

14. It is expected to dispatch 10000 liters of water to a storage system that is 50m high within 20 minutes. The diameter of the steel pipe is 1.5 inch. What is the Reynold number for the water flow? (kinematic viscosity of water is $1 \times 10^{-6} m^2/s$)

15. What is the total dynamic head of the system, if the friction factor for the flow is 0.0135 in problem of previous question?

16. A 400V DC motor-pump is used to pump water to an overhead storage tank. The hydraulic power required by the system is 5 kW. If the pump and motor system are 60% efficient, what is the rated armature current of the DC motor?

10.13. QUESTIONS

17. In a small hydro electric scheme, water is held in a reservoir 200m above the power station. The conversion efficiency from the potential energy to electric energy is 82%. If the mass flow rate is 25kg/s, what is the electric power output?

18. A community has 500 people. Per capita water consumption is 40 litres/day. Borewell depth is 20m and storage tank height is 2m above ground. Find the daily hydraulic energy required.

19. A community has 500 people. Per capita water consumption is 40 litres/day. Borewell depth is 20m and storage tank height is 2m above ground. The motor-pumpset is supplied by a PV array. The incident energy at the place is 14MJ/m^2/day. The motor-pump subsystem efficiency is 45%. Find the power requirement of the PV array.

20. A community has 500 people. Per capita water consumption is 40 litres/day. Borewell depth is 20m and storage tank height is 2m above ground. This motor pumpset is supplied by a PV array. The incident energy at the place is 14MJ/sqm/day. The motor-pump subsystem efficiency is 45%. Find the flow rate of the water being pumped.

Chapter 11

Grid Interaction

11.1 Introduction

The power grid is the main source of power for electrical loads. It is important that the photovoltaic sources interact with the grid in order to make an effective contribution. The photovoltaic source can be used for standalone applications like peltier cooling, water pumping systems, battery charging and other applications. However, the PV source should interact actively with the grid to become a major supplementary source. The roof top photovoltaic systems that are so popular is an example of grid interaction. The PV source can be connected to a single phase grid or a three phase grid. At the home roof top level, the PV sources are connected to the single phase grid. Large PV fields consisting of several arrays of PV modules are connected to three phase grids and provide significant supplementary power to the electrical grid.

11.2 Interconnection principle

PV power is DC power and cannot be directly connected to the AC grid. The PV power has to be processed through an inverter to make it compatible to the AC power before it can be interfaced to the AC grid. Consider the circuit of figure 11.2.1(a). The rhombus shaped voltage source is a controlled voltage source that generates a voltage v_i. The circular shaped voltage source represents the grid having voltage v_g. An inductor L acts as the interfacing impedance between the two sources. The controlled voltage source draws power from a PV source which is processed and converted to AC voltage by means of a DC-AC inverter.

A direct connection of the two voltage sources without the interfacing impedance can be catastrophic. There will always be some difference in voltage between the sources. The voltage sources will have very low to zero impedances, and therefore there will be a large circulating current flowing even for very small voltage differences and cause catastrophic failure. The two voltage sources need to be interfaced by an impedance like an inductance in the simplest form as shown in figure 11.2.1(a). Let the grid voltage be given as

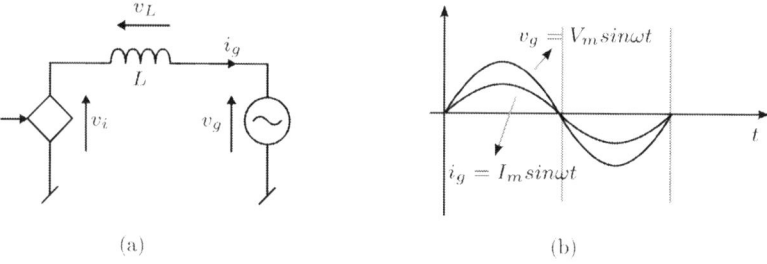

Figure 11.2.1: Interconnection principle

$$v_g = V_m \cdot sin(\omega t) \quad (11.2.1)$$

The current i_g is injected into the grid by drawing power from the controlled voltage source v_i which in-turn will draw power from a PV source. Let the current i_g be injected such that it is in phase with the grid voltage v_g. This means that the current injection into the grid is at unity power factor. Therefore, it is required that the current injected into the grid should be

$$i_g = I_m \cdot sin(\omega t) \quad (11.2.2)$$

A inductor of inductance L is interfacing the controlled voltage source and the grid source. The current i_g flows through the inductor and the voltage across the inductor v_L is given as

$$\begin{aligned} v_L &= L \cdot \frac{di_g}{dt} \\ &= \omega L \cdot I_m \cdot cos(\omega t) \end{aligned} \quad (11.2.3)$$

From equations 11.2.1 and 11.2.3, the voltage that is desired from the controlled voltage source v_i in order to inject a current into the grid at unity power factor is given as

$$\begin{aligned} v_i &= v_L + v_g \\ &= \omega L \cdot I_m \cdot cos(\omega t) + V_m \cdot sin(\omega t) \end{aligned} \quad (11.2.4)$$

Figure 11.2.2 shows the voltage waveforms of the controlled voltage source v_i, the grid source v_g and the voltage across the inductor v_L that will provide unity power factor current injection. Referring to equation 11.2.4, it can be seen that the part $V_m \cdot sin(\omega t)$ will cancel out the grid voltage and essentially the part $\omega L \cdot I_m \cdot cos(\omega t)$ i.e. the voltage across the inductor is the one instrumental in injecting the appropriate value of the current into the grid source. The control block schematic for this grid

11.2. INTERCONNECTION PRINCIPLE

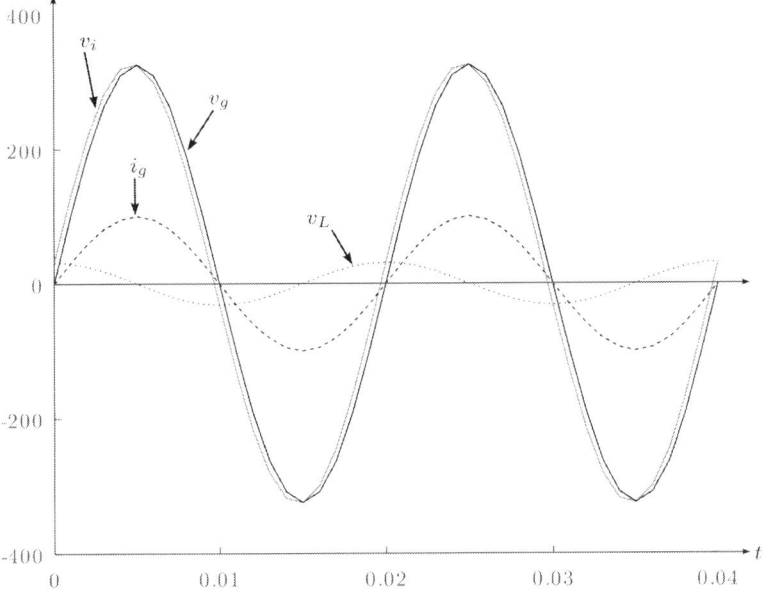

Figure 11.2.2: Voltage waveforms for UPF current injection

connection system is as shown in figure 11.2.3. Changing the value of I_m will control the amplitude of the current injection. As the grid voltage has a fixed rms value, the amount of current injection into the grid is controlled by I_m. The phase angle of the injected current can be adjusted by controlling the value of ϕ. For unity power factor injection, $\phi = 0$. Thus equation 11.2.4 can be generalised as

$$v_i = \omega L \cdot I_m \cdot cos(\omega t + \phi) + V_m \cdot sin(\omega t) \tag{11.2.5}$$

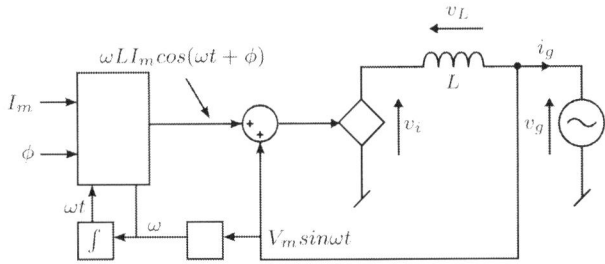

Figure 11.2.3: Block schematic of the grid interaction system

11.3 Controlled inverter source

The block schematic shown in figure 11.2.3 gives a visualisation of the principle of operation. However, the implementation is slightly more complex. The controlled voltage source in practice is not a continuous domain system wherein the output voltage is continuous as shown in figure 11.2.2. The voltage source is built using power semiconductor switches like MoSFETs or IGBTs or similar such components. The power switches are switched either ON or OFF through a pulse width modulation (PWM) strategy. The output of the controlled source will be switching between a high value and a low value. The pulse width modulation will ensure that the average value of the switched output waveform will be equal to the voltage as given in equation 11.2.4.

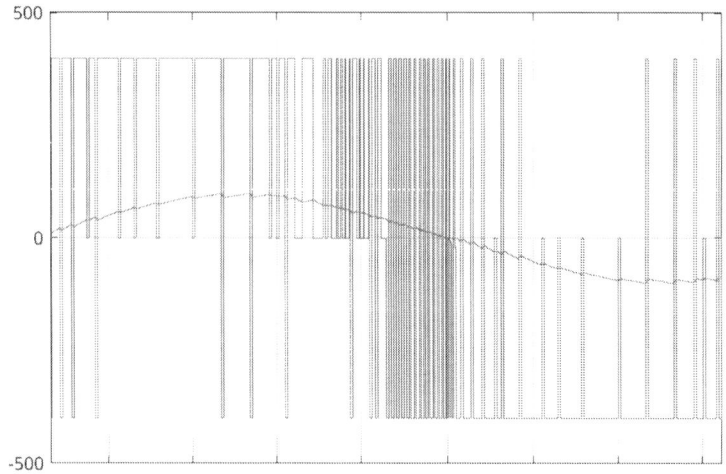

Figure 11.3.1: Inverter voltage and current waveforms

For a controlled voltage source that is a pulse width modulated DC-AC inverter, the inverter output voltage swings between DC positive rail and the negative rail. The inverter output voltage is as shown in figure 11.3.1. The voltage across the interfacing inductor is the difference between the pulse width modulated inverter output voltage and the sinusoidal or cosinusoidal grid voltage. The inductor voltage will have the switching component of the inverter side present. However, the inductor current is the integral of the voltage across the inductor and hence it will be significantly filtered before being injected into the grid. The inductor current waveform is also shown in figure 11.3.1.

The value of the inductor will decide the current ripple of the current that will be injected into the grid. The switching frequency of the inverter switches will also have a bearing on the current ripple component. The inductor value is designed based on the inductor current ripple requirement.

$$L = \frac{v_L}{\left(\frac{\triangle i_g}{\triangle t}\right)} \quad (11.3.1)$$

In equation 11.3.1, v_L is the voltage that appears across the inductor resulting in worst case current ripple. $\triangle i_g$ is the maximum ripple in the inductor current that occurs during the inductor current peak value and $\triangle t$ is the maximum time when the voltage across the inductor is maximum. Let V_{dc} be the dc link voltage of the inverter, m is the modulation index and the worst case $\triangle t$ is the switching period itself, which is T_s when $m = 1$. Then

$$L >= \frac{(V_{dc} - V_m) \cdot m \cdot T_s}{\triangle i_g}$$

$\triangle i_g$ can be considered to be 10% of peak of the maximum grid injection current i.e. 10% of I_m.

For example consider a single phase inverter being interfaced to the grid through an inductor. The dc link voltage is 400V. The grid voltage is sinusoidal in nature with a peak of 325V. A current of 100A peak is to be injected into the grid. The controlled inverter switching frequency is 20kHz and the current ripple due to switching should be less than 3A. The inductor value is

$$L >= \frac{(400 - 325) \cdot (325/400) \cdot 50 \times 10^{-6}}{3}$$
$$>= 1mH$$

The interface between the controlled source and the grid has been discussed with only an inductor acting as the interfacing element. However, one can use a T-network topology as an interface network between grid and the controlled source in order to achieve better filtering properties. The L-C-L T-network configuration is a popular interface filter that will be discussed in the following section.

11.4 T-Network

Consider a generic T-network as shown in figure 11.4.1. It is a 2-port network and there are two impedances that characterise the T-network Z_{10} and Z_{20}. Z_{10} is a driving point impedance which is the impedance looking into the T-network from port 1-0 when an impedance Z_{20} is connected across port 2-0. Likewise, Z_{20} is also a driving point impedance which is the impedance looking into the T-network from port 2-0 when an impedance Z_{10} is connected across port 1-0. Z_{10} and Z_{20} are also called the image impedances.

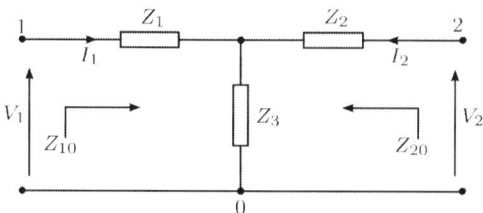

Figure 11.4.1: A general T-network

11.4.1 Characteristic impedance

The driving point impedances or the image impedances can be derived from the T-network circuit of figure 11.4.1. Thus,

$$Z_{10} = Z_1 + \frac{Z_3(Z_2 + Z_{20})}{Z_2 + Z_{20} + Z_3} \qquad (11.4.1)$$

$$Z_{20} = Z_2 + \frac{Z_3(Z_1 + Z_{10})}{Z_1 + Z_{10} + Z_3} \qquad (11.4.2)$$

From equations 11.4.1 and 11.4.2, one obtains

$$Z_{10} = \sqrt{\frac{(Z_1 + Z_3)(Z_1 Z_2 + Z_2 Z_3 + Z_1 Z_3)}{Z_2 + Z_3}} \qquad (11.4.3)$$

$$Z_{20} = \sqrt{\frac{(Z_2 + Z_3)(Z_1 Z_2 + Z_2 Z_3 + Z_1 Z_3)}{Z_1 + Z_3}} \qquad (11.4.4)$$

The driving point impedances or the images impedances can also be obtained from

$$Z_{10} = \sqrt{Z_{1oc} Z_{1sc}} \qquad (11.4.5)$$

where Z_{1oc} is the driving point impedance of port 1-0 when port 2-0 is open circuited and Z_{1sc} is the driving point impedance of port 1-0 when port 2-0 is short circuited.

$$Z_{20} = \sqrt{Z_{2oc} Z_{2sc}} \qquad (11.4.6)$$

where Z_{2oc} is the driving point impedance of port 2-0 when port 1-0 is open circuited and Z_{2sc} is the driving point impedance of port 2-0 when port 1-0 is short circuited.

11.5. L-C-L INTERFACE

For symmetrical T-networks, $Z_1 = Z_2$. Then $Z_{10} = Z_{20} = Z_0$ where Z_0 is called the characteristic impedance. Let $Z_1 = Z_2 = Z$ for a symmetrical T-network. Then from equations 11.4.3 and 11.4.4,

$$Z_0 = \sqrt{Z^2 + 2 \cdot Z \cdot Z_3} \qquad (11.4.7)$$

11.4.2 Propagation constant

When a symmetrical T-network is terminated by impedance Z_0, then

$$Z_0 = \frac{V_1}{I_1} = \frac{V_2}{-I_2}$$

and

$$\frac{V_1}{V_2} = \frac{I_1}{-I_2} = e^\gamma = e^{\alpha + j\beta} = (e^\alpha) \cdot \left(e^{j\beta}\right) \qquad (11.4.8)$$

where γ is the filter propagation constant, α is the attenuation constant and β is the phase constant of the filter. It can be shown that

$$\frac{e^\gamma + e^{-\gamma}}{2} = \cosh\gamma = 1 + \frac{Z_1}{Z_3}$$

and

$$\gamma = \ln\left[1 + \frac{Z_1}{Z_3} + \sqrt{\left(\frac{Z_1}{Z_3}\right)^2 + \frac{2Z_1}{Z_3}}\right] \qquad (11.4.9)$$

For the specific case of low pass filters wherein $Z_1 = Z_2$ are inductive in nature and Z_3 is capacitive in nature, the following relations apply

1. $-1 < \frac{Z_1}{2Z_3} < 0$ in the pass band

2. $\frac{Z_1}{2Z_2} = -1$ at the cut-off frequencies in the region of transition from pass band to stop band.

11.5 L-C-L interface

The simple inductor interface is generally very effective. However, it is rather bulky and heavy as the flux within the inductor core has to operate at grid frequency. An alternate filter topology that is popular is the L-C-L filter. This is a filter that falls in the class of k-derived filters. It is a T-network with two inductors and a capacitor as shown in figure 11.5.1.

Let $L_1 = L_2 = L$.

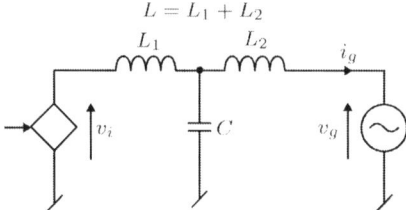

Figure 11.5.1: L-C-L filter interface

For a T-network, the cut-off frequency can be obtained from relation

$$\frac{j\omega_c L}{2 \cdot \frac{1}{j\omega_c C}} = -1$$

$$\frac{\omega_c^2 LC}{2} = 1$$

$$\omega_c = \sqrt{\frac{2}{LC}} \quad (11.5.1)$$

The characteristic impedance of the T-network at any frequency is given from equation 11.4.7 as

$$Z_0 = \sqrt{(j\omega L)^2 + 2 \cdot j\omega L \cdot \frac{1}{j\omega C}} = \sqrt{\frac{2L}{C}\left(1 - \frac{\omega^2 LC}{2}\right)}$$

$$= \sqrt{\frac{2L}{C}\left(1 - \left(\frac{\omega}{\omega_c}\right)^2\right)} \quad (11.5.2)$$

The characteristic impedance at dc is given as

$$Z_{0dc} = \sqrt{\frac{2L}{C}} \quad (11.5.3)$$

The values of L and C are calculated using equations 11.5.3 and 11.5.1. The characteristic impedance as a function of frequency can be plotted using equation 11.5.2.

Consider an example wherein 100A is required to be injected into the grid at 230V rms. The characteristic impedance for the filter is designed for $Z_{0dc} = \frac{230V}{100A} = 2.3\Omega$. The inverter switching frequency is 20kHz. The LCL filter is required to filter 20kHz and its harmonics. Let the cut-off frequency be set at 3kHz. Using equation 11.5.2, the plot of the characteristic impedance over frequency is shown in figure 11.5.2.

From equation 11.5.1,

$$LC = \frac{2}{\omega_c^2} = \frac{2}{4\pi^2 \cdot 3000^2} = 5.63 \times 10^{-9}$$

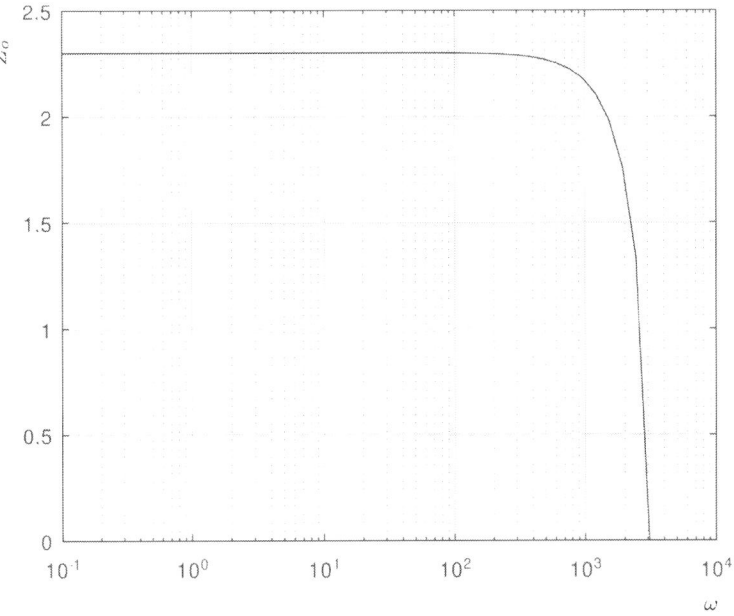

Figure 11.5.2: Characteristic impedance as a function of frequency

and from equation 11.5.3,

$$\frac{L}{C} = \frac{Z_{0dc}^2}{2} = \frac{2.3^2}{2} = 2.645$$

and therefore one obtains
$C^2 = \frac{5.63 \times 10^{-9}}{2.645}$ giving $C = 46\mu F$ and
$L = 2.645C = 0.12mH$

11.6 Transformer-less versus galvanic isolation

Figure 11.6.1 shows the block schematic of a single phase grid interactive system using a simple inductor as the grid interface element. Observe that there is no galvanic isolation. There are several aspects in this scheme that one should be cautious about.

1. The grid neutral is connected to the physical earth through the sub-station grounding or earthing. When a person touches the frame of the PV source, the person establishes a connection to the physical earth and therefore, a leakage current can flow through the path "earth - grid neutral - inverter - DC-DC converter - PV source - capacitance between PV cell interconnects and the metal parts of the frame - person - earth". This can prove to be dangerous for the person who touches the PV panels.

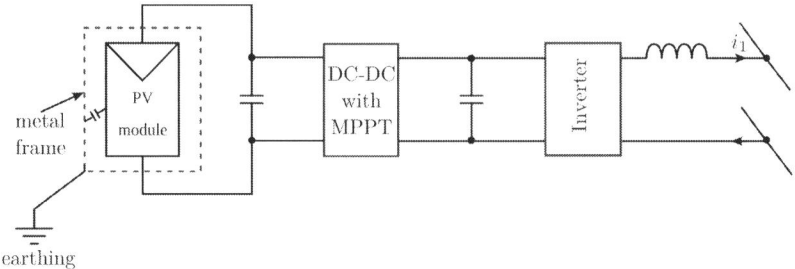

Figure 11.6.1: Schematic of a transformerless grid interaction

Therefore, it is essential that the metal frame parts of the PV panels are earthed, which means that a conductive wire is drawn from all the metal parts of the PV panels and physically connected to earth by preparing an earthing pit. This way the metal parts of the PV panels are at earth potential and even if a person comes in physical contact with the panels, there will be no danger of electric shock to the person.

2. There is a capacitive non-ideality existing between the PV cell interconnects and the PV panel frame. As the PV panel frame is earthed, there is a leakage sneak path that can inject current into the grid through both the live and neutral conductors to earth. This can violate the leakage current limit regulation.

This issue can be addressed by increasing the common mode impedance. The inductor that is used to interface the inverter and the grid can be split into two equal parts wherein one part is placed on the live line and the other part is placed on the neutral line as shown in figure 11.6.2(a). This will increase the common mode impedance without affecting the differential mode impedance. Another way to increase the common mode impedance is by use of a common mode choke between the inverter and the grid as shown in figure 11.6.2(b). This will increase the impedance from line to earth and also neutral to earth. However, the differential impedance will not be affected. The common mode choke consists of two inductors one is along the neutral path and the other along the line path. Both the inductors are wound on the same core with polarity such that the differential currents will cancel out the flux and the common mode currents will face significant impedance.

3. There can be DC current injection due to several non-idealities. The devices in the inverter may have differing characteristics which can reflect as a finite DC injection into the grid. DC injection has detrimental effects on downstream magnetics and needs to be eliminated or at least mitigated. Including galvanic isolation is the solution that will resolve this problem.

The galvanic isolation can be given in two ways. One method is to use a low frequency

11.7. SINGLE PHASE GRID INTERFACE

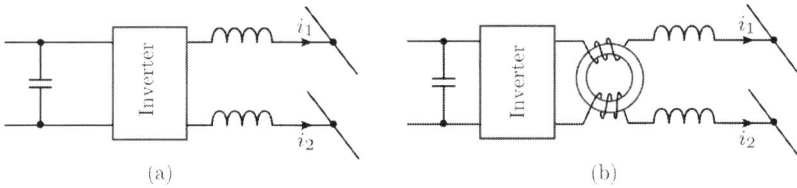

Figure 11.6.2: Leakage current reduction (a) Splitting the inductor (b) common mode choke

transformer and use it as one of the grid interface devices. The other method is to use a high frequency transformer in the DC-DC converter block. Both methods have their advantages and disadvantages. In the case of high frequency transformer in the DC-DC converter block, the transformer core can be designed with high frequency magnetic materials like ferrite which will make the transformer compact and light. However, there will be electronics in the downstream part up to the interface to grid. This would mean that there can be mismatch in characteristics of the semiconductor devices which can result in DC injection.

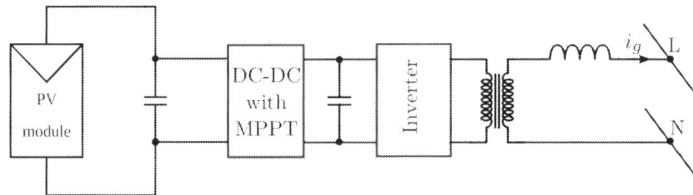

Figure 11.6.3: Grid interaction with galvanic isolation

In the case of the low frequency transformer galvanic isolation, there will not be DC injection. However, the flux in the core has to be designed for the low grid frequency and hence the core will be CRGO silicon steel material which will make the transformer big, heavy and costly. In most cases, one would like to avoid the DC-DC converter block for performing MPPT. It will be shown later that MPPT can be performed with only the inverter stage thereby improving the overall efficiency of the system. If the DC-DC converter block is not present, then the only way to provide galvanic isolation is by low frequency transformer at the grid interface. With careful design, one can use the leakage inductance of the transformer as seen from the grid side to become the abstract grid interface element. This has the advantage of eliminating the physical inductor element.

11.7 Single phase grid interface

Low power home rooftop PV applications in the range of 3 kW to 5 kW typically use single phase grid interface. The block schematic of a single phase grid interactive system is shown in figure 11.7.1. The system comprises of several component blocks which are necessary to make the system work. The following subsections will discuss each of the blocks in detail.

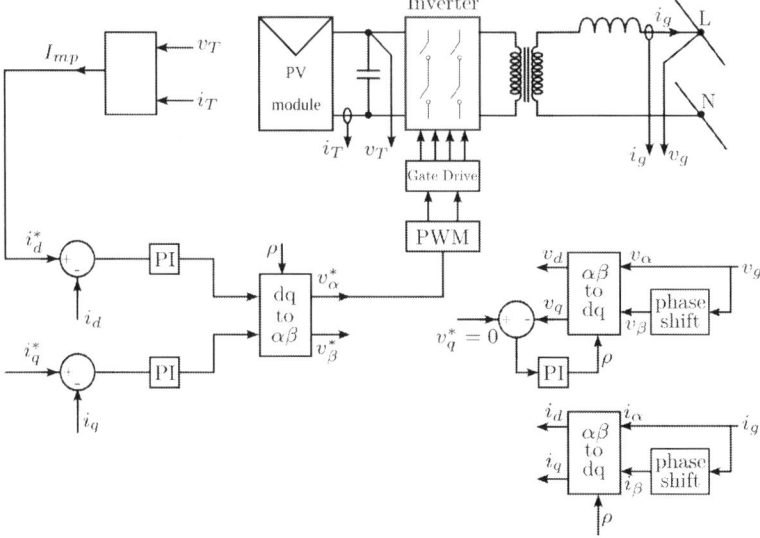

Figure 11.7.1: Single phase grid interaction

11.7.1 Open loop plant

The open loop plant comprises broadly of two parts, viz. (i) the power components that allow the flow of power and interface with the grid, and (ii) the signal part that generates the pulse width modulation signals and the necessary gate drives signals to drive the power semiconductor switches.

The power components comprise of

1. PV source : The PV panels with appropriate series and parallel protection supplies DC power to the DC-link capacitor which is at voltage $V_{dclink} = v_T$.

2. Inverter : A single phase inverter is used to draw switched power from the DC link and deliver pulse width modulated power. The inverter is generally a full-bridge converter with 4 power semiconductor switches with internal body diodes. The inverter used is generally a 2-level inverter or atmost a 3-level inverter wherein each arm of the full-bridge inverter can transact power at V_{dclink}, 0 or $-V_{dclink}$ volts. Multi-level inverter topologies are generally used in high voltage three phase systems. However, one may also use multi-level inverter topologies here too.

3. Grid interface components : The grid interface is done using either an inductor alone to filter the pulse width modulated voltage, or an L-C-L k-derived low pass filter in combination with or without galvanic isolation.

The signal part comprises of the gate drive for the power switches and the PWM generation.

11.7.2 Gate drive

The gate drive is an important circuit that operates the power semiconductor devices ON and OFF. There are several intelligent power modules that consist of power devices in full-bridge or H-bridge connection along with gate drive and protection circuits. There are power modules without gate drives too. In such a case, external gate drive modules need to be used. Several gate drive modules are available. A popular gate drive scheme based on the ISO5451 gate driver IC is given in figure 11.7.2. This gate drive IC incorporates a capacitive isolation between the PWM part of the circuit and the semiconductor device part. The 3.3V power supply is needed for interface with the microcontroller which provides the PWM and fault I/O. This power supply is isolated from the gate drive side power supply. An isolated power supply is used to generate +15V and -15V power supplies which powers up the gate drive part of the circuit. A desat protection is used to protect the semiconductor device against overcurrents. The desat protection circuit also provides a fault information signal which is interfaced to the microcontroller I/O in order to enable/disable the PWM pulses.

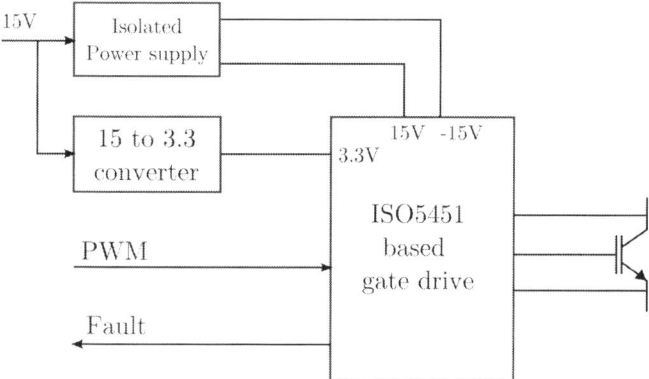

Figure 11.7.2: Gate drive circuit

11.7.3 PWM

Pulse width modulation is the means by which the modulating signal is discretised into ON and OFF signals in order to achieve high efficiency switched power flow. The modulating signal is again re-constructed by filtering which in this case is the grid interface filter. Figure 11.7.3(a) shows a triangular carrier which is used to modulate a low frequency modulating signal like the 50Hz sine wave.

There are two comparators that compare the modulating signal with the triangular carrier. The polarity of the comparison are such that one of the comparator outputs is given to the top switch of a inverter bridge arm and the other output is given to

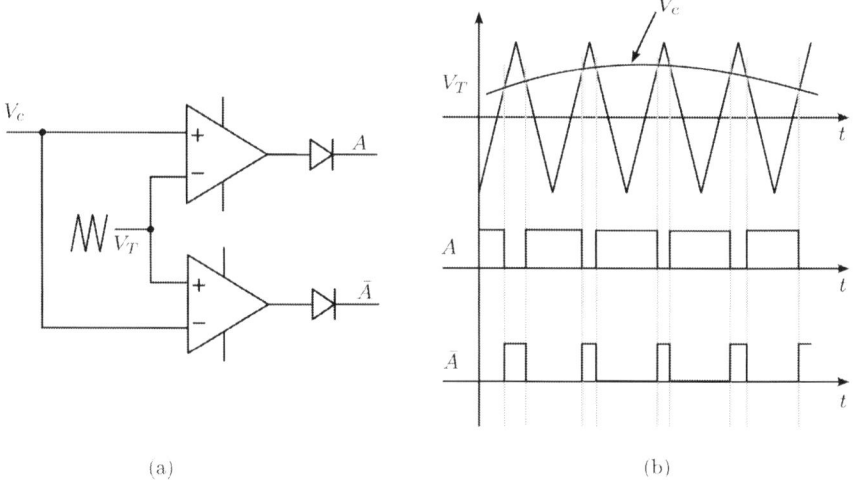

Figure 11.7.3: Simple PWM (a) PWM schematic (b) typical waveforms of the modulating signal, carrier and PWM output

the bottom switch of the inverter bridge arm. The triangle waveform frequency is the carrier frequency which also determines the frequency at which the power semiconductor switches of the bridge arm will operate. In the case of a single phase inverter, there are 2 bridge arms A and B. For the A bridge arm, let a sinusoidal frequency be used for comparing with the triangular carrier. The two comparators will provide the drive signals for the top switch of the A-arm and the bottom switch of the A-arm respectively.

For the B-arm, the modulating signal used for the A-arm is inverted or phase shifted by 180° and used as the modulating signal with the same triangular carrier. This will eliminate the even harmonics in the output. The two comparator outputs are given to the B-arm top and bottom switches. The output is taken out across the centre points of the A-arm and B-arm. The output is a pulse width modulated waveform that contains the low frequency modulating signal. Typical modulating signal, triangular carrier and PWM waveforms are shown in figure 11.7.3(b).

Deadtime : The transition from the top switch of an inverter arm to the bottom switch and vice-versa is a critical time period. One of the switches in transition is moving from ON state to OFF state and the other switch in the arm is moving from OFF state to ON state. There will be a small but finite period of time when both the switches will be conducting and this will result in a current called the "shoot through" current. This can be a large current which flows from the DC link positive rail to the negative rail and is limited only by the conduction resistances of the two semiconductor switches in transition. This shoot through current is not desirable for the health of the switches and must be avoided.

In order to address this problem, a small period of time called the "deadtime" is provided which ensures that the transition of one of the switches of the bridge arm

11.7. SINGLE PHASE GRID INTERFACE

is completed before the other one begins. The deadtime is introduced by a using a monoshot which provides a pulse for the deadtime period. The logic for generating the PWM with deadtime is shown in figure 11.7.4(a). The waveforms at the various nodes are shown in figure 11.7.4(b).

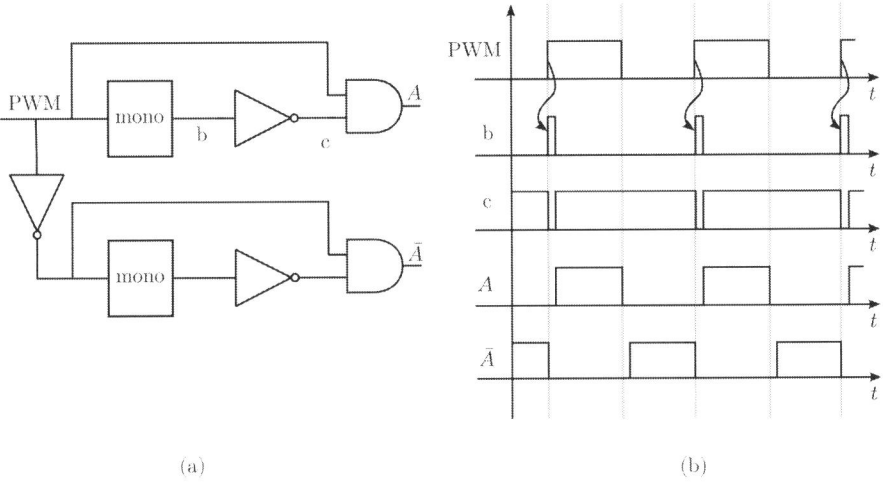

Figure 11.7.4: PWM with deadtime (a) Block schematic (b) Logic waveforms

11.7.4 α, β currents

The open loop plant needs to be controlled through negative feedback. In grid interactive systems, current is being injected into the grid and therefore, most grid interactive systems will sense the inverter output current and control it such that a specific amount of current is injected into the grid. In short, the inverters in grid interactive systems are mostly current controlled systems. As the voltage is set by the grid, the amount of current that is injected is a direct measure of the power that is injected into the grid.

The grid current is AC by nature and therefore continuously varying. The reference current should also be AC current which is compared with the measured or the fed back signal of the actual grid current. As the reference is continuously varying, controller design becomes a tracker problem. For a tracker problem, the controller should have a high dynamic response over a large signal range. In order to improve the dynamics and design a controller such that it needs to handle only small signal deviations in the neighbourhood of the operating point, a set-point control is always better. The controller design problem should be formulated for a set-point control to achieve better performance. In a set-point control, the reference is a constant or DC value and the feed back signal should also be a DC value. Any deviations from the reference becomes the error deviation around the operating point. The controller needs to act for the small signal deviations only and therefore the stress on the controller is much reduced. The important challenge is to convert the AC grid current measurements to DC value that represents the instantaneous value of the grid current. This is achieved through co-ordinate transformation.

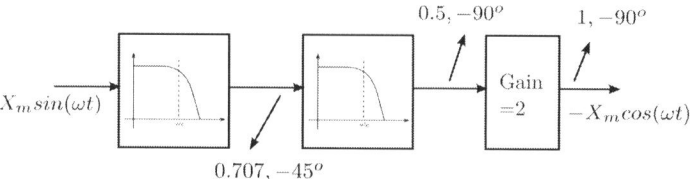

Figure 11.7.5: Quadrature phase shifter

Let the grid current be sinusoidal and is of the form $I_m \cdot \sin\omega t$. There are two essential parts viz. (i) the magnitude part I_m, and (ii) the phase part ωt. The magnitude I_m will give a direct measure of the amount of power that is injected into the grid given that the grid voltage magnitude V_m is fixed by the grid and is constant. The magnitude I_m is a DC value and can be used as the measure parameter for set-point control.

Referring to the block diagram of figure 11.7.1, the injected AC grid current is measured. This measured current is passed through two first order low pass filters which have cut-off frequencies set at grid frequency of 50Hz. At the grid frequency, the attenuation is -3dB which means $\frac{1}{\sqrt{2}}$ for each first order low pass filter and the phase angle is 45° at the cut-off frequency. Therefore, two low pass filters will provide a 90° phase shift and an overall gain of 1/2. The output of the second low pass filter is passed through a gain block and the signal is scaled with gain of 2. This signal along with the measured current signal gives the orthogonal current signals i_α and i_β. These two orthogonal signals can be used to estimate the I_m value of the current at any given instant.

11.7.5 Stationary to rotating frame transformation

The measured single phase grid current can be expressed in the form of two orthogonal currents as follows,

$$i_\alpha = I_m \cdot \cos\omega t$$

and

$$i_\beta = I_m \cdot \sin\omega t$$

The two orthogonal currents and their resultant can be visualised on a cartesian coordinate system called the α-β coordinate system. The axes of the α-β coordinate system is stationary and therefore, it is also called the stationary coordinate system. Consider the coordinate system shown in figure 11.7.6 wherein the x-axis is named as the α-axis and the y-axis is called the β-axis. The current i_α is considered as the α-axis variable and the current i_β is considered as the β-axis variable. The evolution with time of the two orthogonal currents are shown alongside the two axes. Note that

11.7. SINGLE PHASE GRID INTERFACE

i_α is a cosine wave and i_β is a sinusoidal wave. The resultant is a fixed amplitude vector called the space vector, which rotates at frequency ω.

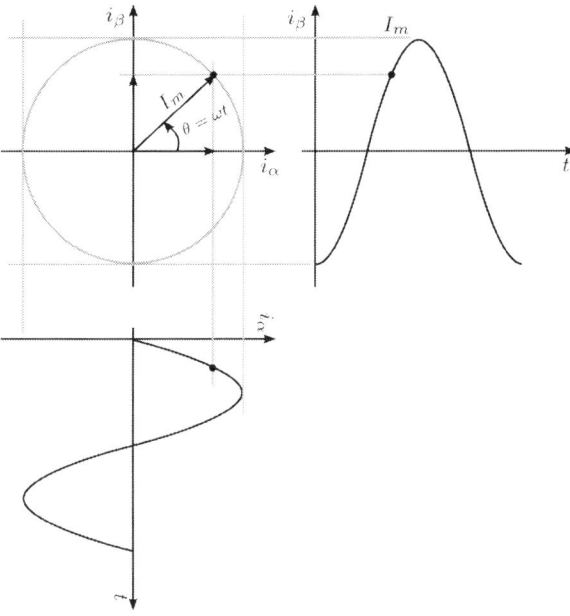

Figure 11.7.6: Currents in the stationary coordinate system

For sinusoidal waveshape current, the amplitude of the grid current space vector is

$$I_g = \sqrt{i_\alpha^2 + i_\beta^2} = \sqrt{(I_m \cdot \cos\omega t)^2 + (I_m \cdot \sin\omega t)^2} = I_m$$

If the current waveshapes contain only the fundamental frequency, then the grid current amplitude is a constant I_m. However, for a general case the grid current amplitude is $\sqrt{i_\alpha^2 + i_\beta^2}$.

Likewise, the grid voltage can also be combined to obtain the grid voltage space vector. It can be seen in figure 11.7.1 that the grid voltage is measured and passed through two first order filters that each have cut-off frequency set at 50Hz. In a similar manner to that described for the current, the voltage is split into the co-sinusoidal part v_α and the sinusoidal part v_β and

$$v_g = \sqrt{v_\alpha^2 + v_\beta^2} = \sqrt{(V_m \cdot \cos\omega t)^2 + (V_m \cdot \sin\omega t)^2} = V_m$$

Consider the diagram shown in figure 11.7.7. The stationary coordinate system is represented by the α-axis and β-axis. The grid voltage space vector and grid current space vector are represented by v_g and i_g respectively. The grid voltage and grid current

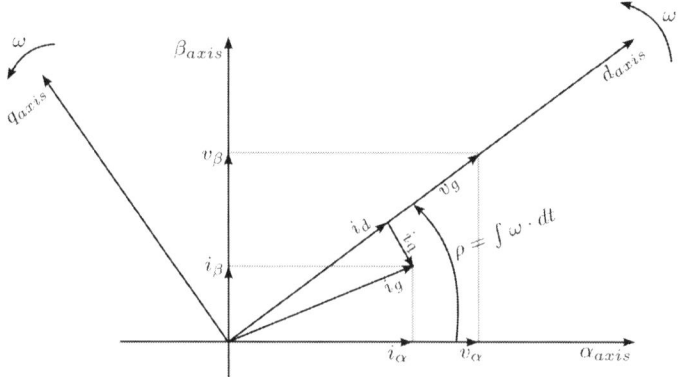

Figure 11.7.7: Grid voltage and current in stationary and rotating co-ordinate systems

can be resolved into orthogonal components along α-axis and β-axis as v_α, i_α and v_β, i_β respectively.

The vectors v_g and i_g are rotating with angular velocity ω. Consider another coordinate system comprising the d-axis and the q-axis such that the entire coordinate system is rotating synchronously with angular velocity ω. As v_g and i_g are also rotating synchronously, they would appear as DC quantities when observed from the rotating d-q coordinate system. Let the d-axis of the d-q coordinate system be aligned along the voltage space vector v_g. The grid current i_g is resolved into i_d along the d-axis and i_q along the q-axis. For sinusoidal grid currents, i_d and i_q are DC quantities. Likewise, for the voltage, v_d is same as v_g and $v_q = 0$ as the d-axis is aligned along the voltage space vector.

The grid current i_g is measured and passed through the phase shifters to obtain the orthogonal current components i_α and i_β. The fundamental frequency of the grid current is ω which is also the synchronous rotating angular velocity of the voltage space vector, current space vector and d-q coordinate system. The components of the current along the d-axis and q-axis are estimated as follows;

Let ρ be the angular displacement of the d-q coordinate system from the stationary coordinate system at a given instant of time. Then at that instant, the grid current can be expressed in the different coordinate systems as

$$i_{g\alpha\beta} = I_g \cdot e^{j\theta} \tag{11.7.1}$$

where θ is the angle of the grid current space vector with respect to the stationary α-axis. This is given as

$$\theta = \cos^{-1}\left(\frac{i_\alpha}{\sqrt{i_\alpha^2 + i_\beta^2}}\right) \tag{11.7.2}$$

11.7. SINGLE PHASE GRID INTERFACE

$$\begin{aligned} i_{gdq} &= I_g \cdot e^{j(\theta-\rho)} \\ &= \left(I_g \cdot e^{j\theta}\right) \cdot e^{-j\rho} \\ &= i_{g\alpha\beta} \cdot e^{-j\rho} \end{aligned} \quad (11.7.3)$$

The space vector equation given in equation 11.7.3 can be resolved using complex quantities and can be written as

$$i_d + ji_q = (i_\alpha + ji_\beta) \cdot (cos\rho - jsin\rho)$$

This can be represented in matrix form as

$$\begin{bmatrix} i_d \\ i_q \end{bmatrix} = \begin{bmatrix} cos\rho & sin\rho \\ -sin\rho & cos\rho \end{bmatrix} \cdot \begin{bmatrix} i_\alpha \\ i_\beta \end{bmatrix} \quad (11.7.4)$$

Equation 11.7.4 gives the transformation from the stationary frame or the $\alpha\beta$ co-ordinate system to the rotating frame or the dq coordinate system.

The reverse transformation from the dq coordinate system to $\alpha\beta$ coordinate system can also be performed. Referring to equation 11.7.3 and re-arranging the terms, one has

$$\begin{bmatrix} i_\alpha \\ i_\beta \end{bmatrix} = \begin{bmatrix} cos\rho & -sin\rho \\ sin\rho & cos\rho \end{bmatrix} \cdot \begin{bmatrix} i_d \\ i_q \end{bmatrix} \quad (11.7.5)$$

Along similar lines, the measured grid voltage can be decomposed into v_α and v_β. The stationary frame signals are transformed to the dq coordinate system and back by the following equations,

$$\begin{bmatrix} v_d \\ v_q \end{bmatrix} = \begin{bmatrix} cos\rho & sin\rho \\ -sin\rho & cos\rho \end{bmatrix} \cdot \begin{bmatrix} v_\alpha \\ i_\beta \end{bmatrix}$$

$$\begin{bmatrix} v_\alpha \\ v_\beta \end{bmatrix} = \begin{bmatrix} cos\rho & -sin\rho \\ sin\rho & cos\rho \end{bmatrix} \cdot \begin{bmatrix} v_d \\ v_q \end{bmatrix}$$

In the d-q domain the current signals i_d and i_q are compared with the DC set-point references i_d^* and i_q^* respectively. If the current needs to be injected into the grid at unity power factor, then this would mean that the current space vector and the grid voltage space vector are in alignment. Under such conditions, only the d-axis current will exist and q-axis current will be zero. Therefore, set $i_q^* = 0$ if unity power factor injection is needed. After comparison, the error signals are passed through controllers like PI controllers. The output of the PI controllers are added with feedforward terms and then transformed from the d-q domain to the stationary coordinate system using equation 11.7.5.

11.7.6 Frequency and angle estimation

In the previous sub-section, the coordinate transformation of signals need the instantaneous angle between the coordinate frames in order to execute the transformation from the stationary to rotating frame and vice-versa. Figure 11.7.8 shows the control block diagram to estimate the frequency and the angle of the rotating coordinate frame with respect to the stationary coordinate frame.

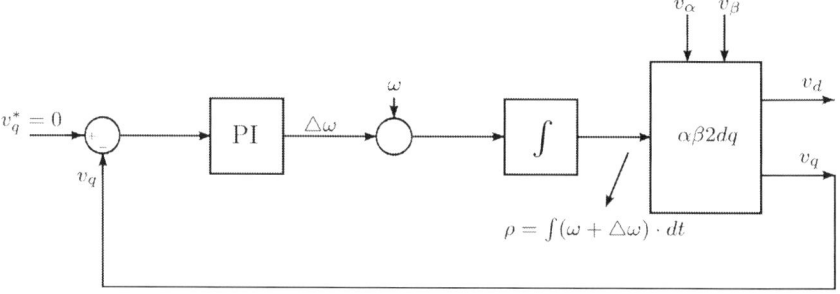

Figure 11.7.8: Phase locked loop control system to estimate frequency and angle

The d-axis of the rotating reference frame is required to be aligned along the grid voltage space vector. Once the d-axis is aligned along the grid voltage space vector, then the direct axis component of the voltage in the d-q coordinate system will be the magnitude of the grid voltage space vector, i.e. $v_d = V_m$ and the quadrature axis component will be zero i.e. $v_q = 0$. This property is used for performing the phase locked loop control. The control system uses a quadrature axis voltage set-point $v_q^* = 0$ which is compared with the v_q which is obtained from the measurement of the grid voltage in the stationary coordinate system and transformed into the rotating coordinate system using the angular displacement ρ. A proportional controller is used to obtain $\triangle \omega$ from the error. The ideal grid frequency $\omega = 2\pi 50$ is used as a constant term that is added to $\triangle \omega$ in order to estimate the correct grid frequency. As it is known that the grid frequency will hover around the ideal 50Hz, the stress on the controller will be lesser as it needs to handle only small signal deviations. Integrating the summed frequency will provide the angular displacement ρ.

11.7.7 Feedforward components

Consider the single phase grid connected system wherein the inverter is interfaced to the grid with the help of an inductor. The voltage drop across the inductor is the difference in voltage between the inverter output and the grid voltage. Thus,

$$L\frac{di_g}{dt} = v_i - v_g$$

where i_g is the grid current space vector, v_g is the grid voltage space vector and v_i is the space vector of the inverter output. Resolving the currents and voltages into orthogonal components in the stationary α-β coordinate system, one obtains,

11.7. SINGLE PHASE GRID INTERFACE

$$L\frac{d(i_\alpha + ji_\beta)}{dt} = (v_{i\alpha} + jv_{i\beta}) - (v_\alpha + jv_\beta)$$

The stationary frame quantities can be transformed to the rotating d-q frame quantities by multiplying with $e^{-j\rho}$ where ρ is the instantaneous angular displacement of the d-q coordinate system from the α-β coordinate system. Thus,

$$L\frac{d(i_\alpha + ji_\beta)e^{-j\rho}}{dt} = (v_{i\alpha} + jv_{i\beta})e^{-j\rho} - (v_\alpha + jv_\beta)e^{-j\rho}$$

Solving and simplifying, one obtains

$$L\frac{di_d}{dt} = (v_{id} - v_d) + \omega L i_q$$

$$L\frac{di_q}{dt} = (v_{iq} - v_q) - \omega L i_d$$

The equation terms are re-arranged in order to express the inverter output voltage in terms of the other parameters. Thus,

$$v_{id} = L\frac{di_d}{dt} - \omega L i_q + v_d \qquad (11.7.6)$$

$$v_{iq} = L\frac{di_q}{dt} + \omega L i_d + v_q \qquad (11.7.7)$$

From equations 11.7.6 and 11.7.7, it can be seen that there are cross coupling terms. $(-\omega L i_q)$ is a quadrature component term that is coupled with the direct axis inverter voltage and $(\omega L i_d)$ is a direct axis term that is coupled with the quadrature axis inverter voltage. These terms are added as feedforward terms in the control loop at a point just after the controller. By this, the stress on the controller will be significantly reduced.

11.7.8 MPPT integration

Referring to figure 11.7.1, one can see that maximum power point tracking is integrated into the control of the grid connnected inverter. The terminal voltage and current of the PV panels are monitored and measured. The terminal parameters v_T and i_T are given as input to the MPPT block. The MPPT block uses any one of the algorithms discussed in chapter 7 in order to achieve the maximum power operating point for the PV panels. The MPPT controller will set the reference for the direct axis component of the grid current.

Consider that the current is injected into the grid at unity power factor. This implies that the quadrature axis component of the grid current space vector $i_q = 0$. In order to achieve this, the quadrature axis current command i_q^* is set to zero. The output of the MPPT controller will be used as set-point for the direct axis current i_d^*. The PI current controllers will provide the appropriate modulating signals to the grid connected inverter such that $i_d = i_d^*$ in the steady state. This would imply that the PV panel is pumping the maximum power for the given insolation.

11.8 Three phase grid interface

The single phase grid connected inverters are used in home applications wherein the roof top solar PV panels are used to pump energy into the single phase grid of residential homes. The power injected is in the range of 3kW to 5kW. However, when large powers are injected into the grid at low voltage or medium voltage levels, invariably three phase grid connected systems are used. Figure 11.8.1 shows the block diagram of a PV source that is feeding into a three phase grid through a three phase inverter and inductive interface.

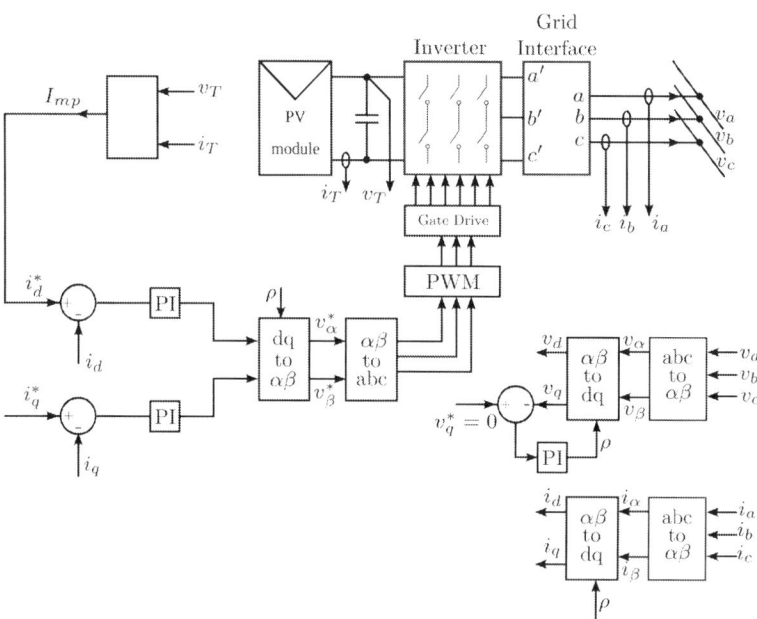

Figure 11.8.1: Three phase grid interaction

The block diagram is similar to the single phase grid injection system. However there are differences which shall be discussed.

1. The inverter is a three phase inverter which implies that there are three bridge arms. One may use the 2-level topology or multilevel topologies. The three

11.8. THREE PHASE GRID INTERFACE

phase inverter is interfaced to the three phase grid either through (i) simple inductors in series or (ii) L-C-L filter or (iii) with galvanic isolation with L or L-C-L filters as shown in figure 11.8.2.

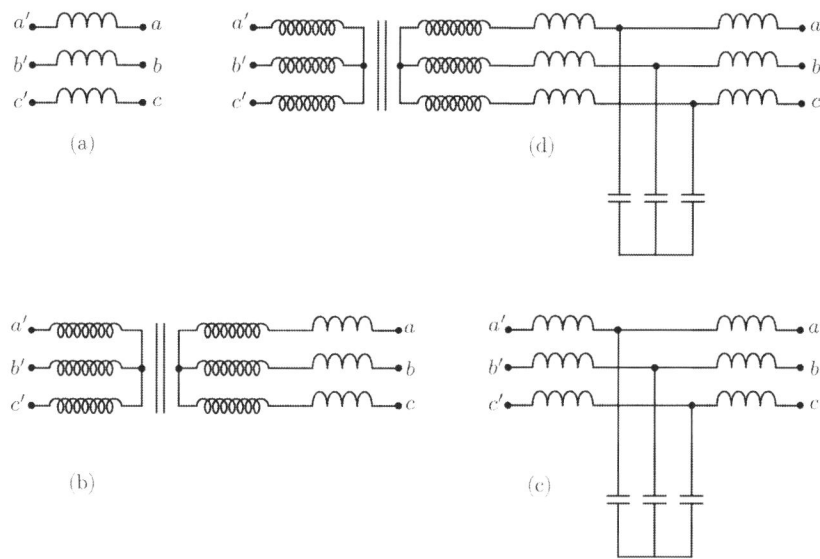

Figure 11.8.2: Three phase grid interaction (a) L interface, (b) Transformer and L interface, (c) L-C-L interface, (d) Transformer and L-C-L interface

2. The three line voltages of the grid and the three line currents from the inverter are measured and fed back for control purpose. The three phase vectors (a, b, c) are converted to two phase orthogonal vectors (α, β). Both are in the stationary coordinate system. Consider the grid current space vector which can be represented as follows,

$$i_g = i_a \cdot e^{j0} + i_b \cdot e^{j\frac{2\pi}{3}} + i_c \cdot e^{j\frac{4\pi}{3}} \qquad (11.8.1)$$

wherein i_a is the space vector of the a-phase which is aligned along the x-axis of the stationary coordinate system. The current i_a being injected into the grid is a sinusoidal current. The current i_b is also a space vector which is the b-phase current and it is spatially at an angular distance of $\frac{2\pi}{3}$ radians from the x-axis. The space vector current i_c is aligned along an axis that has an angular displacement of $\frac{4\pi}{3}$ radians with respect to the x-axis. i_g is the grid current space vector which is the composition of all three phase current space vectors and is given as in equation 11.8.1. This equation can be expanded as

$$i_g = i_\alpha + ji_\beta$$

$$= i_a + i_b \cos\left(\frac{2\pi}{3}\right) + i_c \cos\left(\frac{4\pi}{3}\right) + j\left(i_b \sin\left(\frac{2\pi}{3}\right) + i_c \sin\left(\frac{4\pi}{3}\right)\right)$$

This can be expressed in matrix form as

$$\begin{bmatrix} i_\alpha \\ i_\beta \end{bmatrix} = \begin{bmatrix} 1 & -\frac{1}{2} & -\frac{1}{2} \\ 0 & \frac{\sqrt{3}}{2} & -\frac{\sqrt{3}}{2} \end{bmatrix} \cdot \begin{bmatrix} i_a \\ i_b \\ i_c \end{bmatrix} \qquad (11.8.2)$$

If $i_a = I_m \cos(\omega t)$, $i_b = I_m \cos\left(\omega t - \frac{2\pi}{3}\right)$ and $i_c = I_m \cos\left(\omega t - \frac{4\pi}{3}\right)$, the by applying the above equation 11.8.2, it can be seen that i_α and i_β will be cosinusoidal and sinusoidal with peak magnitude as $\left(\frac{3}{2}I_m\right)$. In order to make the three phase to two phase transformation invariant with respect to magnitude of the space vector, a scaling factor of $\left(\frac{2}{3}\right)$ is included as follows,

$$\begin{bmatrix} i_\alpha \\ i_\beta \end{bmatrix} = \left(\frac{2}{3}\right) \begin{bmatrix} 1 & -\frac{1}{2} & -\frac{1}{2} \\ 0 & \frac{\sqrt{3}}{2} & -\frac{\sqrt{3}}{2} \end{bmatrix} \cdot \begin{bmatrix} i_a \\ i_b \\ i_c \end{bmatrix} \qquad (11.8.3)$$

The transformation matrix given above is equally valid for transforming three phase voltage vectors to two phase voltage vectors in stationary coordinate system.

3. The entire part of the control algorithm in the d-q domain is exactly same as that discussed for the single phase system. The d-axis and q-axis currents are compared with the current commands i_d^* and i_q^*. For unity power factor injection, i_q^* is set to zero. i_d^* is the direct axis current command and it is derived from the output of the MPPT block. The peak power operating point of the PV source at a given insolation will decide the value of i_d^*.

4. The d-q control feed forward term inclusion and conversion from rotating reference frame to stationary reference frame variables are exactly same as that described in the single phase system. However, after conversion of the variable into the α, β coordinate system, one more block is needed to converter the orthogonal two phase quantities to the three phase abc-coordinate axes system. Let the α-axis of the α, β coordinate system and the a-axis of the abc-coordinate system be aligned. Then considering the current space vector and the projections of i_α and i_β on the abc-axes, one obtains,

$$i_a = i_\alpha + 0 \cdot i_\beta$$
$$i_b = -i_\alpha \cos\left(\frac{\pi}{3}\right) + i_\beta \cos\left(\frac{\pi}{6}\right)$$
$$i_c = -i_\alpha \cos\left(\frac{\pi}{3}\right) - i_\beta \cos\left(\frac{\pi}{6}\right)$$

From the above equations, the α, β to abc coordinate transformation in matrix form can be written as

$$\begin{bmatrix} i_a \\ i_b \\ i_c \end{bmatrix} = \begin{bmatrix} 1 & 0 \\ -\frac{1}{2} & \frac{\sqrt{3}}{2} \\ -\frac{1}{2} & -\frac{\sqrt{3}}{2} \end{bmatrix} \cdot \begin{bmatrix} i_\alpha \\ i_\beta \end{bmatrix} \quad (11.8.4)$$

The same matrix relation is also valid for the voltage variables.

5. The phase lock loop control block for estimating the frequency and the angle of the rotating reference frame with respect to the stationary reference frame is also same as that discussed for the single phase system.

6. The pulse width modulator (PWM) is also along similar lines as that discussed for the single phase inverter. The α, β to abc coordinate transformation will result in the modulating signals v_a, v_b and v_c. The triangular carrier will decide the frequency of the switches in the three phase inverter. This type of modulation is called sinusoidal pulse width modulation. If one measures the bridge midpoints with respect to the pole voltage node which is the midpoint of the dc link, then the maximum possible fundamental that one can obtain from the sinusoidal modulation is $\pm \frac{V_{dc}}{2}$. This can be improved by another 14% by using space vector PWM which can result in $\pm 0.57 V_{dc}$ between the bridge midpoint and the pole.

11.9 Questions

1. The following statements indicate the characteristics of the power semiconductor switches that are used in the inverter circuit. Which of the set of characteristics are necessary and sufficient?
   1. Bidirectional current carrying capability
   2. Bidirectional voltage blocking capability
   3. Unidirectional current carrying capability
   4. Unidirectional voltage blocking capability

   a) Both 1 and 2
   b) Both 1 and 4

c) Both 2 and 3
d) Both 3 and 4

2. The output of an inverter is connected to grid via 1:1 transformer. The equivalent circuit of transformer has a leakage reactance of 1 mH at both primary and secondary side. The magnetizing reactance of the transformer is 100 mH. Then the impedance seen between the inverter and grid will be approximately
   a) 100 mH
   b) 101 mH
   c) 2 mH
   d) 1 mH

3. Let $V_r = 10 \cdot sin(100\pi t)$, $V_y = 10 \cdot sin(100\pi t - \frac{2\pi}{3})$ and $V_b = 10 \cdot sin(100\pi t + \frac{2\pi}{3})$ be the voltages in stationary reference frame RYB that have the r-axis, y-axis and b-axis displaced spatially from each other by 120 deg apart. These three voltages are transformed into a dq rotating frame which is rotating at a frequency of 50Hz wherein the q-axis is aligned along the voltage space vector. Then the value of Vd is
   a) 0
   b) 15
   c) -15
   d) 10*sqrt(3)

4. In a 3-phase, 3-leg bridge 2-level inverter, sinusoidal pulse width modulation strategy is employed. Without over-modulation, what is the maximum fundamental phase voltage value that can be obtained with respect to the inverter dc-link voltage V_{dc}?
   a) V_{dc}
   b) $0.5\ V_{dc}$
   c) $0.57\ V_{dc}$
   d) $0.866\ V_{dc}$

5. In a 3-phase, 3-leg bridge 2-level inverter, space vector pulse width modulation strategy is employed. Without over-modulation, what is the maximum fundamental phase voltage value that can be obtained with respect to the inverter dc-link voltage V_{dc}?
   a) V_{dc}
   b) $0.5\ V_{dc}$
   c) $0.57\ V_{dc}$
   d) $0.866\ V_{dc}$

6. Let $V_a = 10 \cdot sin(100\pi t)$, $V_b = 10 \cdot sin(100\pi t - \frac{2\pi}{3})$ and $V_c = 10 \cdot sin(100\pi t - \frac{4\pi}{3})$ be the voltages in a stationary reference frame that have the a-axis, b-axis and c-axis displaced spatially from each other by 120 deg apart. If the three voltages are transformed into a 2-phase stationary frame (alpha-beta), then the loci of voltage in the 2-phase coordinate system will be
   a) a circle with centre at (0,0) and radius of 10

11.9. QUESTIONS

b) a circle with centre at (0,0) and radius of 15
c) an ellipse with centre at (0,0) and with major axis and minor axis lengths of 15 and 10 respectively
d) an ellipse with centre at (0,0) and with major axis and minor axis lengths of 10*sqrt(3) and 10 respectively

7. In a grid connected application, in order to track the frequency and angular position of grid voltage(s), which of the following is done?
   a) q-axis component of grid current reference is set to zero
   b) d-axis component of grid current's reference is obtained from MPPT block
   c) d-axis of the synchronously rotating frame is aligned with the grid voltage space vector
   d) DC-DC converter is interfaced between PV panel and inverter

8. Let $V_a = 10 \cdot sin(100\pi t)$, $V_b = 10 \cdot sin(100\pi t - \frac{2\pi}{3})$ and $V_c = 10 \cdot sin(100\pi t - \frac{4\pi}{3})$ be the voltages in a stationary reference frame that have the a-axis, b-axis and c-axis displaced spatially from each other by 120 deg apart. These three voltages are transformed into a d-q rotating frame that is rotating at a frequency of 20 Hz. The q-axis is aligned along the voltage space vector. Then the frequency of the signal in the d-q axes frame is
   a) dc
   b) 20 Hz
   c) 30 Hz
   d) a50 Hz
   e) 100 Hz

9. An L-C-L filter is to be designed for an inverter that is switching at 20kHz and injecting 50A into a 230 V rms grid. Set the filter cut-off frequency at 2kHz. Calculate the values of L and C.

10. A single phase inverter is being interfaced to the 230 V rms grid through an inductor. The dc link voltage is 400V. A current of 50A peak is to be injected into the grid. The inverter switching frequency is 50kHz and the current ripple due to switching should be less than 1A. Calculate the value of the inductor.

11. For the following signals $V_a = 10 \cdot sin(100\pi t)$, $V_b = 10 \cdot sin(100\pi t - \frac{2\pi}{3})$ and $V_c = 10 \cdot sin(100\pi t - \frac{4\pi}{3})$ in the stationary a-b-c frame, find the value of $|V_\alpha| + |V_\beta|$ at $t = \frac{1}{200}$ s. V_α and V_β are stationary signals in the α-β frame.

12. For the following signals $V_a = 10 \cdot sin(100\pi t)$, $V_b = 10 \cdot sin(100\pi t - \frac{2\pi}{3})$ and $V_c = 10 \cdot sin(100\pi t - \frac{4\pi}{3})$ in the stationary a-b-c frame, find the value of $|V_d| + |V_q|$ at $t = \frac{1}{100}$ s. V_d and V_q are synchronous signals in the d-q frame which is at an angular distance of $\rho = 25\pi t$ from the stationary a-b-c frame.

13. Consider a filter having transfer function of $\frac{a}{s+a}$ where $a = 100\pi$. A sinusoidal signal having amplitude 10 and frequency of 50Hz is passed through this filter. What is the amplitude, frequency and phase of the output signal with respect to the input signal?

14. Let $P_{dq} = K(v_d i_d + v_q i_q)$ be defined as the power in the d-q frame and $P_{abc} = (v_a i_a + v_b i_b + v_c i_c)$ be defined as the power in the a-b-c frame where the currents and voltages are balanced. What is the value of K if power should remain invariant when the variables are transformed across the frames i.e. $P_{dq} = P_{abc}$.

15. A 3-ph inverter interfaces a PV dc bus to a 3 phase grid having 230 Vac rms per phase. If 1000W of power is being injected from the PV into the grid at unity power factor, what are the current reference set point values (Id*, Iq*), if the synchronously rotating d-q reference frame is aligned along the grid voltage space vector?

Chapter 12

Life Cycle Cost

12.1 Introduction

This chapter will discuss a topic that is not related to engineering aspects but nonetheless a very important one. If one needs to compare, bench mark or reference the many PV based systems, the life cycle analysis presented in this chapter will be very useful.

In a PV based system, the PV sources have a lifetime of around 20 to 25 years. The storage batteries may last for around 5 years. This would imply that the batteries need to be replaced if the overall system has to last for more than 5 years. It is so with other sub-systems that make up the overall PV based system. In such a scenario wherein several sub-systems have different life spans, there is a need to evaluate the worth of the overall system to compare and find alternatives.

12.2 Growth models

The value of money and the value of the items under consideration are functions of time. In most cases, the values of money and items may be available at various different time frames. The objective of a meaningful comparison is to bring the corresponding values of money and items to a common reference time frame. The cost comparisons will therefore be based on time frame transformations. For this one must be able to transform the value of an item to a time frame in the past or to a time frame in the future. In order to formulate the algorithm for cost comparison one needs to understand the growth profiles of value of money or item.

Consider an item of interest. Let it be that it had a value yesterday, and it has a different value today and that it will have another different value tomorrow. This item can be bought or sold using money as the medium. This exchange amounts to a transaction. The value of the past, present and future are related by the growth profiles. Growth in the value of an item being bought will imply inflation and growth in the value of money will imply an interest rate.

An item had a value of 100 units of money in the time frame of a year ago. The same item has a value of 150 units of money in the present time frame. There are two process in play. One possibility is that the value of the item has increased or another possibility is that the value of money has decreased or the value of the item and the value of money have changed with time. Time frame transformations will provide a means of accurate comparisons and value estimations.

Interest and inflation have various growth profiles. The following three are the most popular.

1. Linear growth

2. Compound growth

3. Exponential growth

12.2.1 Linear growth

A popular application of linear growth is known as simple interest. Figure 12.2.1 gives a visualisation of this type of growth. Let i be the rate of interest in the interval of time $\triangle t$ which means that the interest at the end of time interval $\triangle t$ is given as $i \cdot \triangle t$. And n is the number of intervals (of duration $\triangle t$) under consideration.

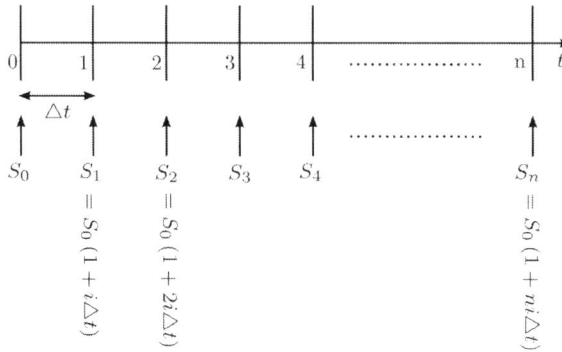

Figure 12.2.1: Linear growth or simple interest

Let S_0 be the initial sum of money at $t = 0$ or time *now*. And S_n is the sum after n intervals of $\triangle t$. Referring to the visualisation of figure 12.2.1, one can see that after 1 interval, the sum of money is given as

$$S_1 = S_0 + S_0 \cdot i\triangle t = S_0(1 + i\triangle t)$$

and after 2 intervals, the initial sum of money has grown to

$$S_2 = S_0 + S_0 \cdot 2i\triangle t = S_0(1 + 2i\triangle t)$$

and likewise after n intervals, the sum of money is given as

12.2. GROWTH MODELS

$$S_n = S_0(1 + ni\triangle t) \quad (12.2.1)$$

It can be seen from equation 12.2.1 that the sum of money grows linearly with time. The plot of sum versus the number of intervals n is shown in the plot shown in figure 12.2.2

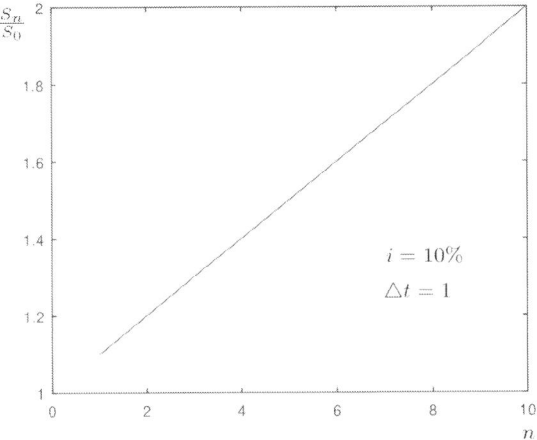

Figure 12.2.2: Plot of $\frac{S_n}{S_0}$ versus n for linear growth

For a specific case, if $\triangle t$ is considered as 1 year, then the interest per annum is i and n is the number of years. Then the sum after n years is given as

$$S_n = S_0(1 + ni) \quad (12.2.2)$$

12.2.2 Compound growth

The popular application of compound growth is called the compound interest. Consider the visualisation given in figure 12.2.3. The rate of interest is i and has the same meaning as discussed for the linear growth model. Likewise the interest at the end of time interval $\triangle t$ is given as $i \cdot \triangle t$.

In this type of growth, not only the interest for the initial principal amount is calculated, the interests gained in the previous intervals is also included in calculation. Therefore, at the end of the first interval, the sum of money is given as

$$S_1 = S_0 + S_0 \cdot i\triangle t = S_0(1 + i\triangle t)$$

and at the end of the 2nd interval, the sum of money is

$$S_2 = S_1 + S_1 \cdot i\triangle t = S_1(1 + i\triangle t) = S_0(1 + i\triangle t)^2$$

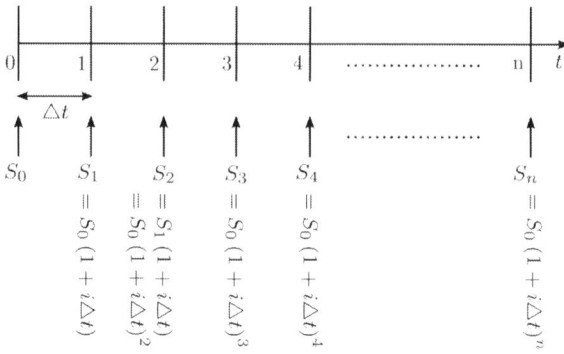

Figure 12.2.3: Compound growth or compound interest

At the end of the nth interval, the sum of money is given as

$$S_n = S_0 (1 + i \triangle t)^n \tag{12.2.3}$$

For the special case when the interval is 1 year, equation 12.2.3 becomes

$$S_n = S_0 (1 + i)^n \tag{12.2.4}$$

As the interest of the pervious intervals is also included in the estimation of the present interest, the interests of previous intervals are compounded giving the compound growth. This can be seen from the plot of sum versus the number of intervals elapsed as given in figure 12.2.4.

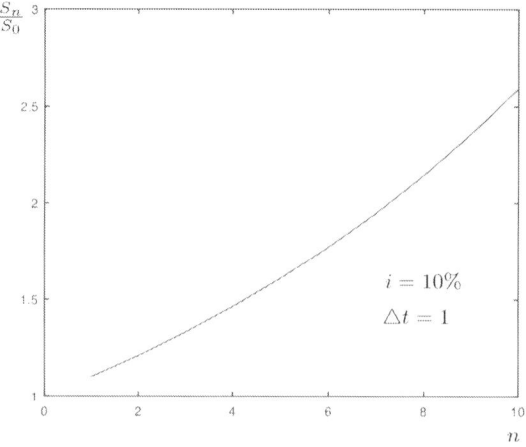

Figure 12.2.4: Plot of $\frac{S_n}{S_0}$ versus n for compound growth

12.2.3 Exponential growth

Compounding on a continuous time scale will result in the exponential growth model. The rate of interest is i and has the same meaning as discussed for the linear growth model. The interval of time $\triangle t$ is further divided into m equal parts as shown in figure 12.2.5. Each sub-interval is of time period $\frac{\triangle t}{m}$. The interest at the end of a sub-interval is $\frac{i \cdot \triangle t}{m}$. In the exponential growth model, the sub-interval time period will tend to 0 and hence $m \to \infty$.

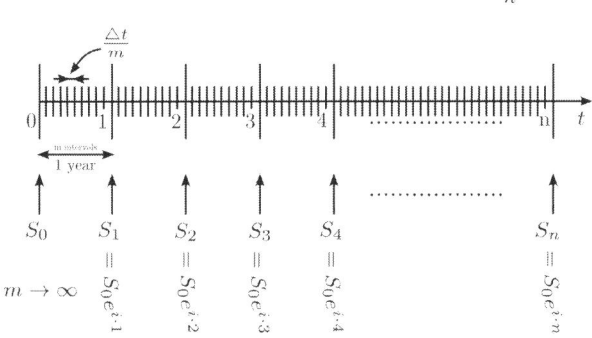

Figure 12.2.5: Exponential growth

Referring to figure 12.2.5 and considering the an interval of period $\frac{\triangle t}{m}$ having mn number of such intervals, then applying the compound growth equation 12.2.3, one obtains,

$$S_n = S_0 \left(1 + i\frac{\triangle t}{m}\right)^{mn} \quad (12.2.5)$$

Equation 12.2.5 is an application of compound growth and may also be called discrete compounding. In order to transform to continuous compounding the following constraints are applied

Let $x = \frac{i \cdot \triangle t}{m}$, and then $m = \frac{i \cdot \triangle t}{x}$ and

$$S_n = S_0 (1+x)^{\left(\frac{i \cdot \triangle t}{x}\right)n}$$

For the continuous compounding case, the sub-interval time period will tend to zero and therefore,

$m \to \infty$ and consequently $x \to 0$

$$S_n = S_0 \left(\lim_{x \to 0} (1+x)^{\frac{1}{x}}\right)^{i \cdot \triangle t \cdot n}$$

$$S_n = S_0 e^{i \cdot \triangle t \cdot n} \quad (12.2.6)$$

wherein $e \triangleq x \underset{}{\overset{lim}{\to}} 0 (1+x)^{\frac{1}{x}}$

For the special case where $\triangle t$ is 1 year, the equation 12.2.6 reduces to

$$S_n = S_0 e^{i \cdot n} \tag{12.2.7}$$

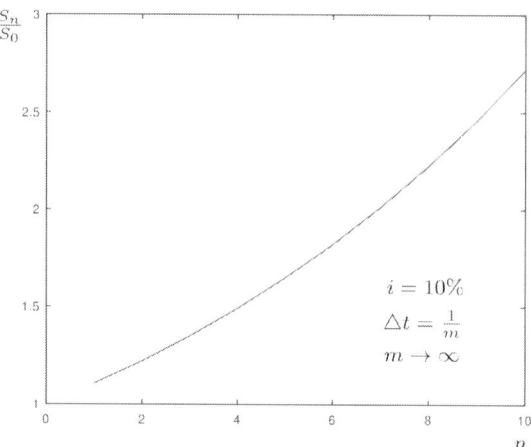

Figure 12.2.6: Plot of $\frac{S_n}{S_0}$ versus n for exponential growth

Figure 12.2.6 gives the plot of the sum versus the number of years. Compare this plot with the plots corresponding to linear growth and compound growth. Exponential growth is also sometimes called continuous compounding as is evident from the discussion above.

12.3 Inflation

The increase in the value of the money over time is reflected in the interest. Likewise, the increase in the value of an item over time is reflected in the inflation. The value of money at the start of the time evolution is S_0. The value of an item at the start of the time evolution is indicated as C_0. Using the more popular compound growth model, one can estimate the value of money after n years as $S_0(1+i)^n$ and the value of the item after n years having inflation rate f is given as $C_0(1+f)^n$.

Let a sum of money S_0 be invested at the start of a time evolution. At the start of the same time evolution, let the cost of an item be C_0. After n years, the value of the money would have appreciated and likewise the value of the item to be purchased would also have appreciated due to inflation. The item can be purchased after n years if the appreciated values of money and item are equal. This means,

$$S_0(1+i)^n = C_0(1+f)^n \tag{12.3.1}$$

12.4. TIME FRAME TRANSFORMATION

Equation 12.3.1 gives the future worth of an item having value C_0 at the start of a given time evolution t_0. This also implies that in order to purchase an item n years after t_0 the sum of money that needs to be set apart at t_0 is S_0 which is given as

$$S_0 = C_0 \left(\frac{1+f}{1+i} \right)^n \qquad (12.3.2)$$

If the starting time of time evolution t_0 is *now*, then in order to purchase an item n years later, which has a future worth of $C_0(1+f)^n$, one needs to set apart a sum of money now according to the estimate equation 12.3.2. This S_0 estimate assumes that the interest rate and the inflation rate are constant for the n years of time. This S_0 estimate is called the present worth of the item that is purchasable n years hence.

Example : An item costs Rs. 10000/- today, and in order to buy the item five years from today, how much should one set aside today. It is expected that the average interest rate is 8 percent and inflation rate is 5% for the next 5 years.

Let S_0 be the sum of money that needs to be set aside. It is the present worth of the item that needs to be purchased 5 years later.

$$S_0 = C_0 \left(\frac{1+f}{1+i} \right)^n$$
$$= 10000 \times \left(\frac{1+0.05}{1+0.08} \right)^5$$
$$= Rs.\, 8686.2$$

12.4 Time frame transformation

One of the important tasks in cost analysis is the estimation of worth at different time frames. The worth of an item may be known at a specific reference time frame. Its equivalent worth at a different time frame should be known in order to perform meaningful comparisons. The equation 12.3.2 is an example of time frame transformation. Consider the sum of money at an arbitrary time frame $t = n$. Let that sum of money be S_n. Let the value of S_n correspond to an equivalent value of S_0 in the time frame $t = 0$. These two values are related by using one of the growth models. Considering the exponential growth model of equation 12.2.7,

$$S_n = S_0 e^{ni}$$

where e^{ni} is the worth transformation factor. Likewise let m be another time frame, then

$$S_m = S_0 e^{mi}$$
$$= S_0 e^{(n-n+m)i}$$
$$= S_0 e^{ni} \cdot e^{(m-n)i}$$
$$S_m = S_n e^{(m-n)i} \quad (12.4.1)$$

Equation 12.4.1 gives the worth transformation relation wherein $e^{(m-n)i}$ is the worth transformation factor. If the exponent is positive then the future estimate of the worth at new time frame m is obtained. Likewise, if the exponent is negative then the past estimate at new time frame m is obtained. Some of the special cases can be listed as follows,

$m = 0$, then $S_0 = S_n e^{-ni}$. This gives the present worth of the value corresponding to $(t = n)$ wherein the new time frame is the "present" or $(t = 0)$.

$n = 0$, then $S_m = S_0 e^{mi}$. This shifts the time frame from present $(t = 0)$ to m years later $(t = m)$. The future worth of S_0 is obtained.

In general the time frame transform equation 12.4.1 can be written as

$$S_m = S_n \cdot W_{nm} \quad (12.4.2)$$

where W_{nm} is the worth transformation factor. The subscript indicates the change in time frame from n^{th} year to m^{th}. Based on the type of growth model used,

$W_{nm} = e^{-(n-m)i}$ for exponential growth model

$W_{nm} = (1+i)^{-(n-m)}$ for compound growth model

$W_{nm} = (1-(n-m)i)$ for linear growth model

12.5 Annual payments

One of the more popular installment payment methods is the equal annual installment payment plan. Consider the time line shown in figure 12.5.1. There is an item that is worth C_0 that needs to be purchased. However, the payment for this item will be made in n equal annual installments over the next n years. Each installment of money has the same value D in the time frame of the payment year. Each of these annual installments are transformed to the present time frame i.e. $(t = 0)$ in order to check if the sum equals the worth C_0 of the item to be purchased.

Applying the time frame transformation equation 12.4.2, one has

$$S_0 = C_0 = D \cdot W_{10} + D \cdot W_{20} + \cdots + D \cdot W_{n0}$$

12.5. ANNUAL PAYMENTS

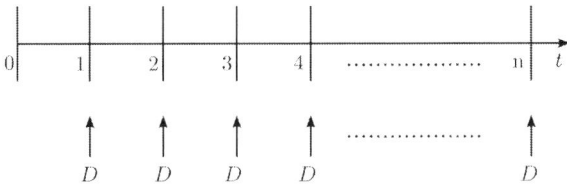

Figure 12.5.1: Annual payment plan

S_0 should be equal to C_0 in order that the item be purchasable. The equal annual installment value can be calculated as

$$D = \frac{C_0}{W_{10} + W_{20} + \cdots + W_{n0}} \qquad (12.5.1)$$

or

$$D = \frac{C_0}{P_w} \qquad (12.5.2)$$

where $P_w = W_{10} + W_{20} + \cdots + W_{n0}$ is called the *present worth factor* that considers all annual installments taken together.

P_w is a geometric progression with inter term factor as W_{10}. The sum to n-terms of a geometric progression can be used to estimate P_w. Alternately, a computer program or a programming environment like MATLAB or Octave can be used to compute P_w.

There is also a variant of the annual installment problem discussed above. Consider an item that is worth C_0 at time frame $t = 0$. This item is being planned for purchase m years later. Equal annual installments will be paid from the mth year on wards till the end of the nth year (where $m < n$). How would one calculate the annual installments?

Let f represent the inflation rate. If f is positive, then it results in inflation of the worth of the item, and if f is negative, then it will result in depreciation in worth of the item. The point of time frame reference for payment of installments is year m. Therefore, all values of items and money should be translated to the time frame where $t = m$. The worth transformation factor using inflation rate as the parameter will be symbolically represented as W_{nm}^f and the worth transformation factor using interest rate as the parameter will be represented as W_{nm}^i. Applying the time frame transformations, the sum of money S_m should be equal to the worth of the item C_m at time frame $t = m$. Thus,

$$\begin{aligned} S_m = C_m &= D + D \cdot W_{(m+1)m}^i + D \cdot W_{(m+2)m}^i + \cdots + D \cdot W_{(n)m}^i \\ &= C_0 \cdot W_{0m}^f \end{aligned}$$

and re-arranging

$$D = \frac{C_0 \cdot W_{0m}^f}{1 + W_{(m+1)m}^i + W_{(m+2)m}^i + \cdots + W_{(n)m}^i} \qquad (12.5.3)$$

or

$$D = \frac{C_0}{P_w}$$

where

$$P_w = \frac{1 + W_{(m+1)m}^i + W_{(m+2)m}^i + \cdots + W_{(n)m}^i}{W_{0m}^f}$$

The worth transformation factors are as described in equation 12.4.2 which is dependent on the type of growth model. Equation 12.5.3 is a more general form for computation of the value of equal annual installments spread over several years.

12.6 Life cycle costing

Life cycle cost (LCC) analyses the cost implications on a system that spans the entire time period of its operation. During the life span of a system, there are several different costs that are incurred at different points in the operational timeline of the system. The costs can be categorised into the following classes,

1. Capital cost (K) : This class of costs are incurred at the initiation or start of the operating life span of the system. This will include procurement of equipments, components, installation and commissioning of the system. In summary all cost incurred till the system begins operation will be classified into this category of cost. The capital cost will occur at time frame $t = 0$ in most systems.

2. Replacement cost (R) : There are several components within the system that will have lifespans that are shorter than the lifespan of the system. These shorter lifespan components need to be replaced. The cost to replace these items need to be time frame transformed and included while planning the life cycle cost of the system.

3. Maintenance cost (M) : Maintenance is another important aspect that will affect the overall cost of the system. Maintenance is generally done at periodic intervals of time during the operating life span of the system. Annual maintenance contract is an example of cost that will be classified in the maintenance category.

4. Energy cost (E) : Several systems need power for its operation. As an example power may be required in electrical form. As a result electrical energy gets consumed due to the operation of the system and electrical energy charges will be incurred either on monthly basis or annual basis. These energy charges should be anticipated and also included in the life cycle cost analysis of the system.

5. Scrap value (S) : At the end of the life span of the system, the system should be decommissioned and dismantled. There will be a scrap value associated with the dismantled residues depending on the recyclability of the dismantled components. Decommissioning, dismantling, recycling charges will fall into the scrap category. A positive scrap value implies that there is good potential for recycling the components and a certain amount of money can be recovered. If the scrap value is negative, then it implies that one must pay to decommission the system and to store the scrap.

The costs from the five categories should be time frame transformed to the time frame when the life span of the system starts i.e. $t = 0$. The sum of all the reflected cost components will give the overall cost or the life cycle cost (LCC) of the system. Thus,

$$LCC = K + R + M + E - S \qquad (12.6.1)$$

There are several systems where one may need to know the annual LCC or the LCC value distributed equally over the entire life span of the system. This will provide a means to compare cost across different systems and also to compare cost of the products obtained from the system. This will provide the unit cost of products like cost of water generated per litre, cost of electricity generated per kWh or per unit, etc. From equation 12.5.3, one can obtain annual LCC (ALCC) as,

$$ALCC = \frac{LCC}{P_w} \qquad (12.6.2)$$

where P_w is the present worth factor.

12.7 Example 1 - Estimation of LCC

Consider a PV based pumping system that has the following data,
 PV array rating - 500 Wp
 Pipe length - 30m
 Well depth - 10m
 Life span of the system - 15 years
 Rate of interest - 10%
 Rate of inflation = 0% (hypothetical)
 Annual maintenance contract - Rs. 1000 per year
 The life and per unit costs of the various components and subsystems are given as

Subsystem	Unit cost	Life
PV array and balance of systems	Rs. 80/Wp	15 years
Motor + pump	Rs. 5/Wp	7.5 years
Miscellaneous like transport	Rs. 3/Wp	-
Pipe cost	Rs. 100/m	5 years
Cost of well	Rs. 300/m	-

Table 12.1: Subsystem cost and life

Based on the above information about the system, the timeline with milestones for the operating lifespan of the system is prepared. The timeline is given in figure 12.7.1. The various cost categories are estimated as follows:

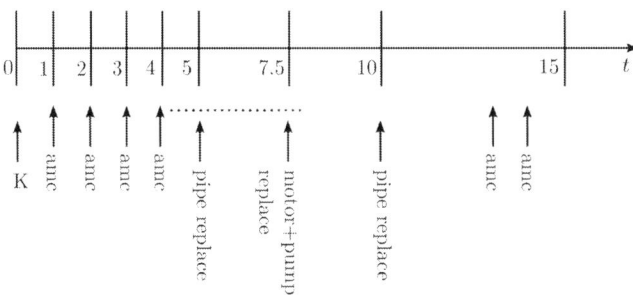

Figure 12.7.1: Timeline with milestones

Capital cost (K)

These costs are at time frame $t = 0$ and therefore do not need time frame transformation.

Item	Cost estimate	Cost
Array cost	$Rs.80 \times 500W_p$	$= Rs.40000.00$
Motor pump cost	$Rs.5 \times 500W_p$	$= Rs.2500.00$
Miscellaneous cost	$Rs.3 \times 500W_p$	$= Rs.1500.00$
Pipe cost	$Rs.100 \times 30m$	$= Rs.3000.00$
Well cost	$Rs.300 \times 10m$	$= Rs.3000.00$
	K	$= Rs.50000.00$

Replacement cost (R)

Here the replacement costs need to be reflected to the $t = 0$ time frame. Referring to the timeline of activities in the operation life span as shown in figure 12.7.1, there is one motor+pump replacement after 7.5 years into the operational life of the system. The value of the motor+pump system at the 7.5 year time frame will be inflated based on the inflation rate. However, inflation rate is considered as zero in this example. Transforming this to the $t = 0$ time frame, one obtains

$$R_1 = 2500 \times \frac{W^f_{0(7.5)}}{W^i_{(7.5)0}}$$

$$= 2500 \times \frac{1}{(1+0.1)^{7.5}} = Rs.1223.2$$

The item pipes have a life of 5 years. This implies that there will be 2 replacements of the pipes during the life space of the system. The value of the pipes at the 5th year and the 10th year will be inflated based in the inflation rate. f is considered as zero for this example. Transforming this replacement cost to the $t = 0$ time frame, one obtains

$$R_2 = 3000 \times \left(\frac{W^f_{0(5)}}{W^i_{(10)0}} + \frac{W^f_{0(10)}}{W^i_{(5)0}} \right)$$

$$= 3000 \times \left(\frac{1}{(1+0.1)^5} + \frac{1}{(1+0.1)^{10}} \right) = Rs.3019.4$$

The total replacement cost add up to $R = R_1 + R_2 = Rs.4242.6$.

Maintenance cost (M)

There is an annual maintenance contract (AMC) that has been put in place for the system for the entire lifespan. This implies that at the end of each year a sum of Rs.1000 will be paid to the maintenance contractor. There will be 15 equal payments made in the entire lifespan of the system. These annual payments should be transformed and reflected in the $t = 0$ time frame. Thus,

$$M = 1000 \cdot P_w$$

where P_w is the present worth factor and is given by

$$P_w = W^i_{(1)0} + W^i_{(2)0} + \cdots + W^i_{(15)0}$$

$$= \frac{1}{(1+0.1)^1} + \frac{1}{(1+0.1)^2} + \cdots + \frac{1}{(1+0.1)^{15}}$$

$$= 7.606$$

Using the estimate of the present worth factor, the maintenance cost is given as
$M = Rs.7606$

Life cycle cost (LCC)

In this example, there are no energy cost and no decommissioning and therefore no scrap value. Therefore the life cycle cost is the sum of the capital cost, replacement cost and the maintenance cost. Thus,

$$LCC = K + R + M$$
$$= 50000 + 4242.6 + 7606$$
$$= Rs.61848.6$$

12.8 Example 2 - Estimation of ALCC

A community has 100 people. The source of water to the community is from borewells supplied by means of hand pumps. 6 hand pumps are installed to meet the requirement of per capita consumption of 50 litres per day. The borewell depth is 20m. The cost of each handpump is Rs.10000. The cost of digging each borewell is at the rate of Rs.800/m. The life of the handpump is given as 10 years. The annual maintenance cost works out to be Rs.1500 per hand pump. If the rate of interest is 8%, and the inflation rate is 5%, what is the unit cost of water for a life cycle period of 20 years?

Based on the above information about the system, the timeline with milestones for the operating lifespan of the system is prepared. The timeline is given in figure 12.8.1. The various cost categories are estimated as follows:

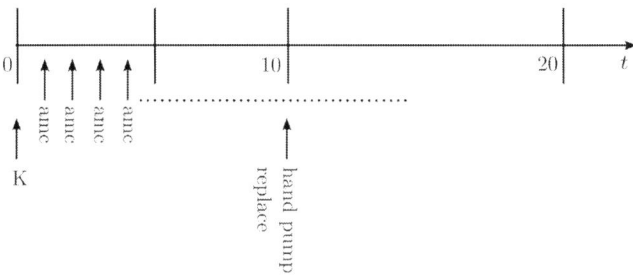

Figure 12.8.1: Timeline with milestones

Capital cost (K)

These costs are at time frame $t = 0$ and therefore do not need time frame transformation.

12.8. EXAMPLE 2 - ESTIMATION OF ALCC

Item	Cost estimate	Cost
Digging of 6 borewells	$Rs.800 \times 20m \times 6$	$= Rs.96000.00$
Hand pump cost	$Rs.10000 \times 6$	$= Rs.60000.00$
	K	$= Rs.156000.00$

Replacement cost (R)

The replacement costs need to be reflected to the $t = 0$ time frame. Referring to the timeline of activities in the operation life span as shown in figure 12.8.1, there is one hand pump replacement after 10 years into the operational life of the system. The value of the hand pump system at the 10 year time frame will be inflated based on the inflation rate. Transforming this cost to the $t = 0$ time frame, one obtains

$$R = 60000 \times \frac{W^f_{0(10)}}{W^i_{(10)0}}$$
$$= 60000 \times \frac{(1+0.05)^{10}}{(1+0.08)^{10}} = Rs.45270$$

Maintenance cost (M)

There is an annual maintenance contract (AMC) that is in place for the system for the entire lifespan. This implies that at the end of each year a sum of Rs.1500 per hand pump will be paid to the maintenance contractor. This means that there will be 20 equal payments made in the entire lifespan of the system. These annual payments should be transformed and reflected in the $t = 0$ time frame. Thus,

$$M = 1500 \times 6 \times P_w$$

where P_w is the present worth factor and is given by

$$P_w = W^i_{(1)0} + W^i_{(2)0} + \cdots + W^i_{(20)0}$$
$$= \frac{1}{(1+0.08)^1} + \frac{1}{(1+0.08)^2} + \cdots + \frac{1}{(1+0.08)^{20}}$$
$$= 9.818$$

Using the estimate of the present worth factor, the maintenance cost is given as $M = Rs.88362$.

Life cycle cost (LCC)

In this example, there are no energy cost and no decommissioning and therefore no scrap value. Therefore the life cycle cost is the sum of the capital cost, replacement cost and the maintenance cost. Thus,

$$LCC = K + R + M$$
$$= 156000 + 45270 + 88362$$
$$= Rs.289632$$

Annual life cycle cost (ALCC)

The value of the life cycle cost spread out equally over each year of the life span is ALCC. This is the annual cost of the system and is given as

$$ALCC = \frac{LCC}{P_w}$$
$$= \frac{289632}{9.818}$$
$$= Rs.29500 \, per \, annum$$

Unit cost of water

Annual water requirement

$$Q = (100 \, people) \cdot \left(\frac{50}{1000} m^3/day\right) \cdot 365 \, days$$
$$= 1825 \, m^3$$

Unit cost of water = $\frac{ALCC}{Q} = \frac{29500}{1825} = Rs.16.16 \, per \, m^3$

12.9 Example 3 - Estimation of break even point

A set of homes are in existence at a remote location. It needs about 10kW of peak power for community. One of the possibilities is to use PV system to supply the community needs. An alternative is to use a micro hydel system to transmit power to the community. The community needs to decide which alternative to select given the following information about the two alternatives,

Micro hydel data

Plant installation cost = Rs. 100000
    Cost of distribution = Rs. 50000
    Distance of plant from community = d_m metres

12.9. EXAMPLE 3 - ESTIMATION OF BREAK EVEN POINT

PV plant data

Cost of PV panels and balance of system = Rs.80 per W peak
   Irradiance at the place, $H_{atmin} = 5\,kWh/m^2/day$
   Distance of plant from community = d_p metres
   The life span of either system is 20 years.
   Cost of transmission line = Rs. 150 per m
   System efficiency = 70%
   The interest rate is 8% and the inflation rate is 5%.

For the purpose of this example let the replacement costs and the maintenance costs be zero. The capital cost can be estimated as a function of the distance. The LCC and ALCC are calculated to estimate the energy generation cost by each alternative for comparison.

Capital cost estimation

For the micro hydel system, the capital cost K_m comprises of the installation cost, distribution cost and the cost of the transmission line. Thus,

$$K_m = 100000 + 50000 + 150 \times d_m$$
$$= 150000 + 150 \cdot d_m$$

For the PV based system, the capital cost K_p comprises of the PV panel cost and the transmission line cost. This is given as

$$K_p = 80 \cdot 10000W + 150 \cdot d_p$$

Life cycle cost

As the replacement and maintenance costs are considered zero in this example, the LCC is simply the capital cost. Thus for the micro hydel system

$$LCC_m = K_m$$

and for the PV based system

$$LCC_p = K_p$$

Annual life cycle cost

The present worth factor is

$$P_w = W^i_{(1)0} + W^i_{(2)0} + \cdots + W^i_{(20)0}$$
$$= \frac{1}{(1+0.08)^1} + \frac{1}{(1+0.08)^2} + \cdots + \frac{1}{(1+0.08)^{20}}$$
$$= 9.818$$

The annual life cycle cost for the micro hydel system is

$$\begin{aligned} ALCC_m &= \frac{LCC_m}{P_w} \\ &= \frac{150000 + 150 \cdot d_m}{9.818} \\ &= 15278 + 15.3 d_m \end{aligned} \quad (12.9.1)$$

The annual life cycle cost for the PV based system is

$$\begin{aligned} ALCC_p &= \frac{LCC_p}{P_w} \\ &= \frac{800000 + 150 \cdot d_p}{9.818} \\ &= 81483 + 15.3 d_p \end{aligned} \quad (12.9.2)$$

Annual energy generation

The annual energy generation for the micro hydel system is

$$\begin{aligned} E_m &= (10kW) \cdot (24hrs) \cdot (365days) \cdot efficiency \\ &= 10 \times 24 \times 365 \times 0.7 \\ &= 61320 kWh \end{aligned}$$

The annual energy generation for the PV based system is

$$\begin{aligned} E_p &= (10kW) \cdot (H_{atmin} hrs) \cdot (365 days) \\ &= 10 \times 5 \times 365 \\ &= 18250 kWh \end{aligned}$$

Cost per kWh

The cost per unit or the cost per kWh is the ratio of the annual LCC to the annual energy generation potential. The cost per unit for the micro hydel system is

$$C_m = \frac{ALCC_m}{E_m}$$
$$= \frac{15278 + 15.3d_m}{61320}$$

and the cost per unit for the PV based system is

$$C_p = \frac{ALCC_p}{E_p}$$
$$= \frac{81483 + 15.3d_p}{18250}$$

Break even point

Equating the cost per unit for micro hydel system with the cost per unit of the PV based system, the condition for break even can be estimated. Thus,

$$\frac{15278 + 15.3d_m}{61320} = \frac{81483 + 15.3d_p}{18250}$$

This can be simplified as

$$d_m = 16896 + 3.36d_p \qquad (12.9.3)$$

Equation 12.9.3 gives the break even criteria for comparing and selecting the type of energy generation system. Consider equation 12.9.3 with respect to an example constraint where in the PV plant is placed within the community home roof tops. Then the distance of the PV plant from the community $d_p = 0$. Under such a condition, if $d_m > 16896$ then the cost per unit of micro hydel plant will be more than the cost per unit of the PV based. This means that if the distance of the micro hydel plant is greater than 16.896 kms away from the community, then the cost of electric energy from the micro hydel plant would be higher than an alternative PV based roof top system.

12.10 Questions

1. A person borrows money from a bank. Which of the growth models will prove to be most expensive for the borrower?
    a) simple growth
    b) compound growth
    c) discrete compound growth
    d) exponential growth

2. Inflation is a term that relates to
   a) increase in value of money with time
   b) decrease in value of money with time
   c) increase in value of item with time
   d) increase in value of time and item with time

3. The present cost of a solar panel is Rs 2000. If the interest rate is 8% and the inflation rate is 5% then how much must one save today in order to purchase the solar panel 5 years from now?

4. A certain sum of money amounts to Rs.80,000 by the end of 3.5 years and amounts to Rs.100,000 by the end of 6 years. Calculate rate of interest per annum by simple growth model.

5. A certain sum of money amounts to Rs.80,000 by the end of 3.5 years and amounts to Rs.100,000 by the end of 6 years. Calculate rate of interest per annum by compound growth model for 2 sub-intervals in a year.

6. Consider a 1kW PV based generating system. The PV array and the balance of system cost is Rs.100/- per peak watt. A life cycle cost analysis is required to be performed for a 25 year period. The rate of interest is 12%. The annual maintenance charge is Rs.1000/- per annum. The minimum daily solar energy incident at the place can be taken as $5 kWh/m^2/day$. Using compound growth model, calculate the unit cost of electric energy produced in Rs./kWh.

7. Consider a situation where one enters into an annual maintenance contract (AMC) for a particular item. The annual maintenance amount is Rs.5000 for a 5 year period. If the rate of interest is 8%, what is the present worth of the AMC?

8. Consider a situation where one enters into an annual maintenance contract (AMC) for a particular item. The annual maintenance amount is Rs.5000 for a 5 year period. If the rate of interest is 8% and the rate of inflation is 5%, what is the present worth of the AMC?

9. Consider a situation where one enters into an annual maintenance contract (AMC) for a particular item. The annual maintenance amount is Rs.5000 for a 5 year period. If the rate of interest is 8% and the rate of inflation is 8%, what is the present worth of the AMC?

10. The life cycle cost of a system is Rs. 10000/- for a life period of 20 years. The rate of interest is 8% and the inflation rate is 5%. What is the annual life cycle cost for the system?

Appendix-A : Typical Datasheet of 308W PV module

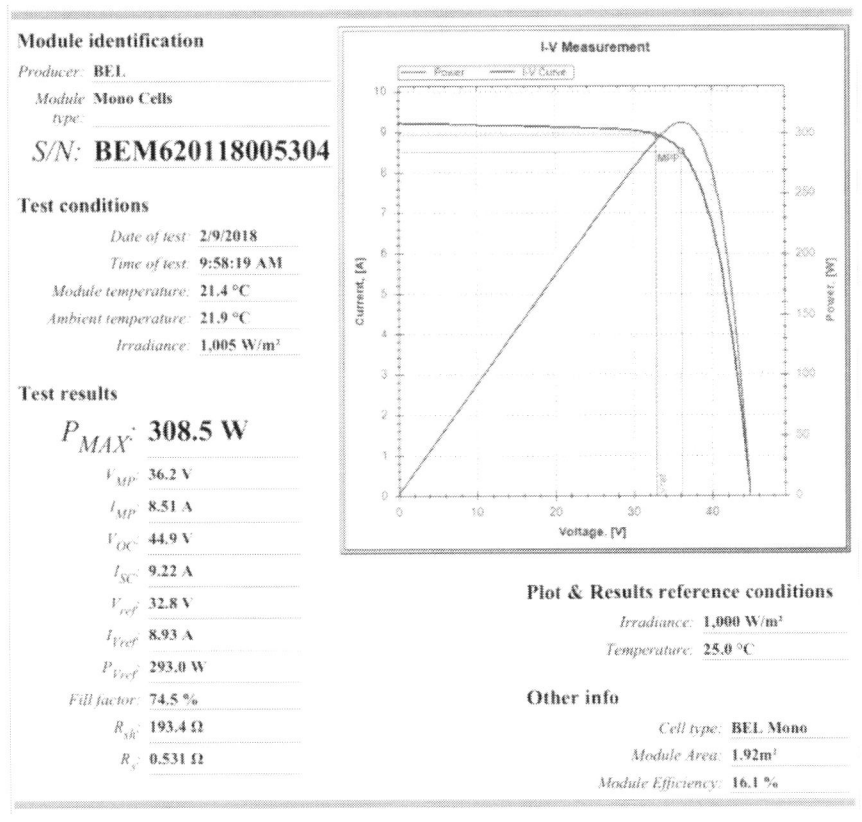

APPENDIX-A : TYPICAL DATASHEET OF 308W PV MODULE

Appendix-B : Typical Datasheet of 240W PV module

Peak Power	240 Wp
Max. Power Voltage (Vmp)	30.18 V
Max. Power Current (Imp)	7.96 A
Open Circuit Voltage (Voc)	36.72 V
Short Circuit Current (Isc)	8.99 A
Cell Efficiency	16.50%
Module Efficiency	14.66%
Maximum System Voltage	1000 V DC
Temp. Coeff. of Isc	+0.045 %/K
Temp. Coeff. of Voc	-0.34 %/K
Temp. Coeff of Pmax	-0.47 %/K
Series Fuse Rating	15 A
Cells	6x10 pieces polycrystalline solar cells series (156mm x 156 mm)
Junction Box	with 3 bypass diodes
Front Glass	toughened safety glass 3.2 mm
Cell Encapsulation	EVA (Ethylene Vinyl Acetate)
Back	composite film
Frame	anodized aluminium profile
Dimensions	1650x992x50mm (LxWxH)
Weight	19.3 kg
Max. Surface Load Capacity	tested up to 5400 Pa; IEC 61215
Hail	maximum diameter of 25 mm with impact speed of 23 m/s
Temperature Range	-40 °C to +85 °C

APPENDIX-B : TYPICAL DATASHEET OF 240W PV MODULE

Appendix-C : Simulation with KiCAD and NgSPICE

This appendix note provides information to setup and simulate an electronic circuit using KiCAD schematic editor and NgSpice. For schematic capture, the KiCAD project includes a schematic editor for GUI based circuit entry. It exports to NgSpice compatible netlist. The generated netlist is further simulated using NgSpice. The steps in this article are tried on a machine that has fedora as the operating system. However, without loss of much generality, the steps are platform independent and should work with other OS too.

Setting up the system

Packages

Install KiCAD packages from respective repositories. Use 'dnf' for fedora and 'apt-get' for ubuntu and use appropriate package manager for the corresponding distribution of linux that you may have. Execute the following commands in the terminal window.

sudo dnf install kicad

sudo dnf install kicad-packages3d

sudo dnf install ngspice

sudo dnf install tclspice

Steps for netlist generation

All KiCAD related files should be stored within a project directory for the given circuit that needs to be simulated. Within the project directory place the model and symbol folders. The model folder contains the library of sub-circuits for various custom components. the symbol directory contains the KiCAD compatible symbol definitions that call up the sub-circuits in the model folder libraries.

KiCAD will create in the project folder default .pcb file, default .pro or project file and a default .sch or schematic file.

Open the .pro project file. Here select the schematic from the view pane and select it.

Within the schematic click on component icon and place the cursor in the schematic sheet and click to open the component dialog box.

Choose the appropriate component and place it on the schematic sheet.

Place all components. Also pick and place a power ground which is the 0-node or the reference node.

Place the cursor on the component, right click an select the rotate function to orient the component properly.

Select the wire icon and interconnect the components and the power ground using the wire mode cursor.

The wheel mouse button can be used to zoom-in and zoom-out

After drawing the circuit, enter the values for the components

Now place a text box anywhere in the sheet. In the text box, within the .control and .endc statements, one can include the various NgSpice analysis and simulation commands

Click on *Tools* menu and annotate the circuit so that the components are sequentially labelled.

From the *File* menu, select *Export* and choose *Netlist*. Select the *Spice Netlist*. It will generate a .cir file that is compatible with NgSpice syntax and ready to be simulated by NgSpice.

Simulate in NgSpice

1. The simulation folder should contain the generated netlist as *fname.cir*.
2. Within the .cir netlist file, the text box content in the schematic is appended at the end as the set of control commands
3. Open a terminal and change directory to the simulation project folder. Here type
4. ¿ ngspice *fname*.cir
5. This will take you to the ngspice environment. In the ngspice environment type
6. ngspice 1¿ run
7. This will execute the simulation commands given in *fname*.cir.

Interactive NgSpice

The interactive NgSpice environment also called nutmeg is a very powerful simulation environment. Refer to the NgSpice manual for detailed description and syntax of the commands. However, given below are some frequently used commands that will help one to get started quickly:

source ¡finame.cir¿ - loads the given circuit file, netlist and commands into the simulator.

run - executes the simulation as specified in the circuit file

listing - shows the netlist and the analysis commands

show ¡device name1¿ ¡device name2¿ ... : ¡parameter name1¿ ¡parameter name2¿ ... - Output the operating point device summary

display - shows the list of vectors in the current plot

setplot [plotname] - select which plot data to be made current

write [file [expr ...]] - write data to a file of the currently active plot data

let varname = expr - assign vector variables

option varname = value - assign value to an option variable

alter varname value - alter the value of a variable

plot $expr_1$ $expr_2$ ... - plot vectors with respect to default x-axis variable

plot $expr_1$ vs $x_{axis-expr}$ $expr_2$ vs $x_{axis-expr}$ ... - plot vectors with specified x-axis variable

set color0 = white - sets the plot background colour to white

set color1 = black - sets the plot foreground colour to black

Model .lib subcircuit examples

```
* LIBRARY OF MY CUSTOM COMPONENTS
*****************************************************************
* Diodes
*****************************************************************
.model diode_def D()
.model diode_zener  D(bv=6.2)
*****************************************************************
*Diode macro model
.SUBCKT diode_pwr 101 102
DX    101 102    diode_model
```

```
Rsh     102 103     10000.0
Csh     103 101     0.01uF
.model  diode_model  D(RS=0.01, CJO=100pF)
.ENDS
*
*****************************************************************
* Switches
*-----------------------------------------------
.model switch_simple SW(RON=0.001)
*-----------------------------------------------
.SUBCKT   switch_pwr   nSp    nSn    nVcp  nVcn
SW        nSp   nDb    nVcp   0     bidir_sw
Dblock    nDb   nSn    d_switch
Dbody     nSn   nSp    d_switch
.model    bidir_sw    SW( RON=0.001 )
.model    d_switch    D()
.ENDS
*****************************************************************
*GY
*        pins   1  2  3  4
.SUBCKT  gy  np  np0  ns  ns0  N=100
rp1 np nrp 0.1
rs1 ns nrs 0.1
rpmid nrp nrpmid 0.1
rsmid nrs nrsmid 0.1
BGp1 nrpmid npmid V=(-1)*i(vs1)*{N}
BGs1 nrsmid nsmid V=i(vp1)*{N}
vp1 npmid np0 0
vs1 nsmid ns0 0
BHs1 nrs nrsmid I=(-1)*v(nrp,np0)/{N}
BHp1 nrp nrpmid I=v(nrs,ns0)/{N}
*Uncomment and give 1% permeance value to include leakage effects
*Clp nrp np0 100nF
*Cls nrs ns0 100nF
rsp1 nrp nrs 100e6
rsm1 np0 ns0 100e6
.ENDS
*****************************************************************
*TF
*     pins   1  2  3  4
.SUBCKT tf np np0 ns ns0 N=1
xGYp np np0 n1 n10 gy N={1}
xGYs n1 n10 ns ns0 gy N={N}
.ENDS
*****************************************************************
```

```
*Electro-Magnetic Transformer with saturation
*       pins   1   2    3   4
.SUBCKT tfx np np0 ns ns0 Np=1000 Ns=1000 C=100uF
xGYp np np0 n1 n10 gy N={Np}
xGYs n2 n10 ns ns0 gy N={Ns}
*   IF Vc   < fsat    THEN {Cper} ELSE   {reducing Cper}
Cper n1 n2 c='v(n1)-v(n2) < {5} ? {C} : ...
                        {0.99*C*exp(-10*(v(n1)-v(n2)))+0.01*C}'
.ENDS
***************************************************************
*FORWARD Xfm
*        pins  1   2   3   4   5   6
.SUBCKT tffor np np0 nd nd0 ns ns0 Np=100 Nd=100 Ns=100 C=100uF
xGYp np np0 n1 n10 gy N={Np}
xGYd n1 n30 nd nd0 gy N={Nd}
xGYs n2 n10 ns ns0 gy N={Ns}
*   IF Vc   < fsat    THEN {Cper} ELSE   {reducing Cper}
Cper n30 n2 c='v(n30)-v(n2) < {5} ? {C} : ...
                        {0.99*C*exp(-10*(v(n30)-v(n2)))+0.01*C}'
.ENDS
***************************************************************
*PUSHPULL Xfm
*        pins     1   2   3   4   5   6
.SUBCKT tfpush np1 npc np2 ns1 nsc ns2 Np=100 Ns=100 C=100uF
xGYp1 np1 npc n1 n10 gy N={Np}
xGYp2 npc np2 n10 n2 gy N={Np}
xGYs1 n3 n30 ns1 nsc gy N={Ns}
xGYs2 n30 n2 nsc ns2 gy N={Ns}
*   IF Vc   < fsat    THEN {Cper} ELSE   {reducing Cper}
Cper n1 n3 c='v(n1)-v(n3) < {5} ? {C} : ...
                        {0.99*C*exp(-10*(v(n1)-v(n3)))+0.01*C}'
.ENDS
***************************************************************
*BRIDGE Xfm
*        pins    1   2   3   4   5
.SUBCKT tfbdg np1 np2 ns1 nsc ns2 Np=100 Ns=100 C=100uF
xGYp  np1 np2 n1 n10 gy N={Np}
xGYs1 n3 n30 ns1 nsc gy N={Ns}
xGYs2 n30 n10 nsc ns2 gy N={Ns}
*   IF Vc   < fsat    THEN {Cper} ELSE   {reducing Cper}
Cper n1 n3 c='v(n1)-v(n3) < {5} ? {C} : ...
                        {0.99*C*exp(-10*(v(n1)-v(n3)))+0.01*C}'
.ENDS
***************************************************************
*---------------------------------------------------------------
```

```
* PV SOURCE
*---------------------------------------------------------------
* connections:    +ve terminal
*                 | -ve terminal
*                 | |
.SUBCKT PVSOURCE 1 2 Isc=1 Vscale=50
*Setting Isc value
VIsc nsc 0 {Isc}
*Setting No. of devices in series
Vnum ndev 0 {Vscale}
*Using PV model eqn. in B-source
*(nVT)=0.05 -- n=2 and VT=0.025
*a conditional statement is used. If current flow is negative
*i.e. PV as sink, then the current should get limited to 0 and
*not allow sink current into PV
Bpv 1 2 i=-(v(nsc)- 1e-7*(exp(v(1,2)/0.05/v(ndev))-1)>0 ? ...
              (v(nsc)- 1e-7*(exp(v(1,2)/0.05/v(ndev))-1)) : 0)
.ENDS
*---------------------------------------------------------------
```

Appendix-D : Octave scripts

Octave script for estimating the model parameters for water vapour content in India

```
# FOURIER SERIES CURVE FITTING FOR WATER VAPOUR CONTENT IN
# ATMOSPHERE FOR INDIA
#
# WATER VAPOUR MODEL:
#
#     w= water vapour content in atmosphere at a given lattitude
#     w = G1 + G2.sint + G3.sin2t + G4.sin3t
#            + G5.cost + G6.cos2t + G7.cos3t
#     where
#         Gi = gi1 + gi2.x + gi3.x^2
#         gi1..gi3 are determined using "wv_india.m" script file
#     and
#             x=(Q-35)
#             t=2.pi(N-80)/365
#
# The program gives the coefficients gi1, .....gi3 for WATER
# VAPOUR CONTENT. The data taken into consideration are N
# (day number of the year), w (water content in the atmosphere)
# and lattitude Q. The data are of 12 locations spread out
# in India
clc
clear
#Day number
N=[15;46;74;105;135;166;196;227;258;288;319;349;
   15;46;74;105;135;166;196;227;258;288;319;349;
   15;46;74;105;135;166;196;227;258;288;319;349;
   15;46;74;105;135;166;196;227;258;288;319;349;
   15;46;74;105;135;166;196;227;258;288;319;349;
   15;46;74;105;135;166;196;227;258;288;319;349;
```

```
    15;46;74;105;135;166;196;227;258;288;319;349;
    15;46;74;105;135;166;196;227;258;288;319;349;
    15;46;74;105;135;166;196;227;258;288;319;349;
    15;46;74;105;135;166;196;227;258;288;319;349;
    15;46;74;105;135;166;196;227;258;288;319;349;
    15;46;74;105;135;166;196;227;258;288;319;349];

#Lattitude in degrees
Q=[23.07;23.07;23.07;23.07;23.07;23.07;
                       23.07;23.07;23.07;23.07;23.07;23.07;
    21.75;21.75;21.75;21.75;21.75;21.75;
                       21.75;21.75;21.75;21.75;21.75;21.75;
    19.12;19.12;19.12;19.12;19.12;19.12;
                       19.12;19.12;19.12;19.12;19.12;19.12;
    22.65;22.65;22.65;22.65;22.65;22.65;
                       22.65;22.65;22.65;22.65;22.65;22.65;
    26.30;26.30;26.30;26.30;26.30;26.30;\
                       26.30;26.30;26.30;26.30;26.30;26.30;
    10.23;10.23;10.23;10.23;10.23;10.23;
                       10.23;10.23;10.23;10.23;10.23;10.23;
    13.00;13.00;13.00;13.00;13.00;13.00;
                       13.00;13.00;13.00;13.00;13.00;13.00;
    21.15;21.15;21.15;21.15;21.15;21.15;
                       21.15;21.15;21.15;21.15;21.15;21.15;
    28.58;28.58;28.58;28.58;28.58;28.58;
                       28.58;28.58;28.58;28.58;28.58;28.58;
    18.53;18.53;18.53;18.53;18.53;18.53;
                       18.53;18.53;18.53;18.53;18.53;18.53;
    08.48;08.48;08.48;08.48;08.48;08.48;
                       08.48;08.48;08.48;08.48;08.48;08.48;
    17.72;17.72;17.72;17.72;17.72;17.72;
                       17.72;17.72;17.72;17.72;17.72;17.72];

#water vapour content
w=[1.67;1.82;2.16;2.73;3.33;4.66;5.37;5.32;4.40;3.27;2.00;1.91;
   1.84;1.85;2.30;2.88;3.42;4.81;5.32;5.16;4.59;3.31;2.29;2.03;
   2.70;2.55;2.90;3.30;3.95;5.17;5.36;5.05;4.78;4.01;2.94;2.93;
   2.10;2.22;2.63;3.49;4.30;5.62;6.28;6.17;5.69;4.55;2.86;2.06;
   1.59;1.60;1.83;2.29;2.86;4.22;5.23;5.25;3.91;2.53;1.71;1.73;
   0.74;0.77;0.77;1.12;1.58;1.86;1.88;1.77;1.82;1.63;1.04;0.91;
   2.73;2.66;2.51;3.25;4.36;5.07;5.09;4.94;4.82;4.64;3.45;3.12;
   1.75;2.09;2.08;2.58;2.99;4.66;5.32;5.41;4.59;3.53;1.99;1.89;
   1.40;1.38;1.66;1.93;2.55;4.01;5.56;5.74;4.19;2.67;1.50;1.42;
   1.77;1.75;2.06;2.45;2.87;3.94;4.26;4.17;3.86;3.15;2.29;2.09;
   3.02;3.16;3.49;4.30;4.47;4.47;4.46;4.29;4.39;4.37;4.02;3.67;
```

```
      2.73;3.21;3.08;3.98;4.59;5.50;5.55;5.48;5.39;4.91;3.18;2.67];
#
x=(Q-35);
t=((2*pi)/365).*(N-80);
#
n=length(x);
c(1:n,1)=1;
c(1:n,2)=x;
c(1:n,3)=x.*x;
#
c(1:n,4)=sin(t);
c(1:n,5)=x.*sin(t);
c(1:n,6)=x.*x.*sin(t);
#
c(1:n,7)=sin(2*t);
c(1:n,8)=x.*sin(2*t);
c(1:n,9)=x.*x.*sin(2*t);
#
c(1:n,10)=sin(3*t);
c(1:n,11)=x.*sin(3*t);
c(1:n,12)=x.*x.*sin(3*t);
#
c(1:n,13)=cos(t);
c(1:n,14)=x.*cos(t);
c(1:n,15)=x.*x.*cos(t);
#
c(1:n,16)=cos(2*t);
c(1:n,17)=x.*cos(2*t);
c(1:n,18)=x.*x.*cos(2*t);
#
c(1:n,19)=cos(3*t);
c(1:n,20)=x.*cos(3*t);
c(1:n,21)=x.*x.*cos(3*t);
#
#
[row,col]=size(c);
for i=1:col
    for j=1:col
        A(i,j)=sum(c(1:n,j).*c(1:n,i));
        b(i)=sum(w.*c(1:n,i));
    end
end
#
cof=A\b';  #inv(A)*b';
k=1;
```

```
for i=1:7
for j=1:3
G(i,j)=cof(k);
k=k+1;
endfor
endfor
G
#*****************************************************************
```

Octave script for estimating the model parameters for clearness index in India

```
# FOURIER SERIES CURVE FIT FOR CLEARNESS INDEX FOR INDIA
#
# CLEARNESS INDEX ESTIMATE MODEL:
#
# Kte = A1 + A2.sint + A3.sin2t + A4.sin3t
#          + A5.cost + A6.cos2t + A7.cos3t
# where
#       Ai = ai1 + ai2.x + ai3.x^2 + ai4.w + ai5.w^2
#       x=(Q-35)
#       t=2.pi(N-80)/365
#       w= water vapour content is estimated from another
#          Fourier curve fit model
#       w = G1 + G2.sint + G3.sin2t + G4.sin3t
#              + G5.cost + G6.cos2t + G7.cos3t
#       and
#           Gi = gi1 + gi2.x + gi3.x^2
#                gi1..gi3 are determined using "wv_india.m"
#                script file
#
# The program gives the coefficients ai1, .....ai5 for
# CLEARNESS INDEX. The data taken into consideration are Kt
# (measured clearness index), N (day number of the year), W
# (water content in the atmosphere) and lattitude Q. The data
# are of 12 locations spread out in India
#
clc
clear
#Kt computed from computed Ho and measured Ha
Kt=[0.7070;0.7164;0.7198;0.7073;0.6958;0.5998;
               0.4553;0.4393;0.6100;0.6952;0.7157;0.7051;
   0.7231;0.7297;0.7242;0.7007;0.6991;0.5693;
```

```
                    0.4330;0.4167;0.5969;0.7041;0.7239;0.7095;
    0.6694;0.6680;0.6766;0.6745;0.6811;0.5305;
                    0.3836;0.3851;0.5229;0.6376;0.6730;0.6661;
    0.6012;0.6410;0.6441;0.6268;0.6366;0.4617;
                    0.4270;0.4101;0.4740;0.5464;0.5924;0.6083;
    0.7394;0.7349;0.7289;0.7087;0.6887;0.6467;
                    0.5592;0.5537;0.6660;0.7269;0.7501;0.7334;
    0.7707;0.7634;0.7009;0.6259;0.5755;0.5086;
                    0.4208;0.4325;0.4565;0.4381;0.5251;0.6662;
    0.6427;0.6882;0.6955;0.6720;0.6467;0.5882;
                    0.5385;0.5574;0.5953;0.5546;0.5453;0.5590;
    0.6906;0.6887;0.6716;0.6611;0.6493;0.5438;
                    0.4021;0.3969;0.5470;0.6779;0.7087;0.6997;
    0.6792;0.6998;0.7108;0.6902;0.6751;0.5934;
                    0.5044;0.5159;0.6504;0.7072;0.7244;0.7034;
    0.6904;0.7098;0.7034;0.6902;0.6972;0.5488;
                    0.4216;0.4315;0.5434;0.6668;0.6888;0.6932;
    0.6809;0.6748;0.6679;0.6165;0.5771;0.5508;
                    0.5108;0.5549;0.6043;0.5683;0.5719;0.6377;
    0.7068;0.6999;0.6770;0.6453;0.6375;0.5026;
                    0.4483;0.4827;0.5643;0.6685;0.7029;0.7030];
#Day number
N=[15;46;74;105;135;166;196;227;258;288;319;349;
    15;46;74;105;135;166;196;227;258;288;319;349;
    15;46;74;105;135;166;196;227;258;288;319;349;
    15;46;74;105;135;166;196;227;258;288;319;349;
    15;46;74;105;135;166;196;227;258;288;319;349;
    15;46;74;105;135;166;196;227;258;288;319;349;
    15;46;74;105;135;166;196;227;258;288;319;349;
    15;46;74;105;135;166;196;227;258;288;319;349;
    15;46;74;105;135;166;196;227;258;288;319;349;
    15;46;74;105;135;166;196;227;258;288;319;349;
    15;46;74;105;135;166;196;227;258;288;319;349;
    15;46;74;105;135;166;196;227;258;288;319;349];

#Lattitude in degrees
Q=[23.07;23.07;23.07;23.07;23.07;23.07;
                    23.07;23.07;23.07;23.07;23.07;23.07;
    21.75;21.75;21.75;21.75;21.75;21.75;
                    21.75;21.75;21.75;21.75;21.75;21.75;
    19.12;19.12;19.12;19.12;19.12;19.12;
                    19.12;19.12;19.12;19.12;19.12;19.12;
    22.65;22.65;22.65;22.65;22.65;22.65;
                    22.65;22.65;22.65;22.65;22.65;22.65;
    26.30;26.30;26.30;26.30;26.30;26.30;
```

```
                             26.30;26.30;26.30;26.30;26.30;26.30;
    10.23;10.23;10.23;10.23;10.23;10.23;
                             10.23;10.23;10.23;10.23;10.23;10.23;
    13.00;13.00;13.00;13.00;13.00;13.00;
                             13.00;13.00;13.00;13.00;13.00;13.00;
    21.15;21.15;21.15;21.15;21.15;21.15;
                             21.15;21.15;21.15;21.15;21.15;21.15;
    28.58;28.58;28.58;28.58;28.58;28.58;
                             28.58;28.58;28.58;28.58;28.58;28.58;
    18.53;18.53;18.53;18.53;18.53;18.53;
                             18.53;18.53;18.53;18.53;18.53;18.53;
    08.48;08.48;08.48;08.48;08.48;08.48;
                             08.48;08.48;08.48;08.48;08.48;08.48;
    17.72;17.72;17.72;17.72;17.72;17.72;
                             17.72;17.72;17.72;17.72;17.72;17.72];

#water vapour content
w=[1.67;1.82;2.16;2.73;3.33;4.66;5.37;5.32;4.40;3.27;2.00;1.91;
    1.84;1.85;2.30;2.88;3.42;4.81;5.32;5.16;4.59;3.31;2.29;2.03;
    2.70;2.55;2.90;3.30;3.95;5.17;5.36;5.05;4.78;4.01;2.94;2.93;
    2.10;2.22;2.63;3.49;4.30;5.62;6.28;6.17;5.69;4.55;2.86;2.06;
    1.59;1.60;1.83;2.29;2.86;4.22;5.23;5.25;3.91;2.53;1.71;1.73;
    0.74;0.77;0.77;1.12;1.58;1.86;1.88;1.77;1.82;1.63;1.04;0.91;
    2.73;2.66;2.51;3.25;4.36;5.07;5.09;4.94;4.82;4.64;3.45;3.12;
    1.75;2.09;2.08;2.58;2.99;4.66;5.32;5.41;4.59;3.53;1.99;1.89;
    1.40;1.38;1.66;1.93;2.55;4.01;5.56;5.74;4.19;2.67;1.50;1.42;
    1.77;1.75;2.06;2.45;2.87;3.94;4.26;4.17;3.86;3.15;2.29;2.09;
    3.02;3.16;3.49;4.30;4.47;4.47;4.46;4.29;4.39;4.37;4.02;3.67;
    2.73;3.21;3.08;3.98;4.59;5.50;5.55;5.48;5.39;4.91;3.18;2.67];
#
x=(Q-35);
t=((2*pi)/365).*(N-80);
#
n=length(x);
c(1:n,1)=1;
c(1:n,2)=x;
c(1:n,3)=x.*x;
c(1:n,4)=w;
c(1:n,5)=w.*w;
#
c(1:n,6)=sin(t);
c(1:n,7)=x.*sin(t);
c(1:n,8)=x.*x.*sin(t);
c(1:n,9)=w.*sin(t);
c(1:n,10)=w.*w.*sin(t);
```

```
#
c(1:n,11)=sin(2*t);
c(1:n,12)=x.*sin(2*t);
c(1:n,13)=x.*x.*sin(2*t);
c(1:n,14)=w.*sin(2*t);
c(1:n,15)=w.*w.*sin(2*t);
#
c(1:n,16)=sin(3*t);
c(1:n,17)=x.*sin(3*t);
c(1:n,18)=x.*x.*sin(3*t);
c(1:n,19)=w.*sin(3*t);
c(1:n,20)=w.*w.*sin(3*t);
#
c(1:n,21)=cos(t);
c(1:n,22)=x.*cos(t);
c(1:n,23)=x.*x.*cos(t);
c(1:n,24)=w.*cos(t);
c(1:n,25)=w.*w.*cos(t);
#
c(1:n,26)=cos(2*t);
c(1:n,27)=x.*cos(2*t);
c(1:n,28)=x.*x.*cos(2*t);
c(1:n,29)=w.*cos(2*t);
c(1:n,30)=w.*w.*cos(2*t);
#
c(1:n,31)=cos(3*t);
c(1:n,32)=x.*cos(3*t);
c(1:n,33)=x.*x.*cos(3*t);
c(1:n,34)=w.*cos(3*t);
c(1:n,35)=w.*w.*cos(3*t);
#
#
[row,col]=size(c);
for i=1:col
    for j=1:col
        A(i,j)=sum(c(1:n,j).*c(1:n,i));
        b(i)=sum(Kt.*c(1:n,i));
    end
end
#
cof=A\b';  #inv(A)*b';
k=1;
for i=1:7
for j=1:5
coeff(i,j)=cof(k);
```

```
k=k+1;
endfor
endfor
coeff
```
#**

Appendix-E : Answers to chapter end questions

Chapter 1

1) c, 2) b, 3) b, 4) a, 5) d, 6) d, 7) a b c, 8) 0.25, 9) 37.9 V, 10) 17%, 11) 3.6 W, 12) 0.73, 13) 9.59 A, 14) 810.35 W/m^2, 15) 70.52 V

Chapter 2

1)b, 2)a, 3)d, 4)e, 5)d, 6)c, 7)10.08kW, 8)57.6W, 9)18.3V, 10)6A, 11)Place an anti-parallel bypass diode across PV cell2 and also PV cell1. This will prevent either cell from acting as a sink, 12)4.735ohms, 13)2.272ohms, 14)80, 15)linearly from Voc of 19V to Isc of 1.9A.

Chapter 3

1)a, 2)b, 3)b, 4)a, 5)b, 6)d, 7)b, 8)c, 9)a, 10)66.15 deg, 11)April 12th, 12)equinox days, 13)8.1 deg, 14)15 deg, 15)1.2, 16)1.1439, 17)1.57 radians, 18)1.12, 19a)1336.88 W/m^2, 19b)May 19th, 19c)0, 20)11.92 hours, 21)15.42 hours and 8.57 hours, 22)11.248 $kWh/m^2/day$, 23)9.585 $kWh/m^2/day$, 24)4.7925 $kWh/m^2/day$, 25)0.5382, 26a)1.385 radians, 26b)6.6346 $kWh/m^2/day$ 26c)9.7799 $kWh/m^2/day$, 26d)1.4741

Chapter 4

1)250 kWh, 2)INR 266.67, 3)5 hours, 4)324 W, 5)125 m^2, 6)8 kWh, 7)144.34 A rms, 8)23.67 kWh, 9)5.7 A, 10)13.16 A

Chapter 5

1)b, 2)a, 3)d, 4)c, 5)a, 6)b, 7)b, 8)b, 9)a, 10)119 Ah, 11)120 A, 12)6.5 ohms, 13)11.334 kWh, 14)415.9 kilolitres, 15)4 times

Chapter 6

1)a, 2)c, 3a)1680 Wh, 3b)1600 Wh, 4)83.3 Ah, 5)2942.9 Wh, 6)3920 Wh, 7)20A and 4.167A, 8)142.86 Ah, 9)C7 or C5, 10)134 Ah and 7.88 m^2

Chapter 7

1)c, 2)b, 3)a, 4)b, 5)a, 6)a, 7)c, 8)d, 9)d, 10)c, 11)a, 12)7.13 ohms, 13)0.1558, 14)288 ohms, 15)0.7, 16)0.9, 17)5A, 18)0.96 ohm, 19)0.307, 20)0.59

Chapter 8

1)a, 2)c, 3)a, 4)c, 5)d, 6)6.5 ohms, 7)-3000 A/s, 8)1 A, 9)-6 A/s, 10)0.576 W

Chapter 9

1)a, 2)b, 3)c, 4)d, 5)c, 6)b, 7)d, 8)b, 9)1.67, 10)1 $^oK/W$, 11)4.74e-5 $^oK/W$, 12)0.0047619 $^oK/W$, 13)8333.3, 14)0.9933 ohm, 15)9.64 W

Chapter 10

1)b, 2)c, 3)c, 4)a, 5)d, 6)c, 7)b, 8)981 joules, 9)10.19 kg, 10)2943 W, 11)5 m, 12)32 m, 13)15320 W, 14)277390, 15)97.3 m, 16)20.833 A, 17)40.22 kW, 18)1.2 kWh, 19)685.7 W, 20)1.428 lt/s

Chapter 11

1)b, 2)c, 3)a, 4)b, 5)c, 6)b, 7)c, 8)c, 9)0.366 mH and 34.6 μF, 10)1.215 mH, 11)42.43, 12)42.43, 13)7.07, 50 Hz and -45°, 14)0.67, 15)2.05 A and 0 A

Chapter 12

1)d, 2)c, 3)Rs. 1737.2, 4)15.38 %, 5)9.12 %, 6)Rs. 7.53/kWh, 7)Rs. 19964.00, 8)Rs. 22922.00, 9)c, 10)Rs. 663.31

Bibliography

1. E. Becquerel, Memoire sur les effets electriques produits sous l'influence des rayons solaires. Comptes Rendus, 1839.
2. W.G. Adams and R.E. Day, The Action of Light on Selenium, Philosophical Trans- actions of the Royal Society of London, 1876.
3. Smith Willoughby, Effect on Light on Selenium during the passage of an Electric Current, Nature, 1873.
4. C.E. Fritts, On a New Form of Selenium Photocell, American Journal of Science, 26:465, 1883.
5. A. Einstein, Uber einem die Erzeugung und Verwandlung des Lichtes betreffenden heuristischen Gesichtspunkt, Annalen der physik, 1905.
6. L. O. Grondahl, The copper-cuprous-oxide rectifier and photoelectric cell, Reviews of Modern Physics, 5:141–168, Apr 1933.
7. D.M. Chapin, C.S. Fuller, and G.L. Pearson, A new silicon p-n junction photocell for converting solar radiation into electrical power. Journal of Applied Physics, 25:676, 1954.
8. M. Halkias, Integrated Electronics. McGraw-Hill electrical and electronic engineering series, Tata McGraw-Hill Publishing Company, 2001.
9. Chenming Hu, Richard M White, Solar Cells : From Basics to Advanced Systems, McGraw-Hill Book Company, 1983.
10. Liu, B.Y.H. and Jordan, R.C., Daily Insolation on Surfaces Tilted towards the Equator, ASHRAE Transactions, 67, 526-541, 1962.
11. S A Klein, Calculation of monthly average insolation on tilted surfaces, Elsevier, Solar Energy, volume 19, Issue 4, pp. 325-329, 1977.
12. A.A.M Sayigh, Solar Energy Engineering, Elsevier, ISBN: 9780124143913, 2012
13. Annamani and Rangarajan S, Solar radiation over India, Allied Publishers pvt. ltd., 1982

14. Ravinder Kumar, L Umanand, Estimation of global radiation using clearness index model for sizing photovoltaic system, Renewable energy, 30 (15), 2221-2233, 2005.

15. L Umanand, Power Electronics: Essentials and Applications, Wiley India Pvt Ltd., 2009.

16. C. Colebrook, Turbulent flow in pipes, Journal of the Inst. Civil Eng., vol. 11, p.133, 1938.

17. J.F. Kreider and F Kreith, Solar Energy Handbook, McGraw Hill, New York, 1981.

18. John W Twidell, Anthony D Weir, Renewable Energy Resources, ELBS, London, 1986.

19. Kitsum K, Switched Mode Powewr Conversion - Basic Theory and Design, Marcel Dekker, Inc. NY, USA, 1984.

20. Leonhard W, Control of Electrical Drives, Springer Verlag, Berlin, 1985.

21. Mohan N, Undeland T M and Robbins W P, Power Electronics: Converters, Applications and Design, John Wiley and Sons, NY, USA, 1989.

22. Ralph E Tarter, Principles of Solid-state Power Conversion, Howard W Sams, 1985.

23. Lander C.W, Power electronics, McGraw-Hill Book Company, NY, USA, 1981.

24. Thorborg, K, Power Electronics, Prentice Hall, NJ, USA, 1988.

25. Jefferson W Tester, Elisabeth M Drake, Michael W Golay, Michael J Driscoll, William A Peters, Sustainable Energy : Choosing among options, PHI Learning Pvt. Ltd., New Delhi, 2009.

26. D A J Rand, R Woods, R M Dell, Batteries for Electric Vehicles, SAE International, Warrendale, Pa., 1998.

27. T R Crompton, Battery Reference Book, 2nd edition, SAE International, Warrendale, Pa., 1996.

Index

Numbers
1MVA generation, 89

A
ACB, 90
ACDB, 90
active equaliser, 162
Active paralleling circuits, 165
airmass, 64
airmass coefficient, 65
albedo, 71
amp-sec balance, 131
annual installment, 260
Annual payments, 260
Atmospheric effects, 62
atmospheric effects, 71
attenuation constant, 231

B
ballast load, 220
Batteries, 151
Batteries in Parallel, 165
Batteries in Series, 160
battery current, 152
battery fires, 165
bi-metallic junction, 171
bidirectional switch, 154
Boltzmann constant, 8
boost converter, 126
borewell, 213
buck converter, 128
buck-boost converter, 130

C
C-rate, 98
capacity, 96
Capital cost, 262
celestial equator, 44
centrifugal pump, 199
characteristic dimension, 186
Characteristic impedance, 230
Charge controller, 154
Charge controller circuit, 155
charge equalisation, 161
Charge pump, 162
charge-discharge cycle, 98
Charging with MPPT, 160
circulating current, 166
clearness index, 66
CoP, 175
Coefficient of performance, 175
coefficient of thermal expansion, 186
cold junction, 172
Colebrook-White formula, 204
common mode impedance, 234
Compound growth, 255
compound interest, 255
compressed air storage, 108
Convection, 184
copper-cuprous oxide, 4
current control, 156
current scaling, 141
current source, 6
cut-off frequencies, 231

D
Daily energy, 49
daily irradiance, 41
Darcy-Weisbach formula, 203
datasheet, 14
Day load, 113
days of autonomy, 115
days of recharge, 116
Deadtime, 238

Declination, 44
delivery head, 201
delivery pipe, 199
depth of discharge, 97
differential mode impedance, 234
diffuse radiation, 71
discharge rate, 202
discharge volume, 202
discrete compounding, 257
dissimilar metals, 171
diurnal, 45
driving point impedances, 230
dynamic head, 201
dynamic pump, 199

E

earth centric, 44
Efficiency, 10
electrodes, 1
Energy cost, 262
Energy script, 74
equal annual installments, 260
equatorial plane, 43
Estimation of ALCC, 266
Estimation of break even point, 268
Exponential growth, 257
extra-terrestrial radiation, 42

F

Feedforward components, 244
Fill factor, 13
fill factor, 33
flat plate collector, 47
flux walking, 136
flyback active parallel circuit, 167
Flyback converter, 133
Flywheel storage, 105
forced convection, 188
Forced cooling, 181
Fourier law, 183
Free convection, 186
Frequency and angle estimation, 244
friction factor, 203
friction head, 201
Full bridge converter, 137

G

Gate drive, 237
gear box, 220
geometric progression, 261
gravimetric energy density, 101
gravimetric power density, 102
gravitational acceleration, 186
Grid tied system, 82
Growth models, 253

H

Half bridge converter, 136
heat equation, 172
heat flow, 177
heat flow rate equation, 183
heat pump, 172
Heatsink, 179
Hill climbing method, 146
horizon plane, 46
horizontal flat plate collector, 47
hot junction, 172
hydraulic energy, 202
hydraulic power, 202
hydraulic system, 200
hydro generator, 217
hysteresis comparator, 155

I

I-V Characteristics, 8
IGC, 220
Identical cells in parallel, 23
Identical cells in series, 19
image impedances, 230
incomer, 81
incremental resistance method, 143
induction generator, 220
induction generator controller, 220
inertia of the flywheel, 106
Inflation, 258
input capacitor, 129
input resistance, 126
input resistance control, 138
Insolation, 40
Insolation based emulation, 33
internal series resistance, 152
irradiance, 40

Irradiance simulation, 55

K
kinematic viscosity, 186

L
L-C-L interface, 231
LCC, 262
Large tilt generalisation, 57
latitude angle, 43
Life cycle cost, 262
Linear growth, 254
Lithium polymer, 103
load distribution, 114
load line, 20
locale centric, 46

M
MCCB, 90
MPPT, 125
MPPT integration, 245
MPPT with shading, 147
Maintenance cost, 262
Mass transport, 194
Maximum power point, 10
maximum power point, 21
maximum power point tracking, 125
Measurement setup, 30
meridional axis, 45
metallization, 5
Model, 7
model for battery, 151
monocrystalline, 4
Moody Chart, 204

N
net zero energy, 82
NgSPICE, 15
ngSpice SUBCKT, 29
Night load, 113
Non-identical cells in parallel, 24
Non-identical cells in series, 21
Nusselt's number, 185

O
Open circuit voltage, 10

open delta, 220
open loop plant, 236
Optimum fixed tilt, 60
outcomer, 82
overhead tank, 199

P
p-n junction, 5
p-type substrate, 5
PV collector, 43
PV panel current, 152
PV source, 7
PV source emulation, 32
parallel combination, 23
parking lot, 87
pass band, 231
peak power point, 125
peak sun, 40
peak sun hours, 43
Peltier cooling, 177
Peltier effect, 171
peltier element, 180
peltier elements, 171
Peltier refrigeration, 189
penstock, 216
permeance, 133
phase constant, 231
phase locked loop, 244
photocurrent, 6
photons, 4
photovoltaic cell, 5
photovoltaic effect, 1
Pico hydel system, 218
planck's constant, 3
polycrystalline, 4
positive displacement, 199
Power slope, 143
present worth factor, 261
Pressure heads, 200
pressure source, 200
primary cell, 96
prime mover, 199
Propagation constant, 231
protection for parallel connection, 25
protection for series connection, 25
pulse phase method, 145

Pulse width modulation, 237
pumped hydro, 216
Pumped hydro storage, 106
Pushpull converter, 135
pyranometer, 66

Q
Quadrature phase shifter, 240
quantum of action, 3
quantum physics, 3

R
Rayleigh number, 186
reciprocating pump, 199
Reference cell method, 139
regenerative type, 166
relays, 155
Replacement cost, 262
reservoir, 164
Resistive equaliser, 161
reverse saturation current, 8
Reynolds number, 188
roof top PV modules, 82
rotating co-ordinate systems, 242
rotating frame, 240
roughness ratio, 204

S
SLI, 97
SPICE model for a PV, 27
Sampling method, 142
schmitt comparator, 145
Scrap value, 263
secondary cell, 96
selenium, 2
series combination, 19
series connection, 19
Short circuit current, 9
silicon solar cell, 4
silver bromide, 1
silver chloride, 1
simple interest, 254
single phase grid, 225
Single phase grid interface, 235
Slope Compensation, 157
Sodium Ion, 103

solar constant, 40
Solar geometry, 44
solar insolation, 40
solar power per unit area, 10
solar radiation spectrum, 42
Specific power, 182
spectral irradiance, 40
Spherical coordinate, 44
standard insolation, 11, 40
stationary coordinate system, 241
submersible pump, 213
suction head, 201
suction pipe, 199
sump, 200
sunrise angle, 51
sunset angle, 51
switching time period, 127

T
T-Network, 229
tariff, 82
Temperature coefficients, 12
Temperature effects, 11
temperature gradient, 172
terminals, 6
thermal coefficient, 183
thermal conductivity, 183
thermal design, 182
thermal diffusivity, 186
thermal domain, 177
thermal nomographs, 181
thermal resistance, 177
thermal resistivity, 183
thermal runaway, 165
thin film selenium, 3
three phase grid, 225
Three phase grid interface, 246
threshold comparison, 145
Tilt factor, 71
tilted flat plate, 53
Time frame transformation, 259
torque-speed characteristic, 219
Transformer-less, 233
transformerless grid interaction, 234
tropic of cancer, 44
tropic of capricorn, 44

turns ratio, 136

U
unity power factor, 226

V
vertically placed collectors, 76
vitreous selenium, 2
volt-sec balance, 131
voltage scaling, 139
voltage source, 6
volumetric energy density, 101
volumetric power density, 102

W
Water pumping, 199
water turbine, 216
worth transformation factor, 261

Z
zener regulators, 155
zenith angle, 51
zero net cost, 83

Made in the USA
Columbia, SC
15 September 2024

85480c22-c27a-4b31-afdb-6209fc237106R01